FRANCES JOHNSON 1975-6
D.W. 70 UNIVERSITY HALL
KT-550-330

Longman Mathematical Texts

Set theory and abstract algebra

Logic 309

DIC

Longman Mathematical Texts

Edited by Alan Jeffrey and Iain Adamson

Random variables L. E. Clarke
Electricity C. A. Coulson and T. J. M. Boyd
Continuum mechanics A. J. M. Spencer
Elasticity R. J. Atkin and N. Fox
Set theory and abstract algebra T. S. Blyth
Elementary mathematical analysis I. T. Adamson
Optimization A. H. England and W. A. Green
Functions of a complex variable E. G. Phillips
The theory of ordinary differential equations J. C. Burkill

Longman Mathematical Texts

Set theory and abstract algebra

T. S. Blyth

Reader in Pure Mathematics in the University of St Andrews

Longman
London and New York

Longman Group Limited
London
Associated companies, branches and representatives throughout the world.

Published in the United States of America by Longman Inc., New York.

© Longman Group Limited 1975

All rights reserved. No part of this publication may be reproduced, stored in a retrieval system, or transmitted in any form or by any means, electronic, mechanical, photocopying, recording, or otherwise, without the prior permission of the Copyright owner.

First published 1975

Library of Congress Cataloging in Publication Data
Blyth, Thomas Scott
Set theory and abstract algebra
(Longman mathematical texts)
Includes index.
1. Set theory. 2. Algebra, Abstract. I. Title
QA248.B62 512′.02 74-77989
ISBN 0 582 44284 2

Printed in Great Britain by
Adlard & Son Ltd., Bartholomew Press, Dorking

Preface

When the author was an undergraduate (1956–60 for your information!) there were very few texts available on abstract algebra. In those days, set theory was dismissed in half a lecture and, though it was considered quite something to teach a little group theory and some linear algebra, modern topics such as modules, homological algebra, categories, algebraic topology, etc., were considered too sophisticated and quite unsuitable. Nowadays the situation is vastly different: the inclusion of more and more algebra in undergraduate curricula (to the detriment of geometry, alas, say some), mainly in the 1960s, has given rise to a proliferation of texts, so much so that they can now be bought "by the yard" to fill your bookshelves. At the introductory level there are already many splendid texts. Some, however, are perhaps too elementary in that they contain no results of any depth; and others, at the other end of the spectrum, are often too encyclopaedic. The latter are often declared to be the best, however, since they are enjoyed by readers who have already learned the basic framework of the subject from more elementary expositions. The object of this text (and the excuse for producing yet another on abstract algebra) is to provide something which lies between these extremes, namely a text which is sophisticated enough not only to prepare the site and lay a concrete foundation but also to build thereon a splendid house or two.

We have divided the text into two main chapters, each of which is in several sections. In the first of these chapters we deal with the (naive) theory of sets, covering all the standard theory found in most first-year courses. Proceeding beyond this we develop, through the notion of equipotent sets, the theory of cardinals and use this to give a description of the natural numbers. More difficult topics (such as general cartesian products) are kept to the end of a section so that they may easily be omitted if the reader's prescribed course does not include them. Likewise, though we make free use of Zorn's axiom at this level, we have restricted its use in Chapter 1 to the section on infinite cardinals. This section may easily be assigned for supplementary reading if so desired. However, it is the author's opinion and experience that the extra effort involved in giving a more sophisticated approach to topics such as set theory is repaid by a much greater interest and understanding on the part of the student; and a description of the natural numbers via the theory of cardinals (albeit difficult in places) is an excellent illustration of the theory at work and one which even beginners find fascinating (and a refreshing change from the calculus which they all claim to have previously done at school!). In Chapter 2 we begin algebra proper and develop the basic machinery (semigroups, groups, rings, fields) contained in most first- and second-year curricula. One of the main (and indeed the prettiest) outcomes is a development of the number system. This is rarely covered in detail at this level.

Perhaps the greatest kick one gets out of mathematics is in solving problems; for it is here that the tools of the trade are put to use. It is a pity that many good texts are let down by their exercises. All too often the student who is unable even to begin a problem will give up and spend the time doing something else instead. In this text we have supplied a large number of varied exercises. Some are easy and some are difficult – and there are lots of hints. Indeed, even though some of the hints actually tell the reader how to do the entire problem, we feel that they are justified; for doing problems is an essential part of the learning process and a significant part of the examination process!

There are two notable types of mathematician: those who are somewhat pedantic and those who are skilled in the art of hand-waving. Whilst we can wave as well as the next man when the need arises, we have taken the view that in a formal introductory text such as this it is wise to be as pedantic as is necessary to maintain clarity throughout. This may well come as a source of annoyance to some mathematicians; but then it is not for them that this text is written! In particular, we mention that we shall doggedly maintain a distinction between a mapping $f: E \to F$ and the induced mappings $f^{\rightarrow}$, $f^{\leftarrow}$ on the power sets (these mappings usually being written as f, f^{-1} which is a source of confusion to some members of every class). Likewise, when dealing with laws of composition we prefer to use Bourbaki's $\top$ sign and, with the odd exception for typographical reasons, do so until we meet the notion of a ring.

Our approach to the topics covered is as modern as can be expected, by which we mean that we use universal-type definitions whenever possible. This has the pedagogical advantage of throwing everything back on to the foundation of sets and mappings. It also makes a course on the theory of categories a worthwhile inclusion at the honours level after a course on modules and linear algebra. Commutative diagrams will therefore be found throughout the text and many other important concepts will be found in the exercises.

It is a pleasure to record here my thanks to Dr I. T. Adamson of the University of Dundee for all his help and encouragement in preparing this text. It has grown out of many discussions with him over the past few years and is by far the better for his reading of the manuscript. I would also like to thank my colleagues Dr M. J. Beetham and Dr E. F. Robertson for their assistance with the proof reading and for checking the exercises. The traditional free copy is small recompense for their labour. Finally, I would like to thank the publishers for including this volume in their new series (or is it a sequence?) of mathematical texts.

T. S. B.
St Andrews,
January 1974.

Contents

1

Set theory and the natural numbers

§1. Sets; inclusion; intersection; union; complementation; number systems

A mathematical theory is essentially a deductive theory and as such is presented in the form of a sequence of definitions and theorems. In such a presentation, each definition is expressed in terms of previously defined concepts and each theorem is proved using previously established results. However, to begin such a theory we must have certain undefined **primitive terms** and unproved **primitive theorems** or **axioms**. These terms and axioms form the **intuitive basis** of the theory under consideration.

Modern mathematics is founded on a rigorous axiomatisation of **logic** and **set theory**. It would be completely out of the question at this stage, however, to attempt an axiomatisation of such topics and so, as a workable compromise, we shall accept without formal definition the intuitive concept of a **set**, that of **element belonging to a set** and the "obvious" axioms (or "common sense") of our everyday logic.

Set theory is the fundamental language of mathematics. Coupling it with logic, from which it is really inseparable, we obtain a language endowed with an extremely high degree of precision. In using this language the reader will become more and more aware of the close arguments involved in giving mathematical proofs and will also realise that, in order to proceed into any mathematical discipline, he has to develop an insistence on being precise in expressing what he means.

Intuitively, we think of a **set** as a collection of **objects**, these objects being anything we care to imagine, each of which is said to **belong** to the set.

Remark. This is, of course, not a proper mathematical definition since the words "set" and "collection" are synonymous. However, our intuition suffices to grasp the idea we have in mind.

We usually denote sets by capital letters and objects in the sets by lower case letters, though this is not to be taken as a universal convention since there is no reason why we should not consider a set as an object itself, say an object in a set whose objects are themselves sets! If E is a given set and x is a given object, we assume (what is surely in accordance with our intuition) that either x belongs to E or x does not belong to E. If x belongs to E we shall write $x \in E$ and if x does not belong to E we shall write $x \notin E$. Equivalent ways of reading $x \in E$ are: "x is an element of E", "x is contained in E" or simply "x is in E". We shall say that sets E, F are **equal** and write $E=F$ if they consist of the same objects; otherwise we shall say that they are **distinct** (or **unequal**) and write $E \neq F$.

Sometimes it is possible to display the elements of a set. In so doing we shall use what is known as the "curly brackets" or "braces" notation. For example, if E is the set consisting of the objects a, b, c then we write

$$E=\{a,b,c\}.$$

Note that, from the definition of equality for sets, we have

$$\{a,b,c\}=\{a,a,b,c,a,c,c,b,a\}=\{c,b,a\}=\text{etc.}$$

If a set E consists of a single object x then we write $E=\{x\}$; such a single-element set is often called a **singleton**. Note that every object x determines a singleton, namely $\{x\}$; moreover, we have $x \in \{x\}$.

We say that a set E is a **subset** of a set F and write $E \subseteq F$ if every element of E is an element of F. We also describe this situation by saying that E is **included** in F or that E is **contained** in F, though the latter terminology is better avoided. For notational convenience, we introduce another way of saying this, using the logical signs $\Rightarrow$ and $\Leftrightarrow$. We shall use the sign $\Rightarrow$ between two statements as an abbreviation for "implies" (or, equivalently, "if . . . then . . ."); and the sign $\Leftrightarrow$ between two statements as an abbreviation for "implies and is implied by" (or, equivalently, "if, and only if"). Thus, for example, we write

$$E \subseteq F \Leftrightarrow \begin{cases} \text{for all objects } x, \\ x \in E \Rightarrow x \in F. \end{cases}$$

In a similar way, we can express the fact that E, F are equal by

$$E=F \Leftrightarrow \begin{cases} \text{for all objects } x, \\ x \in E \Leftrightarrow x \in F. \end{cases}$$

Remark. In connection with the usage of "if and only if" it should be noted that, in making mathematical definitions, it is an almost universally accepted convention to use "if" when one really means "if and only if".

The two principal properties of the logical symbols $\Rightarrow$ and $\Leftrightarrow$ are the following: each is

(a) **reflexive**, by which we mean that, for every statement X, we have $X \Rightarrow X$ and $X \Leftrightarrow X$;
(b) **transitive**, by which we mean that, for all statements X, Y, Z we have

$$\text{if } X \Rightarrow Y \text{ and } Y \Rightarrow Z \text{ then } X \Rightarrow Z;$$

$$\text{if } X \Leftrightarrow Y \text{ and } Y \Leftrightarrow Z \text{ then } X \Leftrightarrow Z.$$

Property (b) may of course be abbreviated further, namely that for all statements X, Y, Z we have

$$(X \Rightarrow Y \text{ and } Y \Rightarrow Z) \Rightarrow (X \Rightarrow Z);$$

$$(X \Leftrightarrow Y \text{ and } Y \Leftrightarrow Z) \Rightarrow (X \Leftrightarrow Z).$$

Note that each of these is of the form "if . . . then . . ." and not of the form "if and only if".

The principal properties of $\subseteq$ (called the relation of **set inclusion**) follow from the above properties of $\Rightarrow$ and $\Leftrightarrow$ and are as follows.

Theorem 1.1 *The relation* $\subseteq$ *of set inclusion is*

(a) **reflexive**: *for every set* E, $E \subseteq E$;
(b) **transitive**: *for all sets* E, F *and* G,

$$(E \subseteq F \text{ and } F \subseteq G) \Rightarrow E \subseteq G;$$

(c) **anti-symmetric**: *for all sets* E *and* F,

$$(E \subseteq F \text{ and } F \subseteq E) \Rightarrow E = F.$$

Proof. (a) Let E be any set and let x be any object. Then we have $x \in E \Rightarrow x \in E$ by the reflexivity of $\Rightarrow$ and so $E \subseteq E$.

(b) Let E, F, G be any sets with $E \subseteq F$ and $F \subseteq G$. Then for all objects x we have $x \in E \Rightarrow x \in F$ and $x \in F \Rightarrow x \in G$. The transitivity of $\Rightarrow$ now yields $x \in E \Rightarrow x \in G$ whence we have $E \subseteq G$.

(c) Let E, F be any sets with $E \subseteq F$ and $F \subseteq E$. Then for all objects x we have $x \in E \Rightarrow x \in F$ and $x \in F \Rightarrow x \in E$. Consequently, $x \in E \Leftrightarrow x \in F$ and so $E = F$.

If E, F are sets with $E \subseteq F$ and $E \neq F$ then we shall write $E \subset F$ and call E a **proper** subset of F; we also say that E is **properly included** (or **properly contained**) in F. It should be noted that $\subset$ (called the relation of **strict inclusion**) is transitive but is neither reflexive nor anti-symmetric. If E, F are such that E is not a subset of F then we write $E \nsubseteq F$. Note that $E \nsubseteq F$ and $F \subset E$ are not the same in general; for example, if $E = \{1,2\}$ and $F = \{1,3\}$ then we have $E \nsubseteq F$ and $F \nsubseteq E$.

In the logic which underwrites the whole of mathematics, a **proposition** (or **statement**) is taken intuitively to be a meaningful sentence which may be classified as either true or false. For example, if E is any set and x is any object then either x is in E or x is not in E; so the sentence "x is in E" is either true or false. By an **open sentence** (or **formula**) we shall mean a sentence, written $S(x, y, z, \ldots)$, involving symbols $x, y, z, \ldots$ such that when the names of given objects $a, b, c, \ldots$ are substituted for $x, y, z, \ldots$ respectively at every occurrence of $x, y, z, \ldots$ we obtain a proposition concerning $a, b, c, \ldots$. For example, each of the following is an open sentence:

x is divisible by 4;
$x - y$ is a multiple of 5;
x is a professor;
x is a sister of y;
$x^2 + y^2 = z^2$;
$x^2 - 1 = (x+1)(x-1)$.

Remark. Again, this is not a proper mathematical definition since symbols are themselves objects. However, our intuition again suffices to grasp the idea we have in mind.

Given an open sentence of the form $S(x)$, we shall denote by $\{a; S(a)\}$ the set consisting of those objects a for which the proposition $S(a)$ is true. Other common notations are $\{a : S(a)\}$ and $\{a \mid S(a)\}$. If $E = \{a; S(a)\}$ then we shall say that the open sentence $S(x)$ is a **characteristic property** of E. For example, if M denotes the set of all the mathematicians in the world and if I denotes the set of all the people who can play a musical instrument then

$$\{x;\ x \in M \text{ and } x \in I\}$$

describes the set of those people who are mathematicians and who can play a musical instrument. This set can, of course, be described also as the set of mathematicians who can play a musical instrument, so we can regard this as equivalent to

$$\{x \in M;\ x \in I\}.$$

We shall now extend our intuitive idea of a set to allow as a set a "collection consisting of no objects". The courtesy of regarding this as a set has several advantages. This conventional **empty set** will be denoted by the Norwegian letter Ø; some authors, however, prefer to use the symbol □. In allowing Ø the status of a set, we gain the advantage of being able to talk about a set without knowing at the outset whether or not it has any elements. We shall agree that a set E which has elements will be called **non-empty** and shall denote this by writing $E \neq \varnothing$.

Consider now, for any set E and any object x, the statement

$$x \in E \text{ and } x \notin E.$$

By the logical principle of non-contradiction, this statement is false. Roughly speaking, therefore, there is no object x such that $x \in E$ and $x \notin E$. The concept of an empty set usefully describes this situation: given any set E we can write

$$\varnothing = \{x;\ x \in E \text{ and } x \notin E\}.$$

For any set E we have $\varnothing \subseteq E$; for the statement "every element of Ø is an element of E" is true simply because Ø has no elements at all. In the case where E is non-empty we have $\varnothing \subset E$.

Given any sets E, F we define the **intersection** of these sets to be the set, denoted by $E \cap F$, given by

$$E \cap F = \{x;\ x \in E \text{ and } x \in F\}.$$

Thus, if x is any object we have

$$x \in E \cap F \Leftrightarrow (x \in E \text{ and } x \in F).$$

It is of course possible to have $E \cap F = \varnothing$; for example, if E is the Greek alphabet and $F = \{♭, ♮, ♯\}$ is a set of musical symbols then $E \cap F$ is empty. In this case, we say that E and F are **disjoint**.

We define the **union** of the sets E, F to be the set, denoted by $E \cup F$, given by

$$E \cup F = \{x;\ x \in E \text{ or } x \in F\}.$$

Thus, if x is any object we have

$$x \in E \cup F \Leftrightarrow (x \in E \text{ or } x \in F).$$

Note that here (and indeed always in mathematics) the word "or" is taken in its *inclusive* sense, so that by "$x \in E$ or $x \in F$" we mean "$x \in E$ or $x \in F$ *or both*".

It is clear from the above definitions that for all sets E, F we have

$E \cap F \subseteq E$ and $E \cap F \subseteq F$; and similarly $E \subseteq E \cup F$ and $F \subseteq E \cup F$. Moreover, for every set E we have $\emptyset \cap E = \emptyset$ and $\emptyset \cup E = E$.

If E is any set and A is a subset of E then we define the **complement of A in E** to be the set $\complement_E(A)$ given by

$$\complement_E(A) = \{x \in E;\ x \notin A\}.$$

Thus, if x is any object we have

$$x \in \complement_E(A) \Leftrightarrow (x \in E \text{ and } x \notin A).$$

It is clear that $\complement_E(E) = \emptyset$ and $\complement_E(\emptyset) = E$ for any set E.

Given subsets A, B of a set E we define the **difference set** $A \backslash B$ (some authors write $A - B$) by

$$A \backslash B = \{x \in E;\ x \in A \text{ and } x \notin B\}.$$

The notation $A \backslash B$ is read "A cut down by B" or "A slash B". Clearly we have $A \backslash B = A \cap \complement_E(B)$ and, in the particular case where B is a subset of A, we have $A \backslash B = \complement_A(B)$.

It is sometimes convenient to represent sets by means of **Venn diagrams,** the interpretation of which is as follows. Since we are assuming that a given object either belongs to a given set or does not, we can represent a set by the interior of a closed contour and if an object belongs to the set then we represent it by a point inside the contour, whereas if it does not belong to the set then we represent it by a point outside the contour.

For example, the Venn diagrams depicting $E \cap F$ and $E \cup F$ are the following:

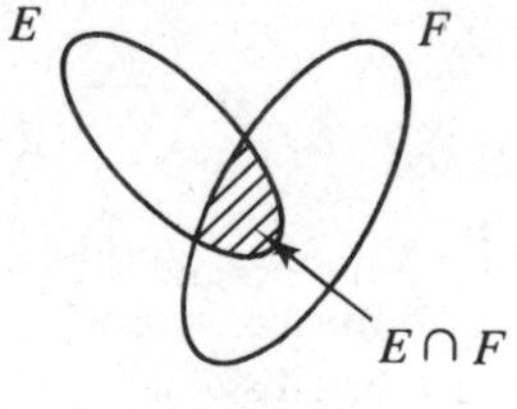

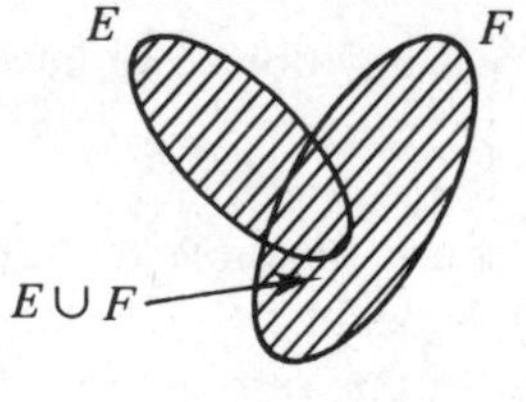

Likewise, the Venn diagrams depicting complements and difference sets are the following:

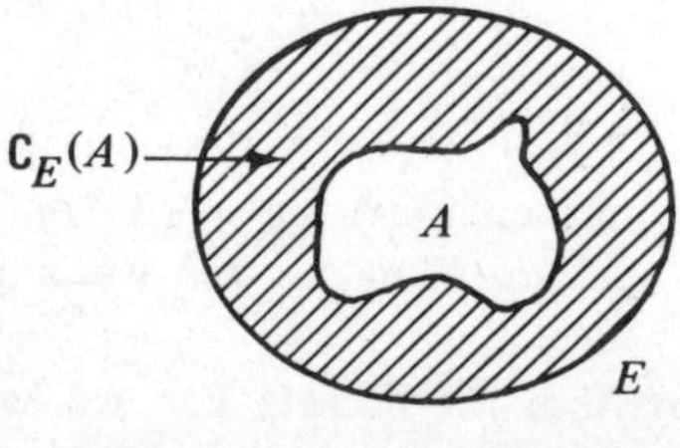

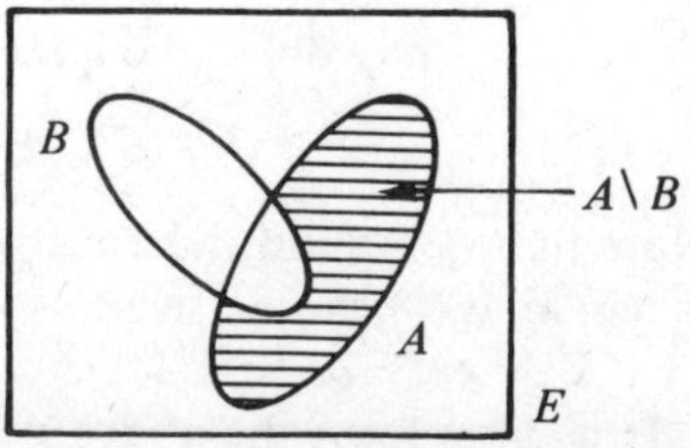

We now list the fundamental properties of intersection, union and complementation. To avoid repetition, we assume in *(a)*→*(g)* below that E, F, G are any given sets and in *(h)*→*(k)* that E is any set and that A, B are any subsets of E.

(a) $\boldsymbol{E \cap E = E = E \cup E}$.

This property is usually referred to by saying that $\cap$ and $\cup$ are **idempotent.** The proof is immediate from the definitions.

(b) $\boldsymbol{E \cap F = F \cap E; \quad E \cup F = F \cup E}$.

This property is referred to by saying that $\cap$ and $\cup$ are **commutative.** The proof is again immediate.

(c) $\boldsymbol{E \cap (F \cap G) = (E \cap F) \cap G; \quad E \cup (F \cup G) = (E \cup F) \cup G}$.

This property is referred to by saying that $\cap$ and $\cup$ are **associative**. The proof is once more immediate. Note that by this result we can write, without ambiguity, each side of the first equality as simply $E \cap F \cap G$ (i.e., omitting the brackets) and likewise each side of the second by $E \cup F \cup G$.

(d) $\boldsymbol{E \cap (F \cup G) = (E \cap F) \cup (E \cap G); \quad E \cup (F \cap G) = (E \cup F) \cap (E \cup G)}$.

This property is referred to by saying that each of $\cap$, $\cup$ is **distributive** over the other. We prove the first equality, the proof of the second being entirely similar. We have, for every object x,

$$\begin{aligned} x \in E \cap (F \cup G) &\Leftrightarrow x \in E \text{ and } x \in F \cup G \\ &\Leftrightarrow x \in E \text{ and } (x \in F \text{ or } x \in G) \\ &\Leftrightarrow (x \in E \text{ and } x \in F) \text{ or } (x \in E \text{ and } x \in G) \\ &\Leftrightarrow x \in E \cap F \text{ or } x \in E \cap G \\ &\Leftrightarrow x \in (E \cap F) \cup (E \cap G), \end{aligned}$$

so the result follows from the definition of equality for sets.

(e) $\boldsymbol{E \subseteq F \Rightarrow E \cap G \subseteq F \cap G; \quad E \subseteq F \Rightarrow E \cup G \subseteq F \cup G}$.

This property is referred to by saying that $\cap$ and $\cup$ are **inclusion-preserving**. The proof is almost immediate; for example, if $E \subseteq F$ then for any object x we have

$$x \in E \cap G \Rightarrow (x \in E \text{ and } x \in G) \Rightarrow (x \in F \text{ and } x \in G) \Rightarrow x \in F \cap G$$

and so $E \cap G \subseteq F \cap G$.

(f) $\boldsymbol{E \subseteq F \Leftrightarrow E \cap F = E \Leftrightarrow E \cup F = F}$.

Note how this property relates $\cap$ and $\cup$ to $\subseteq$. Suppose first that $E \subseteq F$. Then by (a), (e) and (b) above we have $E = E \cap E \subseteq F \cap E = E \cap F$. Since we always have $E \cap F \subseteq E$ it follows by Theorem 1.1(c) that $E \cap F = E$. Conversely, if $E \cap F = E$ then since we always have $E \cap F \subseteq F$ it follows that $E \subseteq F$. This then establishes $E \subseteq F \Leftrightarrow E \cap F = E$. Similarly we can show that $E \subseteq F \Leftrightarrow E \cup F = F$.

(g) $\boldsymbol{(E \subseteq F}$ **and** $\boldsymbol{E \subseteq G) \Rightarrow E \subseteq F \cap G}$; $\boldsymbol{(E \subseteq G}$ **and** $\boldsymbol{F \subseteq G) \Rightarrow E \cup F \subseteq G}$.

This property may be interpreted as follows. If E, F are sets such that $E \subseteq F$ then we often say that E is **smaller** than F or, equivalently, that F is **greater** than E. Now $F \cap G$ is smaller than both F and G, so the first statement of (g) says that $F \cap G$ is the largest set included in both F and G. Likewise, the second statement says that $E \cup F$ is the smallest set which includes both E and F. To prove the first statement, let $E \subseteq F$ and $E \subseteq G$. Then by (e) and (f) above we have $E = E \cap G \subseteq F \cap G$. The second statement is proved similarly.

(h) $\boldsymbol{A \cap \complement_E(A) = \varnothing}$; $\boldsymbol{A \cup \complement_E(A) = E}$.

It is essentially from this property that the terminology "complement" comes. The statements are immediate from the fact that, for every object x, either $x \in A$ or $x \notin A$.

(i) $\boldsymbol{\complement_E(\complement_E(A)) = A}$.

This is sometimes referred to by saying that the operation of taking complements is **involutive**. The proof is immediate from the fact that, for every $x \in E$, $x \notin \complement_E(A) \Leftrightarrow x \in A$.

(j) $\boldsymbol{\complement_E(A \cap B) = \complement_E(A) \cup \complement_E(B)}$; $\boldsymbol{\complement_E(A \cup B) = \complement_E(A) \cap \complement_E(B)}$.

These properties are known as the **de Morgan laws**; note the change from $\cap$ to $\cup$ and from $\cup$ to $\cap$. To prove the first we note that, for every object x,

$$\begin{aligned} x \in \complement_E(A) \cup \complement_E(B) &\Leftrightarrow (x \in E \text{ and } x \notin A) \text{ or } (x \in E \text{ and } x \notin B) \\ &\Leftrightarrow x \in E \text{ and } (x \notin A \text{ or } x \notin B) \\ &\Leftrightarrow x \in E \text{ and } x \notin A \cap B \\ &\Leftrightarrow x \in \complement_E(A \cap B). \end{aligned}$$

The first therefore follows from the definition of equality for sets. The second may be established similarly. Alternatively, the second may be deduced from the first by using (i).

(k) $A \subseteq B \Leftrightarrow \complement_E(B) \subseteq \complement_E(A)$.

This follows from the following observations:

$$\begin{aligned} A \subseteq B &\Leftrightarrow A \cap B = A && \text{[by (f)]} \\ &\Leftrightarrow \complement_E(A \cap B) = \complement_E(A) && \text{[by (i)]} \\ &\Leftrightarrow \complement_E(A) \cup \complement_E(B) = \complement_E(A) && \text{[by (j)]} \\ &\Leftrightarrow \complement_E(B) \subseteq \complement_E(A) && \text{[by (f)].} \end{aligned}$$

If we agree to write $B \supseteq A$ as an alternative to $A \subseteq B$ then (k) may be written $A \subseteq B \Leftrightarrow \complement_E(A) \supseteq \complement_E(B)$. For this reason, property (k) is often referred to by saying that the operation of taking complements is **inclusion-inverting.**

We conclude this section with some remarks on number systems. Particularly important sets in mathematics are:

(a) the set $\mathbf{N} = \{0, 1, 2, 3, \ldots\}$ of **natural numbers**;
(b) the set $\mathbf{Z} = \{\ldots, -2, -1, 0, 1, 2, \ldots\}$ of **integers**;
(c) the set **Q** of **rational numbers**;
(d) the set **R** of **real numbers**;
(e) the set **C** of **complex numbers.**

Remark. It should be noted that some mathematicians prefer not to include 0 in **N**.

The reader will be familiar with most of these sets, recognising **Q** as the set of numbers of the form m/n where m,n are integers with $n \neq 0$; **R** as the (rather vague) set of numbers we have been using since our early schooldays, in particular in measurement and in the calculus; and **C** as the (rather mysterious) set of numbers of the form $a+ib$ where $a,b \in \mathbf{R}$ and i is "something that satisfies the property $i^2 = -1$". These number systems are related by the following chain of strict inclusions:

$$\mathbf{N} \subset \mathbf{Z} \subset \mathbf{Q} \subset \mathbf{R} \subset \mathbf{C}.$$

We shall appeal to the reader's familiarity with these number systems in the exercises and illustrative examples throughout this text. Concerning the problem of just what a number is, we shall be able to use the set theory we are developing to give the reader a much deeper idea as to what a natural number is. Having laid this foundation, we shall then develop some algebraic machinery which we shall use, together with **N**, to give a construction of **Z**, **Q**, **R** and **C**.

Exercises for §1

1. Let A, B, C be the subsets of **N** given by $A=\{2, 4, 6, \ldots\}$, $B=\{1, 3, 5, \ldots\}$, $C=\{1, 2, 3, 4\}$. Describe the sets

$$C \cap A; \quad A \cup (B \cap C); \quad B \cup (A \cap C); \quad (A \cup B) \cap C.$$

2. For each positive integer n let $n\mathbf{Z}=\{nm; m \in \mathbf{Z}\}$ be the set of all integer multiples of n. Determine

$$2\mathbf{Z} \cap 7\mathbf{Z}; \quad (3\mathbf{Z} \cap 6\mathbf{Z}) \cup 18\mathbf{Z}; \quad 6\mathbf{Z} \cap 15\mathbf{Z}.$$

3. If E, F are given sets, establish the **absorption laws**

$$E \cap (E \cup F)=E=E \cup (E \cap F).$$

Hence show that if A, B, C are subsets of a set E such that $A \cap B= A \cap C$ and $A \cup B=A \cup C$ then $B=C$. By considering appropriate Venn diagrams, or otherwise, show that neither of these conditions by itself is sufficient to ensure that $B=C$.

4. Prove that $\{n \in \mathbf{N}; n \text{ is even}\}=\{n \in \mathbf{N}; n^2 \text{ is even}\}$.

[*Hint*. For the implication $\Leftarrow$ show that if $n \in \mathbf{N}$ is odd then n^2 is also odd.]

5. Let Π be a plane, let A, B, C be three points of Π and let $r \in \mathbf{R}$. Describe the sets

(a) $\{P \in \Pi; PA=r\}$;
(b) $\{P \in \Pi; PA=PB\}$;
(c) $\{P \in \Pi; PA=PB=PC\}$;
(d) $\{P \in \Pi; PA+PB=2r\}$.

[Make sure that your answers include all possible cases.]

6. Let Σ be three-dimensional space, let A, B be two points of Σ and let $r \in \mathbf{R}$. Describe the sets

(a) $\{P \in \Sigma; PA \leqslant r\}$;
(b) $\{P \in \Sigma; \text{angle } PAB=90°\}$;
(c) $\{P \in \Sigma; \text{angle } PAB=60°\}$;
(d) $\{P \in \Sigma; \text{angle } APB=90°\}$.

7. List the elements of the set

$$\{x \in X; (x+5)(x+1)(x-2)(3x-7)(x^2-2)(x^2+1)=0\}$$

for the cases where X is **C**, **R**, **Q**, **Z**, **N**.

8. For a finite set E (i.e., a set having only a finite number of elements) let $\#(E)$ denote the number of elements in E. Prove that if A, B are finite sets then $\#(A \cup B) = \#(A) + \#(B) - \#(A \cap B)$. Deduce that if A, B, C are finite sets then $\#(A \cup B \cup C) = \#(A) + \#(B) + \#(C) - \#(A \cap B) - \#(A \cap C) - \#(B \cap C) + \#(A \cap B \cap C)$.
[*Hint*. $A \cup B \cup C = (A \cup B) \cup C$.]

9. In a certain examination, all the candidates attempted at least one of questions 1, 2, 3. Question 1 was attempted by forty candidates, question 2 by forty-seven and question 3 by thirty-one. Nine candidates attempted at least questions 1 and 2, fifteen attempted at least questions 1 and 3, eleven attempted at least questions 2 and 3, and six candidates attempted all three questions. How many candidates wrote the examination?

[*Hint*. Use exercise 8.]

10. In a group of 75 students, each of whom studied at least one of the subjects mathematics, physics and chemistry, 40 studied mathematics, 60 studied physics and 25 studied chemistry. Only 5 studied all three. Show that
(a) at least 25 studied mathematics and physics;
(b) at least 10 studied physics and chemistry;
(c) at most 20 studied mathematics and chemistry.

11. Let A, B, C be subsets of a set E. Prove that
(a) $A \cup B = E \Leftrightarrow \complement_E(A) \subseteq B$;
(b) $A \cap B = \emptyset \Leftrightarrow B \subseteq \complement_E(A)$;
(c) $(A \cup B = E \text{ and } A \cap B = \emptyset) \Leftrightarrow B = \complement_E(A)$;
(d) $(A \cap B) \cup C = A \cap (B \cup C) \Leftrightarrow C \subseteq A$;
(e) $(A \cap C) \cup (B \cap \complement_E(C)) = \emptyset \Leftrightarrow B \subseteq C \subseteq \complement_E(A)$.

12. If A, B, C are subsets of a set E, simplify

$$(A \cap B) \cup (C \cap A) \cup \complement_E(\complement_E(A) \cup \complement_E(B)).$$

13. Let E, F, G be sets. Prove that

$$(E \cap F) \cup (F \cap G) \cup (G \cap E) = (E \cup F) \cap (F \cup G) \cap (G \cup E).$$

14. Given subsets A, B of a set E define the **symmetric difference set** $A \triangle B$ by $A \triangle B = A \backslash B \cup B \backslash A$. Show that $A \triangle B = (A \cup B) \backslash (A \cap B)$ and illustrate by a Venn diagram. Show that $\triangle$ is commutative $[A \triangle B = B \triangle A]$ and associative $[A \triangle (B \triangle C) = (A \triangle B) \triangle C]$. Show also that $\cap$ is distributive over $\triangle$ $[A \cap (B \triangle C) = (A \cap B) \triangle (A \cap C)]$.

15. If A, B, C are subsets of a set E prove that

$$(A \backslash B) \backslash C \subseteq A \backslash (B \backslash C).$$

Find a necessary and sufficient condition for this inclusion to be equality and give an example in which the inclusion is strict.

16. If A, B are subsets of a set E define $A \mid B = \complement_E(A \cap B)$. Show that $\complement_E(A) = A \mid A$ and that $A \cap B = (A \mid B) \mid (A \mid B)$. Express $\cup$ in terms of $\mid$ alone.

§2. Sets of sets

Very often in mathematics one has to deal with **sets of sets**; that is, sets whose elements are themselves sets. For example, given a set E, the set which consists of all the subsets of E is called the **power set** of E and is denoted by $\mathbf{P}(E)$. In particular, if $E = \{x, y, z\}$ then we have

$$\mathbf{P}(E) = \{\emptyset, \{x\}, \{y\}, \{z\}, \{x, y\}, \{x, z\}, \{y, z\}, E\}.$$

In situations such as this, one has to be careful how one uses the symbol $\in$. For example, if $x \in A \subseteq E$ then we have $x \in A$ and $A \in \mathbf{P}(E)$. As a consequence we have $\{x\} \subseteq A$ so that $\{x\} \in \mathbf{P}(A) \subseteq \mathbf{P}(E)$, and $\{A\} \subseteq \mathbf{P}(E)$ so that $\{A\} \in \mathbf{P}(\mathbf{P}(E))$. Care should be taken to note that for every object x we have $x \in \{x\}$ and $x \neq \{x\}$. In particular, we have $\emptyset \in \{\emptyset\}$ and $\emptyset \neq \{\emptyset\}$; in other words, the set consisting of the empty set is not empty. [If we had used the word "bag" instead of "set", this would read "a bag containing an empty bag is not empty"!]

> *Remark*. Despite one's intuitive inclination to reject the possibility, it *is* indeed possible, given a set A and an object x, to have both $x \in A$ and $x \subseteq A$; in other words, it is possible for an element of a set to be a subset of the set. By way of illustration, let $A = \{1, 2, \{1, 2\}\}$; then clearly $\{1, 2\} \in A$ and $\{1, 2\} \subseteq A$. Note also that $\complement_A\{1, 2\} = \{\{1, 2\}\}$ is both an element and a subset of $\mathbf{P}(A)$.

Our aim in this section is to point out a paradoxical difficulty which can arise from too naive an approach to set theory. This is not a serious problem since fortunately most sets encountered in mathematics are of a harmless nature. Without complicating the issue unduly, let us show that it is undesirable to talk of the "*set* of all sets". In fact, if such a *set* U exists then, being a set, it must be an element of itself: $U \in U$. Consider now the subset V of U which consists of all sets which are not elements of themselves (this was first considered by Bertrand Russell in 1901). If we suppose that $V \notin V$ then V is not an element of itself and so, by the definition of V, we must have $V \in V$. On the other hand, if we suppose that $V \in V$ then, again from the definition of V, we have $V \notin V$. How does one avoid such an impasse? The only really satisfactory way is to revert to the system of axioms on which the theory of sets is founded and include in the system an axiom which will remove the possibility of such impasses arising. One way of helping the situation is to restrict the idea of a set in such a way that "very large collections" such as V are not counted as sets. It is very fortunate that most of the properties dealt with in mathematics (and indeed all the properties we shall deal with) are "set forming" in the sense that there is a set whose elements satisfy the property in question. Matters lying outside this framework belong properly to the realm of the mathematical logician.

Exercises for §2

1. Given the sets $A=\{1, 2\}$, $B=\{1, \{2\}\}$, $C=\{\{1\}, \{2\}\}$ and $D=\{\{1\}, \{2\}, \{1, 2\}\}$, determine the sets

$$A \cap C; \quad (B \cap D) \cup A; \quad (A \cap B) \cup D; \quad (A \cap B) \cup (C \cap D).$$

2. Determine the power set of each of the following sets:

$$\emptyset; \quad \{\emptyset\}; \quad \{\emptyset, \{\emptyset\}\}; \quad \{\emptyset, \{\emptyset\}, \{\{\emptyset\}\}\}.$$

3. Construct an infinite set, *every* element of which is a subset.

[*Hint*. Look at the sets in Exercise 2.]

4. For the set $E=\{1, 2, \{1, 2\}\}$ determine $\mathbf{P}(E)$ and $E \cap \mathbf{P}(E)$.

5. If E and F are given sets, show that

(a) $\mathbf{P}(E \cap F)=\mathbf{P}(E) \cap \mathbf{P}(F)$;
(b) $\mathbf{P}(E \cup F) \supseteq \mathbf{P}(E) \cup \mathbf{P}(F)$.

Give an example to show that the inclusion in (b) can be strict.

§3. Ordered pairs; cartesian product sets

Given any two objects x, y we define the **ordered pair** (x, y) by

$$(x, y)=\{\{x\}, \{x, y\}\}.$$

The most important (and, indeed, the only) property of ordered pairs is given in the following result.

Theorem 3.1 *The ordered pairs* (x, y) *and* (x^*, y^*) *are equal if and only if* $x=x^*$ *and* $y=y^*$.

Proof. Suppose first that $y \neq x$ and $y^* \neq x^*$. Then we have

$$\begin{aligned}(x, y)=(x^*, y^*) &\Leftrightarrow \{\{x\}, \{x, y\}\}=\{\{x^*\}, \{x^*, y^*\}\}\\ &\Leftrightarrow \{x\}=\{x^*\} \text{ and } \{x, y\}=\{x^*, y^*\}\\ &\Leftrightarrow x=x^* \text{ and } y=y^*,\end{aligned}$$

so the result holds in this case.

Suppose now that $y=x$. Then we have

$$(x, y)=(x, x)=\{\{x\}, \{x, x\}\}=\{\{x\}, \{x\}\}=\{\{x\}\}$$

and so, in this case,

$$\begin{aligned}(x, y)=(x^*, y^*) &\Leftrightarrow \{\{x\}\}=\{\{x^*\}, \{x^*, y^*\}\}\\ &\Leftrightarrow \{x\}=\{x^*\}=\{x^*, y^*\}\\ &\Leftrightarrow x^*=y^*=x\ [=y].\end{aligned}$$

This therefore establishes the result in this case. The only other case to consider is the case where $y^*=x^*$; this is similar to the case $y=x$.

Corollary. *If* $x \neq y$ *then* $(x, y) \neq (y, x)$.

Remark. What is important in the above result is not the way an ordered pair is defined but rather the result itself; indeed, all we require is a construction of (x, y) from x and y such that Theorem 3.1 holds and the above definition gives such a construction.

In the ordered pair (x, y) we call x the **first coordinate** and y the **second coordinate.** The motivation for this is a geometric one which we shall now see.

If E, F are sets then we define the **cartesian product set** $E \times F$ by

$$E \times F=\{(x, y);\ x \in E,\ y \in F\}.$$

It is clear that $E \times \emptyset=\emptyset=\emptyset \times E$ for every set E and that if E and F are non-empty sets then $E \times F \neq F \times E$ unless $E=F$. We represent the cartesian product set $E \times F$ pictorially as follows:

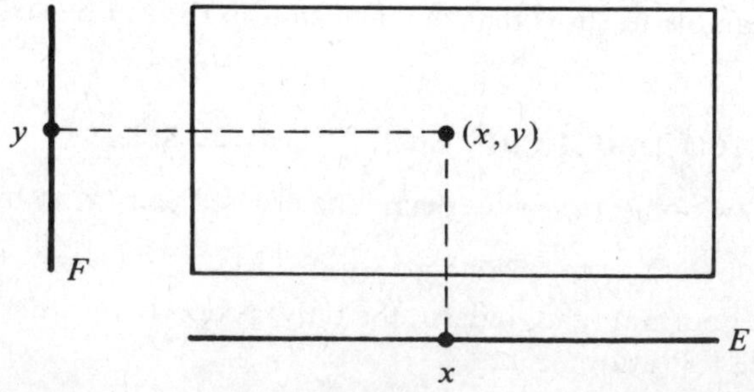

For the geometric motivation behind the above notion, consider the set **R** of real numbers: $\mathbf{R} \times \mathbf{R}$ is the set consisting of all ordered pairs of real numbers; i.e., the set which represents the plane in cartesian geometry. Thus, for example, plane curves can be regarded as subsets of $\mathbf{R} \times \mathbf{R}$. By way of illustration, the set

$$E = \{(x, y) \in \mathbf{R} \times \mathbf{R};\ y^2 = x^2\}$$

gives an exact description of what is loosely called "the line pair $y^2 = x^2$". The corresponding diagram for this set is

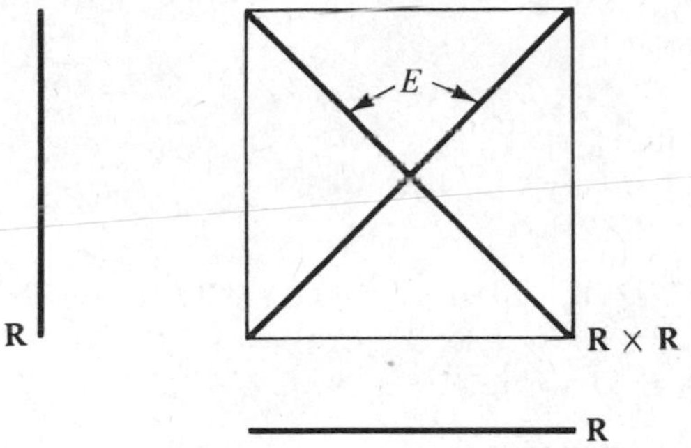

As we shall see in the sections to follow, the notion of cartesian product set is of fundamental importance.

Using the concept of an ordered pair we can define an ordered triple: given any three objects x, y, z we define the **ordered triple** (x, y, z) by

$$(x, y, z) = ((x, y), z).$$

In other words, it is an ordered pair, the first coordinate of which is an ordered pair. Given sets E, F, G we define

$$E \times F \times G = \{(x, y, z);\ x \in E,\ y \in F,\ z \in G\}.$$

It is clear that in this way we can represent three-dimensional space by $\mathbf{R} \times \mathbf{R} \times \mathbf{R}$.

A word of caution at this point: our notation (a, b) for an ordered pair, although standard, should not be confused with a similar notation often used to represent the open interval of real numbers $\{x \in \mathbf{R};\ a < x < b\}$. We avoid such a clash of notation by defining, once and for all, the following subsets of **R**:

$$[a, b] = \{x \in \mathbf{R};\ a \leqslant x \leqslant b\};$$
$$]a, b[= \{x \in \mathbf{R};\ a < x < b\};$$
$$[a, b[= \{x \in \mathbf{R};\ a \leqslant x < b\};$$
$$]a, b] = \{x \in \mathbf{R};\ a < x \leqslant b\}.$$

For convenience we also introduce the following notation:

$$[a, \rightarrow] = \{x \in \mathbf{R};\ a \leqslant x\};$$
$$]a, \rightarrow] = \{x \in \mathbf{R};\ a < x\};$$
$$[\leftarrow, a] = \{x \in \mathbf{R};\ x \leqslant a\};$$
$$[\leftarrow, a[= \{x \in \mathbf{R};\ x < a\}.$$

Exercises for §3

1. Mark on a coordinate plane the following subsets of $\mathbf{R} \times \mathbf{R}$:

(1) $\{(x, y);\ x=0\}$;
(2) $\{(x, y);\ x>y\}$;
(3) $\{(x, y);\ x^2+y^2 \leqslant 1\}$;
(4) $\{(x, y);\ x \in [0, 1], y \in [2, 3]\}$;
(5) $\{(x, y);\ x \in [\leftarrow, 1], y \notin [2, 4[\}$;
(6) $\{(x, y);\ x^2+y^2 \geqslant 25\} \cap \{(x, y);\ 4y \geqslant 3x\}$.

2. Let A, B, C, D be given by $A=\{x \in \mathbf{R};\ 1<x \leqslant 2\}$, $B=\{x \in \mathbf{R};\ 1 \leqslant x<2$ or $x=3\}$, $C=\{x \in \mathbf{R};\ 2 \leqslant x$ or $x=5\}$, $D=\{0, 3, 5\}$. Determine the sixteen sets $E \times F$ where $E, F \in \{A, B, C, D\}$.

3. If A, B, C are sets prove that

(1) $A \times (B \cap C)=(A \times B) \cap (A \times C)$;
(2) $A \times (B \cup C)=(A \times B) \cup (A \times C)$;
(3) $(A \times B) \cap (B \times A)=(A \cap B) \times (A \cap B)$;
(4) $(A \cup B) \times (C \cup D)=(A \times C) \cup (A \times D) \cup (B \times C) \cup (B \times D)$.

4. Prove that for any objects x, y, z, x^*, y^*, z^*,

$$(x, y, z)=(x^*, y^*, z^*) \Leftrightarrow x=x^*, y=y^*, z=z^*.$$

5. Prove that for any sets A and B,

$$A \times B=B \times A \Leftrightarrow A=\emptyset,\ B=\emptyset \text{ or } A=B.$$

6. Prove that, for every object x, $\{x\} \times \{x\}=\{\{\{x\}\}\}$.

§4. Relations; functional relations; mappings

Definition. Let E and F be sets. Then by a **relation between** E **and** F we shall mean a pair $(E \times F, S(x, y))$ whose first coordinate is $E \times F$ and whose second coordinate is an open sentence $S(x, y)$. The subset G of $E \times F$ consisting of those pairs $(a, b) \in E \times F$ for which $S(a, b)$ is true is written $G=\{(a, b) \in E \times F;\ S(a, b)\}$ and is called the **graph** of the relation.

Example 4.1. By way of illustration, let $A=\{1, 2, 3, 4\}$ and $B=\{1, 2, 3\}$. Then in the diagram

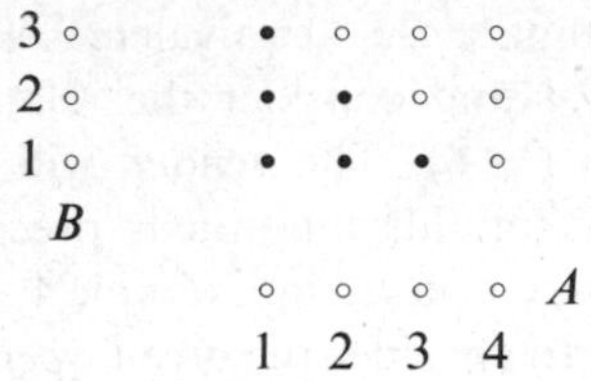

which represents $A \times B$, the set of pairs marked • forms a subset G which may be described by

$$G=\{(x, y) \in A \times B;\ x+y \leqslant 4\}.$$

Thus $(A \times B,\ x+y \leqslant 4)$ is a relation between A and B with graph G.

Example 4.2. Given any set E, consider the relation $(E \times E,\ x=y)$. This is called the **relation of equality** on E. Its graph is the set

$$D=\{(x, y) \in E \times E;\ x=y\}=\{(x, x);\ x \in E\}.$$

For obvious reasons, this is called the **diagonal** of $E \times E$:

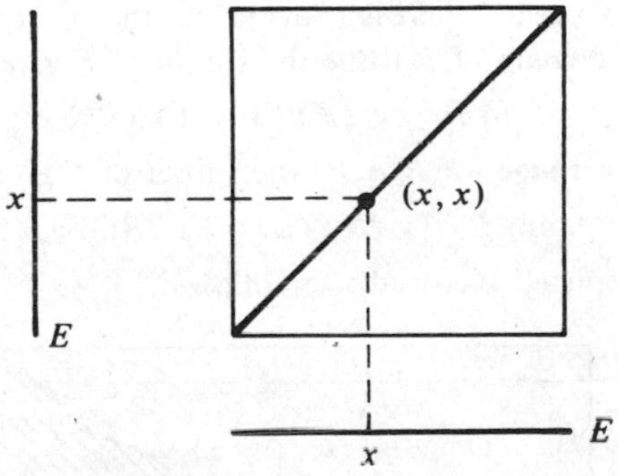

Example 4.3. $(\mathbf{R} \times \mathbf{R};\ x$ is a cow) is a relation between $\mathbf{R}$ and $\mathbf{R}$ whose graph is $\emptyset$.

If E, F are sets and R denotes a relation between E and F with associated graph G then we shall read the statement $(x, y) \in G$ as "x and y satisfy the relation R" which we shall henceforth abbreviate to xRy.

It should be noted that although we have defined a relation as an ordered pair $(E \times F,\ S(x, y))$ we cannot use Theorem 3.1 to define the notion of *equality* for relations, for we have no way of expressing what we mean by equality for open sentences. Despite this, however, we do regard (in an intuitive way) the open sentences $x+y \leqslant 4$ and $x(y+1) \leqslant 6$, for example, as being "different". On the other hand, there may well be a specific situation where we might wish to consider these two particular open sentences as, in some sense, "equivalent". For example, let A and B be as in Example 4.1 and consider the relation between A and B given by $(A \times B,\ x(y+1) \leqslant 6)$. The reader will have no difficulty in verifying that the graph of this relation is precisely the graph of the relation $(A \times B,\ x+y \leqslant 4)$ discussed in Example 4.1; so here is a situation in which we would like to consider the given open sentences as "equivalent". We obviate these difficulties by defining the notion of equivalence for *relations* as follows.

Definition. Let $(E \times F,\ S(x, y))$ and $(A \times B,\ T(x, y))$ be given relations. Then we say that these relations are **equivalent** if

(1) $E = A$; (2) $F = B$; (3) their graphs are equal.

Given a relation R between the sets E and F, we shall now introduce important subsets of E and F. To facilitate our presentation, we shall make use of the symbol $\exists$ (called the **existential quantifier**) which we shall take as an abbreviation for "there exists". Thus, for example, $(\exists x \in \mathbf{N})\ x+2=5$ is read "there is a natural number x such that $x+2=5$".

We define the **domain** of R to be the subset of E given by

$$\text{Dom } R = \{x \in E;\ (\exists y \in F)\ xRy\}$$

and the **image** (or **range**) of R to be the subset of F given by

$$\text{Im } R = \{y \in F;\ (\exists x \in E)\ xRy\}.$$

These sets are depicted pictorially as follows:

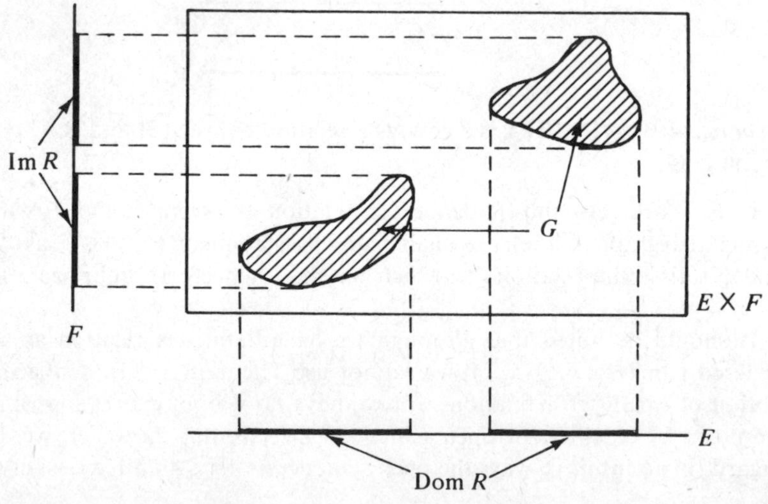

We note that if R_1 and R_2 are relations between E and F such that R_1 and R_2 are equivalent according to the above definition then the fact that their graphs are equal implies that Dom R_1 = Dom R_2 and Im R_1 = Im R_2. The converse of this is not true in general; in other words, if R_1 and R_2 are relations between E and F such that Dom R_1 = Dom R_2 and Im R_1 = Im R_2 then R_1 and R_2 are not equivalent in general. For example, the relations

$$R_1 = (\mathbf{R} \times \mathbf{R},\ x^2 + y^2 \leqslant 1)$$

$$R_2 = (\mathbf{R} \times \mathbf{R},\ x \in [-1, 1] \text{ and } y \in [-1, 1])$$

are such that Dom R_1 = Dom $R_2 = \{(x, 0);\ x \in [-1, 1]\}$ and Im R_1 = Im $R_2 = \{(0, y);\ y \in [-1, 1]\}$ but they are not equivalent since the graph of R_1 is the unit disc whereas that of R_2 is the square region $[-1, 1] \times [-1, 1]$.

There are many important types of relation. We shall devote the rest of this section to studying one of these and shall meet others in subsequent sections.

Definition. Let E and F be sets. A relation R between E and F is said to be **functional** if, for every $x \in$ Dom R, the set $\{y \in F;\ xRy\}$ is a singleton.

The graph corresponding to a functional relation is therefore of the following typical form.

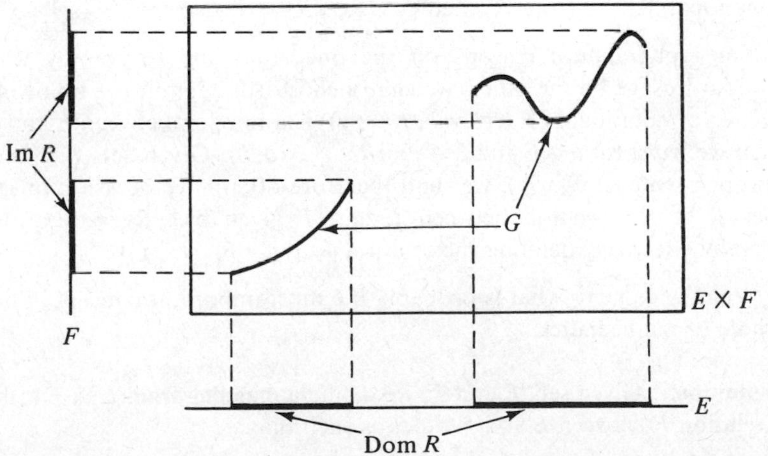

Example 4.4. The relation described in Example 4.1 is not functional.

For example, we have $2 \in \operatorname{Dom} R$ and

$$\{y \in B;\ 2Ry\} = \{y \in B;\ 2+y \leqslant 4\} = \{1, 2\},$$

which is not a singleton.

Example 4.5. The relation of equality in Example 4.2 is clearly a functional relation.

Example 4.6. Consider the relation $R = (\mathbf{R} \times \mathbf{R},\ x^2 + y^2 = 1 \text{ and } y \geqslant 0)$. The graph of R is depicted in the following diagram (the picture on the right is the abbreviated sketch normally given):

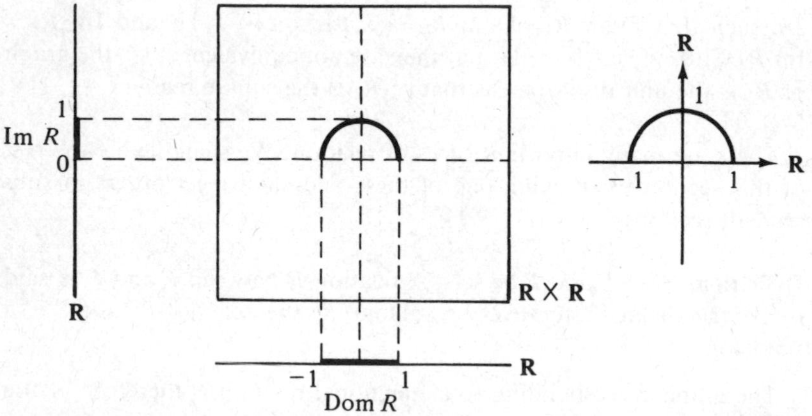

It is clear that $\operatorname{Dom} R = [-1, 1]$ and $\operatorname{Im} R = [0, 1]$. To see that R is functional, let $x \in \operatorname{Dom} R$; then $\{y \in \mathbf{R};\ xRy\} = \{z\}$ where $z = \sqrt{(1 - x^2)}$.

For typographical reasons on the one hand and to comply with standard usage on the other, we shall henceforth commit the following abuses of notation. If $R = (E \times F,\ S(x, y))$ is a relation between E and F then we have, for $a \in E$ and $b \in F$, $aRb \Leftrightarrow S(a, b)$. Given sets E, F and an open sentence $S(x, y)$, we shall therefore often have occasion to say "let R be the relation between E and F given by $xRy \Leftrightarrow S(x, y)$", thereby effectively defining the relation $R = (E \times F,\ S(x, y))$.

We now come to what is probably the most important concept in the whole of mathematics.

Definition. Given sets E and F, we define a **mapping from** E **to** F to be a relation R between E and F which is such that

(1) $\operatorname{Dom} R = E$;
(2) R is functional.

Note that these conditions may be re-stated in the form of the single condition

(3) for every $x \in E$ there is a unique (i.e., precisely one) $y \in F$ such that xRy.

Example 4.7. For any set E the relation of equality on E (see Example 4.2) is functional and has domain E. Hence it is a mapping from E to E, called the **identity mapping** on E.

Example 4.8. For every $r \in \mathbf{R}$ let $[r]$ denote the greatest integer which is less than or equal to r; i.e. $[r]$ is defined by the conditions $[r] \in \mathbf{Z}$ and $[r] \leqslant r < [r]+1$. Consider the relation between $\mathbf{R}$ and $\mathbf{Z}$ given by

$$xRy \Leftrightarrow [x]=y.$$

It is readily seen that the graph of the relation R is as follows (in which the sign •→ indicates a constant value throughout an interval of the form $[n, n+1[$):

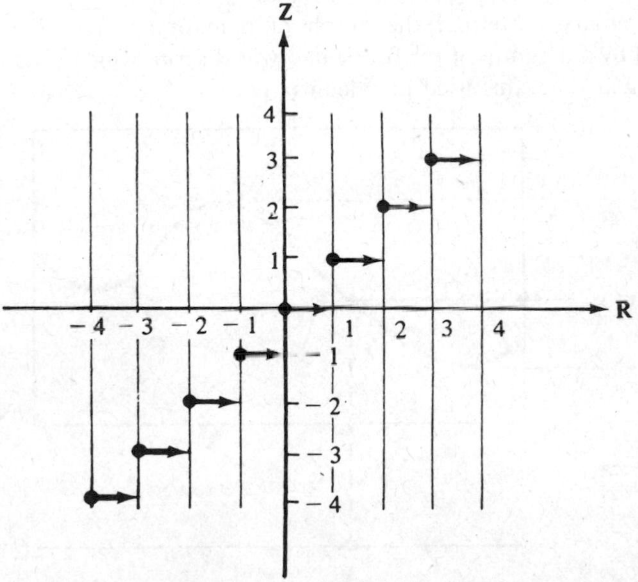

It is clear from this graph that Dom $R=\mathbf{R}$. Moreover, R is functional since, given any $t \in \mathbf{R}$, we have $tR[t]$ and

$$\{y \in \mathbf{Z};\ tRy\} = \{y \in \mathbf{Z};\ y=[t]\} = \{[t]\},$$

so that for each $t \in \mathbf{R}$ there is a unique $y \in \mathbf{Z}$ (namely $y = [t]$) such that tRy. Hence R is a mapping from $\mathbf{R}$ to $\mathbf{Z}$.

It is customary practice to denote mappings by lower case letters and we shall do so henceforth. There is also a universally accepted notation for mappings: in describing a mapping f from a set E to a set F we write $f: E \to F$ or $E \xrightarrow{f} F$. The unique $y \in F$ which corresponds to a given $x \in E$ [see (3), above] is called the **image of** x **under** f and will be denoted by $f(x)$. With this notation, the graph of f is then given by

$$G = \{(x, y) \in E \times F;\ y = f(x)\}.$$

For a mapping $f: E \to F$ the set E is called the **domain** of f and F is called the **codomain** of f. We shall sometimes call E the **departure set** and F the **arrival set** of f. When no confusion can arise over E and F we shall often write a given mapping $f: E \to F$ in the form $x \mapsto f(x)$.

For example, the identity mapping on E can be written as $f: E \to E$ where $f(x) = x$ for all $x \in E$. This mapping is of especial importance; we usually denote it by id_E. Likewise, in Example 4.8, we have a mapping $f: \mathbf{R} \to \mathbf{Z}$ given by $f(r) = [r]$ for every $r \in \mathbf{R}$.

In summary therefore, the graph of a mapping $f: E \to F$ may be depicted by a diagram of the following typical form, which is consistent with similar diagrams used in calculus.

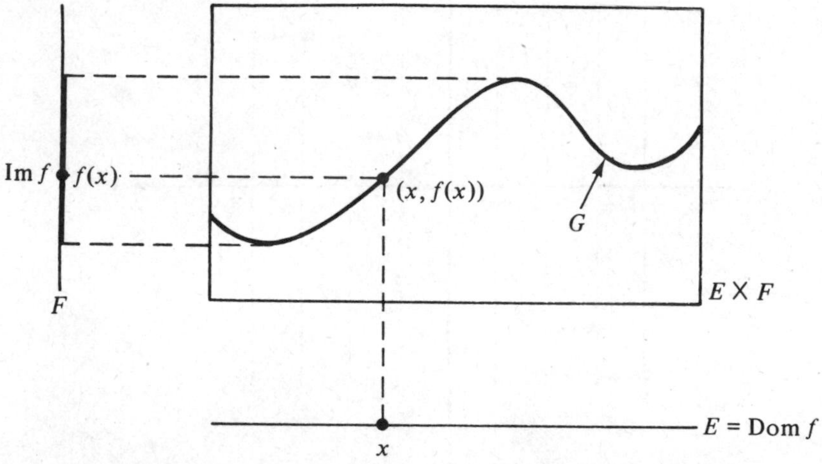

In what follows we shall make use of the symbol $\forall$ (called the **universal quantifier**) which we shall take as an abbreviation for "for all" or "for every". For example, the above mapping $f: \mathbf{R} \to \mathbf{Z}$ can be described by $(\forall r \in R)\ f(r) = [r]$.

Since mappings are in particular relations we have that given mappings $f: E \to F$ and $g: A \to B$ are equivalent if

(1) $E = A$;
(2) $F = B$;
(3) the graphs of f, g are equal.

Now it is clear that, in this case, if (3) holds then (1) holds and, for every element x of the common domain, $f(x) = g(x)$. Conversely, if (1) holds and $f(x) = g(x)$ for every element x of the common domain then it is equally clear that (3) holds. We can therefore replace the above conditions by the equivalent set of conditions

(a) $E = A$;
(b) $F = B$;
(c) $(\forall x \in E)\, f(x) = g(x)$.

We shall henceforth therefore agree to say that the mappings $f: E \to F$ and $g: A \to B$ are **equal** if (a), (b), (c) hold.

In connection with mappings, the reader should note carefully the following remarks.

Remark 1. The word *function* is often used instead of mapping [especially in analysis, where image sets are usually subsets of **R** or of **C** and $f(x)$ is often called the *value* of f at the point x]. However, care should be taken to avoid such inaccuracies as "the function $f(x)$"; in fact, if $f: E \to F$ then $f(x) \in F$ for every $x \in E$ so that $f(x)$ is a function only if $\operatorname{Im} f$ is a set of functions. Thus, for example, we should not talk about "the function $\cos x$" but rather "the function cos" or even "the function $x \mapsto \cos x$" though in this last case we should really also specify the departure set and the arrival set.

Remark 2. Many authors identify a mapping with its graph; i.e. they define a mapping $f: E \to F$ to be a subset of $E \times F$ such that for every $x \in E$ there is a unique $y \in F$ such that $(x, y) \in f$. We shall not adopt this practice since, as we shall see later, it can lead to confusion.

Remark 3. Many authors write their mappings on the right; i.e., where we write $f(x)$ they write xf. We cannot say that we shall definitely not do this, for we often do so without thinking! For example, consider the mapping $\complement_E: \mathbf{P}(E) \to \mathbf{P}(E)$ given by $A \mapsto \complement_E(A)$; here we have written the mapping on the left. But, when the set E with respect to which we are taking complements is always the same, we often economise in writing by letting A' denote the complement of A in E. In so doing we have a mapping $':\mathbf{P}(E)\to\mathbf{P}(E)$ given by $A\mapsto A'$; and here the mapping appears on the right!

Remark 4. Note that in the definition of equality for mappings we insist that the codomains (arrival sets) be equal. Thus, for example, if $f:E\to F$ is a mapping described by $x\mapsto f(x)$ then we distinguish between this mapping and the mapping $g:E\to\operatorname{Im} f$ also described by $x\mapsto f(x)$. In particular, we distinguish between the mapping $f:\mathbf{Z}\to\mathbf{Z}$ described by $n\mapsto n^2$ and the mapping $g:\mathbf{Z}\to\mathbf{N}$ also described by $n\mapsto n^2$. At first sight this may appear somewhat pedantic but we shall see later that not to do so can lead to confusion.

Exercises for §4

1. For every $x\in\mathbf{R}$ define

$$|x|=\begin{cases} x & \text{if } x\geqslant 0;\\ -x & \text{if } x\leqslant 0.\end{cases}$$

Show that the relation between $\mathbf{R}$ and $\mathbf{R}$ defined by $xRy \Leftrightarrow y=|x|$ is functional and determine its domain and image.

2. Consider in turn each of the following relations between $\mathbf{R}$ and $\mathbf{R}$. Determine their domains and images and say which are functional:

(a) $x^2<y$; (b) $\sin x=y$; (c) $x=\sin y$;

(d) $x^3=y$; (e) $x^2=y^2$; (f) $e^x=y$;

(g) $\log x=y$; (h) $x^2+y^2=1$; (i) $x^2-y^2=1$;

(j) $xy=1$; (k) $y=\dfrac{x+3}{x-4}$; (l) $x^2y=1$;

(m) $|y|\leqslant|x|\leqslant 1$; (n) $y=|x|-[x]$; (o) $x+y=1$;

(p) $|x|+y=1$; (q) $x+|y|=1$; (r) $|x|+|y|=1$;

(s) $x-2\leqslant y\leqslant x+1$; (t) $1/(x^2-y^2)\in\mathbf{N}$;

(u) $y=\begin{cases}1 & \text{if } x\in\mathbf{N},\\ 0 & \text{if } x\in\mathbf{R}\backslash\mathbf{N};\end{cases}$ (v) $y=\begin{cases}x & \text{if } x\in\mathbf{Q},\\ -x & \text{if } x\in\mathbf{R}\backslash\mathbf{Q}.\end{cases}$

3. Give an example of mappings $f, g:\mathbf{R}\to\mathbf{R}$ such that $\operatorname{Im} f=\operatorname{Im} g$ but $f\neq g$.

4. Let E, F be sets and let R be a relation between E and F. Define the **converse** of R to be the relation R^t between F and E given by

$$yR^tx \Leftrightarrow xRy.$$

Determine the domain, image and graph of R^t in terms of those of R.

5. Draw the graph of the mapping $f:\mathbf{R}\to\mathbf{R}$ given by

$$f(x)=\begin{cases}2x-1 & \text{if } x\geqslant 0,\\ x+1 & \text{if } x<0.\end{cases}$$

6. Illustrate on a sketch the subset S of $\mathbf{R}\times\mathbf{R}$ given by

$$S=\{(x,y);\ |x|+|y|\geqslant 1 \text{ and } x^2+y^2\leqslant 1\}.$$

If the mapping $f:S\to\mathbf{R}$ is given by the prescription

$$f(x,y)=\sqrt{(x^2+y^2)},$$

what is $\operatorname{Im} f$?

§5. Induced mappings; composition; injections; surjections; bijections

In this section we shall discuss a variety of important types of mapping. We begin by asking the reader to recall the definition of $\mathbf{P}(E)$ as the set of all subsets of E. We shall show that any mapping $f:E\to F$ gives rise, in a natural way, to mappings from $\mathbf{P}(E)$ to $\mathbf{P}(F)$ and from $\mathbf{P}(F)$ to $\mathbf{P}(E)$. Given $f:E\to F$, let us consider, for every subset X of E, the subset $f^{\to}(X)$ of F given by

$$f^{\to}(X)=\begin{cases}\{f(x);\ x\in X\} & \text{if } X\neq\varnothing;\\ \varnothing & \text{if } X=\varnothing.\end{cases}$$

It is clear that the assignment $X\mapsto f^{\to}(X)$ yields a mapping from $\mathbf{P}(E)$ to $\mathbf{P}(F)$. We denote this mapping by $f^{\to}$.

> *Remark*. It should be noted that most mathematicians write $f(X)$ for $f^{\to}(X)$. Now it is quite clear that the mappings f and $f^{\to}$ are not the same, so we shall retain the notation $f^{\to}$ to avoid confusion. However, in order to agree with standard terminology, we make the following definition. We call $f^{\to}(X)$, which is the image of X under $f^{\to}$, the **direct image of** X **under** f. Note in particular that if $f:E\to F$ then $f^{\to}(E)=\operatorname{Im} f$.

Given $f:E\to F$, consider now, for every subset Y of F, the subset

$f^{\leftarrow}(Y)$ of E given by

$$f^{\leftarrow}(Y)=\begin{cases}\{x\in E; f(x)\in Y\} & \text{if } \operatorname{Im} f\cap Y\neq\emptyset;\\ \emptyset & \text{if } \operatorname{Im} f\cap Y=\emptyset.\end{cases}$$

It is clear that the assignment $Y\mapsto f^{\leftarrow}(Y)$ defines a mapping from $\mathbf{P}(F)$ to $\mathbf{P}(E)$. We denote this mapping by $f^{\leftarrow}$.

> *Remark.* Most mathematicians write $f^{-1}(Y)$ instead of $f^{\leftarrow}(Y)$. As this is often confused with a notation which we shall meet later, we shall retain the notation $f^{\leftarrow}$ to avoid confusion. For every $Y\in\mathbf{P}(F)$ we call $f^{\leftarrow}(Y)$ the **pre-image of** Y **under** f.

An important point to note about the mappings $f^{\rightarrow}$ and $f^{\leftarrow}$ is that, although $f^{\rightarrow}(X)=\emptyset$ if and only if $X=\emptyset$, it is possible to have $f^{\leftarrow}(Y)=\emptyset$ with $Y\neq\emptyset$; for example, $\sin^{\leftarrow}\{2\}=\emptyset$.

We shall say that the mappings $f^{\rightarrow}$ and $f^{\leftarrow}$ are the mappings which are **induced** on the power sets by the mapping f. Immediate properties of these induced mappings are given in the following result.

Theorem 5.1 *If f is a mapping from a set E to a set F then, for all subsets E_1, E_2 of E and all subsets F_1, F_2 of F,*

(i) $E_1\subseteq E_2\Rightarrow f^{\rightarrow}(E_1)\subseteq f^{\rightarrow}(E_2)$; (j) $F_1\subseteq F_2\Rightarrow f^{\leftarrow}(F_1)\subseteq f^{\leftarrow}(F_2)$;

(ii) $f^{\rightarrow}(E_1\cup E_2)=f^{\rightarrow}(E_1)\cup f^{\rightarrow}(E_2)$; (jj) $f^{\leftarrow}(F_1\cup F_2)=f^{\leftarrow}(F_1)\cup f^{\leftarrow}(F_2)$;

(iii) $f^{\rightarrow}(E_1\cap E_2)\subseteq f^{\rightarrow}(E_1)\cap f^{\rightarrow}(E_2)$; (jjj) $f^{\leftarrow}(F_1\cap F_2)=f^{\leftarrow}(F_1)\cap f^{\leftarrow}(F_2)$;

(iv) $f^{\rightarrow}(\complement_E(E_1))\supseteq\complement_{\operatorname{Im} f}(f^{\rightarrow}(E_1))$; (jw) $f^{\leftarrow}(\complement_F(F_1))=\complement_E(f^{\leftarrow}(F_1))$;

(v) $f^{\rightarrow}(E_1)\subseteq F_1\Leftrightarrow E_1\subseteq f^{\leftarrow}(F_1)$.

Proof. (i), (j): These properties may be expressed by saying that $f^{\rightarrow}$ and $f^{\leftarrow}$ are **inclusion-preserving**. To prove (i) we observe that if $E_1\subseteq E_2$ then every object of the form $f(x)$ where $x\in E_1$ is of the form $f(x)$ where $x\in E_2$; consequently we have (i). To prove (j) we let $F_1\subseteq F_2$ and verify (j) in each of the cases (a) $F_1\cap\operatorname{Im} f=\emptyset=F_2\cap\operatorname{Im} f$; (b) $F_1\cap\operatorname{Im} f=\emptyset\neq F_2\cap\operatorname{Im} f$; (c) $F_1\cap\operatorname{Im} f\neq\emptyset\neq F_2\cap\operatorname{Im} f$. In case (a) we have $f^{\leftarrow}(F_1)=\emptyset=f^{\leftarrow}(F_2)$; in case (b) we have $f^{\leftarrow}(F_1)=\emptyset\subset f^{\leftarrow}(F_2)$; and in case (c) we have $f^{\leftarrow}(F_1)\subseteq f^{\leftarrow}(F_2)$ since, for every element x of E,

$$x\in f^{\leftarrow}(F_1)\Rightarrow f(x)\in F_1\Rightarrow f(x)\in F_2\Rightarrow x\in f^{\leftarrow}(F_2).$$

(ii), (jj): These properties may be expressed by saying that $f^{\rightarrow}$ and $f^{\leftarrow}$ are **$\cup$-preserving**. To prove (ii) we note first that $E_1, E_2\subseteq E_1\cup E_2$ and so from (i) we obtain $f^{\rightarrow}(E_1)\cup f^{\rightarrow}(E_2)\subseteq f^{\rightarrow}(E_1\cup E_2)$. For the reverse inclusion, it suffices to observe that every object of the form $f(x)$ with $x \in E_1 \cup E_2$ is necessarily of the form $f(x)$ with $x \in E_1$ or $x \in E_2$. The proof of (jj) is similar.

(iii), (jjj): Note that we have $\subseteq$ in (iii) and $=$ in (jjj). The proof of (iii) is immediate from (i) using the fact that $E_1 \cap E_2 \subseteq E_1, E_2$. A similar observation together with (j) gives $f^{\leftarrow}(F_1 \cap F_2) \subseteq f^{\leftarrow}(F_1) \cap f^{\leftarrow}(F_2)$. To obtain equality in (jjj), it suffices to observe that if we have $x \in f^{\leftarrow}(F_1) \cap f^{\leftarrow}(F_2)$ then $f(x) \in F_1$ and $f(x) \in F_2$ so that $f(x) \in F_1 \cap F_2$ and consequently $x \in f^{\leftarrow}(F_1 \cap F_2)$.

(iv), (jw): To establish (iv), we observe that if x is an element of the right hand side then $x = f(y)$ where $y \notin E_1$; and consequently x is an element of the left hand side. As for (jw), this follows immediately from (jj) and (jjj) on taking $F_2 = \complement_F(F_1)$ and using the fact that $f^{\leftarrow}(\emptyset) = \emptyset$ and $f^{\leftarrow}(F) = E$.

(v): This is immediate from the fact that for every $x \in E_1$ we have $f(x) \in F_1 \Leftrightarrow x \in f^{\leftarrow}(F_1)$.

Remark. Note that equality does not hold in general in (iii) and (iv); see Exercises 5.1, 5.2 and 5.3.

We shall now investigate how some mappings can be combined to form new mappings; in particular, this will explain the awful term "function of a function" found in older books on analysis.

Suppose that we have a diagram of non-empty sets and mappings of the form

$$A \xrightarrow{f} B \xrightarrow{g} C.$$

Given an arbitrary element x of A, let us follow its "path" through the diagram. Its image under f is the element $f(x)$ of B and this element has in turn as image under g the element $g[f(x)]$ of C. Now it is clear that we can define a mapping from A to C by the assignment $x \mapsto g[f(x)]$. This assignment effectively allows us to by-pass B and the mapping so defined is such that its effect on x is the same as first applying f to obtain $f(x)$ and then applying g to $f(x)$ to obtain $g[f(x)]$. This process is known as **composing** the mappings f and g and a formal definition is as follows.

Definition. Given the diagram $A \xrightarrow{f} B \xrightarrow{g} C$ of non-empty sets and mappings, the **composite** of these mappings is defined to be the mapping denoted by $g \circ f: A \to C$ and given by the prescription

$$(\forall x \in A) \quad (g \circ f)(x) = g[f(x)].$$

A pictorial representation of this definition is as follows:

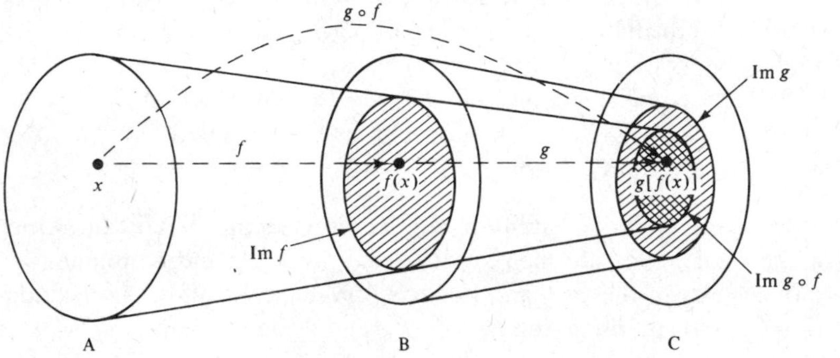

Remark 1. It is extremely important to observe that in the above definition the mapping $g \circ f$ is defined only when the codomain (arrival set) of f is the same as the domain (departure set) of g.

Remark 2. Note that if $g \circ f$ is defined then $f \circ g$ may not be defined. In fact, for both to be defined we require a diagram of the form $A \xrightarrow{f} B \xrightarrow{g} A$.

Remark 3. Note also that even when both $f \circ g$ and $g \circ f$ are defined we need not have $f \circ g = g \circ f$. In fact in the diagram $A \xrightarrow{f} B \xrightarrow{g} A$ we have $\text{Dom } g \circ f = A$ and $\text{Dom } f \circ g = B$ so when $A \neq B$ we have $f \circ g \neq g \circ f$.

Remark 4. It is also extremely important to note the order in which f and g appear in the definition of $g \circ f$. In fact, if we were to write our mappings on the right then such composites would appear in the opposite order, for then we would have $(xf)g = x(f \circ g)$.

The most important property of composition of mappings is that it is **associative**, by which we mean the following.

Theorem 5.2 *If $A \xrightarrow{f} B \xrightarrow{g} C \xrightarrow{h} D$ is a diagram of sets and mappings then $h \circ (g \circ f) = (h \circ g) \circ f$.*

Proof. We note first that the various composites appearing in the equality to be established are defined. Moreover, $h \circ (g \circ f)$ and $(h \circ g) \circ f$

each have departure set A and arrival set D. Now for every $x \in A$ we have

$$[h \circ (g \circ f)](x) = h[(g \circ f)(x)] = h(g[f(x)]);$$

$$[(h \circ g) \circ f](x) = (h \circ g)[f(x)] = h(g[f(x)]).$$

The result therefore follows from the definition of equality for mappings.

Corollary. *If* $h \circ g \circ f : A \to D$ *is given by* $(h \circ g \circ f)(x) = h(g[f(x)])$ *then* $h \circ (g \circ f) = h \circ g \circ f = (h \circ g) \circ f$.

Let us now consider the following general situation. Let A, B, C be sets and let f, g be mappings from A to B and A to C respectively. We illustrate this by the diagram

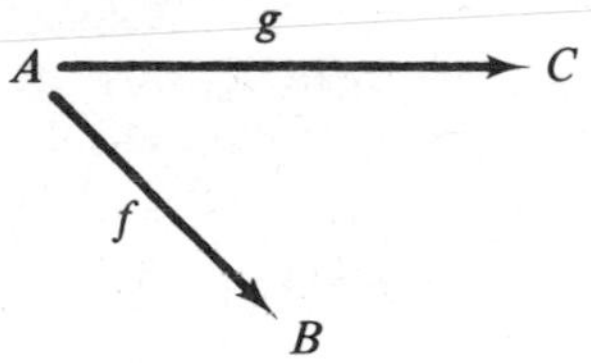

It is clear that there may be a multitude of mappings from B to C; but under what conditions does there exist a mapping $h : B \to C$ such that $h \circ f = g$? If h is such a mapping, we say that the augmented diagram

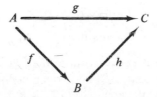

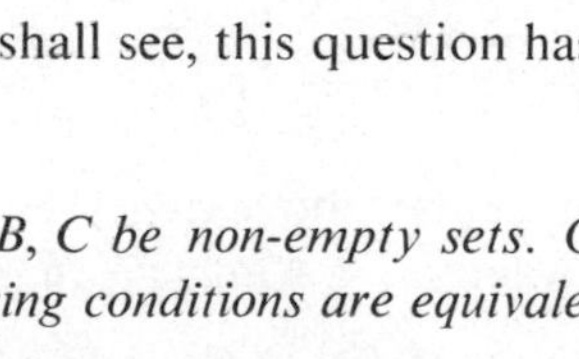

is **commutative**. As we shall see, this question has far-reaching algebraic consequences.

Theorem 5.3 *Let A, B, C be non-empty sets. Given mappings $f : A \to B$ and $g : A \to C$, the following conditions are equivalent:*

(1) *there exists a mapping $h : B \to C$ such that $h \circ f = g$;*

(2) $(\forall x, x^* \in A) \quad f(x) = f(x^*) \Rightarrow g(x) = g(x^*)$.

Proof. (1) $\Rightarrow$ (2): If $h : B \to C$ exists such that $h \circ f = g$ then for all $x, x^* \in A$ we have

$$f(x) = f(x^*) \Rightarrow h[f(x)] = h[f(x^*)] \Rightarrow g(x) = g(x^*).$$

(2) $\Rightarrow$ (1): Suppose now that (2) holds and consider the subset G of $\text{Im} f \times C$ defined by

$$G=\{(y, z); (\exists x \in A)\ y=f(x),\ z=g(x)\}.$$

We observe that $G \neq \emptyset$; for, given any $x \in A$ we have $(f(x), g(x)) \in G$. We shall show first that G is the graph of a mapping $t: \text{Im} f \to C$. For this purpose, we have to show that for every $y \in \text{Im} f$ there is a unique $z \in C$ such that $(y, z) \in G$. Now the existence of at least one such z is clear: simply choose $x \in A$ such that $y = f(x)$ and set $z=g(x)$. To establish the uniqueness of such an element z, suppose that we have $(y, z) \in G$ and $(y, z^*) \in G$. Then by the definition of G there exist $x, x^* \in A$ such that

$$y=f(x)=f(x^*), \quad z=g(x), \quad z^*=g(x^*).$$

The standing hypothesis that (2) holds now gives $g(x)=g(x^*)$ and so $z=z^*$. This then shows that G is the graph of a mapping, t say, from $\text{Im} f$ to C. Moreover, since $(f(x), g(x)) \in G$ for every $x \in A$ we have $g(x)=t[f(x)]$ for every $x \in A$. We now construct a mapping $h: B \to C$ by the prescription

$$h(x)=\begin{cases} t(x) & \text{if } x \in \text{Im} f; \\ \text{any } \alpha \in C & \text{if } x \in B \setminus \text{Im} f. \end{cases}$$

Then for every $x \in A$ we have

$$(h \circ f)(x)=h[f(x)]=t[f(x)]=g(x).$$

It now follows that $h \circ f=g$ as required.

> *Remark.* It is instructive to note that we did not write $g=t \circ f$ in the above. In fact, $t \circ f$ may not be defined; for the departure set of t is $\text{Im} f$ and the arrival set of f is B. If we let $f^+: A \to \text{Im} f$ be the mapping given by $f^+(x)=f(x)$ for every $x \in A$ then what we do have is $g=t \circ f^+$.

As a first application of the above general result, we consider the important special case where $C=A$ and $g=id_A$, the identity mapping on A.

Theorem 5.4 *Let A, B be non-empty sets. Given any mapping $f:A \to B$, the following conditions are equivalent*:

(1) *there exists $g:B \to A$ such that $g \circ f=id_A$*;

(2) $(\forall x, x^* \in A) \quad f(x)=f(x^*) \Rightarrow x=x^*$;

(3) *for every non-empty set C and all mappings h, k:C→A,*

$$f \circ h = f \circ k \Rightarrow h = k.$$

Proof. (1) ⇔ (2): Applying the previous theorem to the diagram

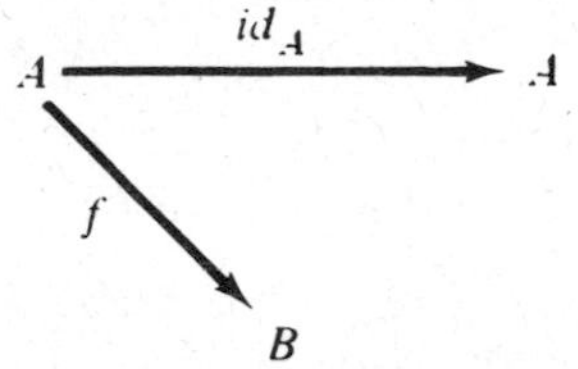

we see that the existence of $g: B \to A$ such that $g \circ f = id_A$ is equivalent to the condition $(\forall x, x^* \in A)$ $f(x) = f(x^*) \Rightarrow id_A(x) = id_A(x^*)$. Since $id_A(x) = x$ for every $x \in A$, the equivalence of (1) and (2) now follows.

(2) ⇒ (3): If (2) holds and $h, k: C \to A$ then

$$\begin{aligned} f \circ h = f \circ k &\Rightarrow (\forall y \in C) \quad (f \circ h)(y) = (f \circ k)(y) \\ &\Rightarrow (\forall y \in C) \quad f[h(y)] = f[k(y)] \\ &\Rightarrow (\forall y \in C) \quad h(y) = k(y) \qquad \text{[by (2)]} \\ &\Rightarrow h = k. \end{aligned}$$

(3) ⇒ (2): Suppose that (2) does not hold. Then for some $x, x^* \in A$ with $x \neq x^*$ we have $f(x) = f(x^*)$. Let C be any non-empty set and let $h, k: C \to A$ be the "constant mappings" given by

$$(\forall y \in C) \qquad h(y) = x, \quad k(y) = x^*.$$

Then clearly $h \neq k$ and

$$(\forall y \in C) \qquad (f \circ h)(y) = f[h(y)] = f(x) = f(x^*) = f[k(y)] = (f \circ k)(y),$$

so that $f \circ h = f \circ k$. Thus if (2) does not hold then neither does (3). This is logically equivalent to the implication (3) ⇒ (2).

Let us now examine the "dual" situation to the above: i.e., where all the arrows are reversed. Specifically, given a diagram

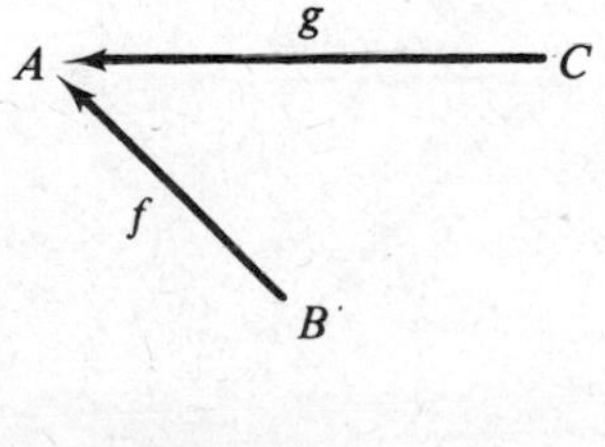

when does there exist $h:C \to B$ such that $f \circ h = g$? This question is slightly easier to answer.

Theorem 5.5 *Let A, B, C be non-empty sets. Given mappings $f:B \to A$ and $g:C \to A$, the following conditions are equivalent*:

(1) *there exists $h:C \to B$ such that $f \circ h = g$;*

(2) $\operatorname{Im} g \subseteq \operatorname{Im} f$.

Proof. (1) $\Rightarrow$ (2): If (1) holds then we have

$$(\forall x \in C) \qquad g(x) = (f \circ h)(x) = f[h(x)] \in \operatorname{Im} f,$$

from which it follows that $\operatorname{Im} g \subseteq \operatorname{Im} f$.

(2) $\Rightarrow$ (1): Suppose now that (2) holds. Then for every $x \in C$ we have $g(x) \in \operatorname{Im} f$, so for every $x \in C$ there exists $y \in B$ such that $g(x) = f(y)$. Given any $x \in C$ label y_x any one of the elements $y \in B$ such that $g(x) = f(y)$; i.e., associate with every $x \in C$ a chosen element y_x in the subset $f^{\leftarrow}\{g(x)\}$ of B. Then we can define a mapping $h:C \to B$ by the prescription

$$(\forall x \in C) \qquad h(x) = y_x.$$

This mapping h then has the property that

$$(\forall x \in C) \qquad (f \circ h)(x) = f[h(x)] = f(y_x) = g(x)$$

and so satisfies (1).

We shall now consider a particularly important special case of Theorem 5.5 which is dual to the result of Theorem 5.4.

Theorem 5.6 *Let A, B be non-empty sets. Given any mapping $f:A \to B$, the following conditions are equivalent*:

(1) *there exists $g:B \to A$ such that $f \circ g = id_B$;*

(2) $\operatorname{Im} f = B$;

(3) *for every non-empty set C and all mappings h, $k:B \to C$,*

$$h \circ f = k \circ f \Rightarrow h = k.$$

Proof. (1) $\Leftrightarrow$ (2): Applying Theorem 5.5 to the diagram

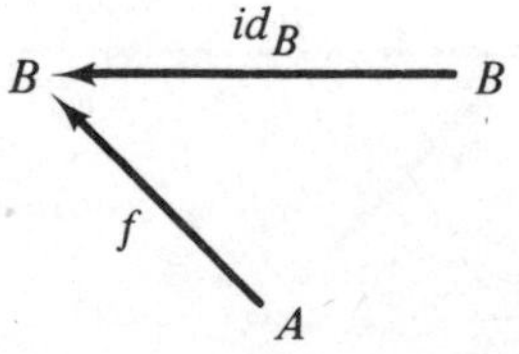

we see that the existence of $g: B \to A$ such that $f \circ g = id_B$ is equivalent to the condition $\operatorname{Im} id_B \subseteq \operatorname{Im} f$. But we know that $\operatorname{Im} id_B = B \supseteq \operatorname{Im} f$ and so this second condition is equivalent to $B = \operatorname{Im} f$.

(1) $\Rightarrow$ (3): Suppose that (1) holds, so that there exists $f^*: B \to A$ such that $f \circ f^* = id_B$. Then from $h \circ f = k \circ f$ we deduce, using the associativity of $\circ$ (Theorem 5.2), that

$$h = h \circ id_B = h \circ (f \circ f^*) = (h \circ f) \circ f^* = (k \circ f) \circ f^* = k \circ (f \circ f^*) = k \circ id_B = k.$$

(3) $\Rightarrow$ (2): Suppose now that f satisfies (3). We shall suppose that B contains at least two distinct elements α, β. [If not, then B is necessarily a singleton and automatically $\operatorname{Im} f = B$.] Let $h, k: B \to B$ be given by the prescriptions

$$h(x) = \begin{cases} x & \text{if } x \in \operatorname{Im} f; \\ \alpha & \text{otherwise,} \end{cases} \qquad k(x) = \begin{cases} x & \text{if } x \in \operatorname{Im} f; \\ \beta & \text{otherwise.} \end{cases}$$

Then we have

$$(\forall y \in A) \qquad h[f(y)] = f(y) = k[f(y)]$$

and consequently $h \circ f = k \circ f$. Applying (3), we deduce that $h = k$. Now, if $B \neq \operatorname{Im} f$ then we have $\operatorname{Im} f \subset B$ and so there exists $x \in B$ with $x \notin \operatorname{Im} f$. For such an element x we then have $\alpha = h(x) = k(x) = \beta$, a contradiction to the hypothesis that α, β are distinct. We conclude, therefore, that $B = \operatorname{Im} f$ which is (2).

The types of mapping introduced in Theorems 5.4 and 5.6 will prove to be of immense importance in what follows. For this reason they have special names. If $f: A \to B$ satisfies any of the mutually equivalent properties of Theorem 5.4 then f is said to be **injective** (or an **injection**); and if f satisfies any of the mutually equivalent properties of Theorem 5.6 then f is said to be **surjective** (or a **surjection**).

Of the properties listed in Theorem 5.4, the most useful in practice is property (2). Thus $f: A \to B$ is injective if

$$(\forall x, x^* \in A) \qquad f(x) = f(x^*) \Rightarrow x = x^*.$$

Note that this is logically equivalent to the property

$$(\forall x, x^* \in A) \qquad x \neq x^* \Rightarrow f(x) \neq f(x^*)$$

which may be described by saying that f maps distinct objects into distinct objects. For this reason, the terminology **one-one mapping** is sometimes used instead of injection. Property (3) of Theorem 5.4 is often described by saying that f is **left cancellative** with respect to composition of mappings; we shall often make use of the fact that this is equivalent to f being injective.

Of the properties listed in Theorem 5.6, the most useful in practice is property (2). Thus $f: A \to B$ is surjective if the image of f is the codomain (arrival set) B. For this reason, the terminology **onto mapping** is sometimes used instead of surjection. Property (3) of Theorem 5.6 is often described by saying that f is **right cancellative** with respect to composition of mappings; we shall often make use of the fact that this is equivalent to f being surjective.

> *Remark.* Note that if we were to disregard arrival sets in the definition of equality for mappings then all mappings would be surjective!

Example 5.1. Let $S=\{a, b, c\}$ and $T=\{x, y\}$. Then the mapping $f: S \to T$ described by setting $f(a)=f(b)=x$ and $f(c)=y$ is surjective (every element of T is the image of some element of S). This mapping is not injective, however, since $a \neq b$ but $f(a)=f(b)$.

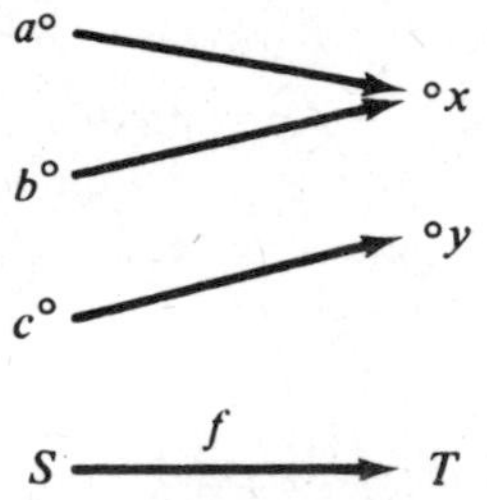

Example 5.2. The mapping $f: \mathbf{N} \to \mathbf{N}$ given by the prescription $f(n)=n+1$ is not surjective (there is no element n such that $f(n)=0$) but it is injective since if $f(m)=f(n)$ then $m+1=n+1$ and so $m=n$.

Example 5.3. For any non-empty set E the identity mapping id_E is both injective and surjective.

Example 5.4. Let E, F be non-empty sets. If $f: E \to F$ is any mapping and A is a non-empty subset of E then we define the **restriction of** f **to** A to be the mapping $f_A: A \to F$ given by $(\forall x \in A)\, f_A(x)=f(x)$. A particularly important case arises when $F=E$ and $f=id_E$. The restriction of id_E to the subset A is written ι_A. The mapping $\iota_A: A \to E$ is then given by $\iota_A(x)=x$ for every $x \in A$ and is clearly injective; it is called the **canonical injection of** A **into** E. It is also sometimes called the **canonical inclusion of** A **in** E.

Example 5.5. If E_1, E_2 are non-empty sets then the **first** and **second projections** from $E_1 \times E_2$ are defined to be, respectively, the mappings

$pr_1: E_1 \times E_2 \to E_1$ and $pr_2: E_1 \times E_2 \to E_2$ given by the prescriptions $pr_1(e_1, e_2) = e_1$ and $pr_2(e_1, e_2) = e_2$. Clearly, both pr_1 and pr_2 are surjective.

Concerning the induced mappings $f^{\to}$ and $f^{\leftarrow}$ introduced at the beginning of this section and their relationship to composition of mappings, we have the following two results.

Theorem 5.7 *Given the diagram* $A \xrightarrow{f} B \xrightarrow{g} C$ *of non-empty sets and mappings, we have* $(g \circ f)^{\to} = g^{\to} \circ f^{\to}$ *and* $(g \circ f)^{\leftarrow} = f^{\leftarrow} \circ g^{\leftarrow}$.

Proof. Given any non-empty subset S of A we have

$$\begin{aligned}(g^{\to} \circ f^{\to})(S) = g^{\to}[f^{\to}(S)] &= \{g(x);\ x \in f^{\to}(S)\} \\ &= \{g(x);\ (\exists y \in S)\quad x = f(y)\} \\ &= \{g[f(y)];\ y \in S\} \\ &= (g \circ f)^{\to}(S).\end{aligned}$$

Since moreover $(g^{\to} \circ f^{\to})(\emptyset) = \emptyset = (g \circ f)^{\to}(\emptyset)$ we deduce that $g^{\to} \circ f^{\to} = (g \circ f)^{\to}$.

Similarly, for every subset T of C we have

$$(f^{\leftarrow} \circ g^{\leftarrow})(T) = \begin{cases} \emptyset & \text{if } \operatorname{Im} f \cap g^{\leftarrow}(T) = \emptyset; \\ \{x \in A;\ f(x) \in g^{\leftarrow}(T)\} & \text{otherwise,} \end{cases}$$

$$(g \circ f)^{\leftarrow}(T) = \begin{cases} \emptyset & \text{if } \operatorname{Im} g \circ f \cap T = \emptyset; \\ \{x \in A;\ g[f(x)] \in T\} & \text{otherwise.} \end{cases}$$

The result will follow, therefore, if we can show that

$$\operatorname{Im} f \cap g^{\leftarrow}(T) = \emptyset \Leftrightarrow \operatorname{Im} g \circ f \cap T = \emptyset.$$

Using Theorem 5.1 (jw), (v) this is established as follows:

$$\begin{aligned}\operatorname{Im} f \cap g^{\leftarrow}(T) = \emptyset &\Leftrightarrow \operatorname{Im} f \subseteq \complement_B(g^{\leftarrow}(T)) = g^{\leftarrow}(\complement_C(T)) \\ &\Leftrightarrow \operatorname{Im} g \circ f = g^{\to}(\operatorname{Im} f) \subseteq \complement_C(T) \\ &\Leftrightarrow \operatorname{Im} g \circ f \cap T = \emptyset.\end{aligned}$$

Theorem 5.8 *Let A, B be non-empty sets and $f: A \to B$ any mapping. If A_1 is any subset of A and B_1 is any subset of B we have*

$$A_1 \subseteq (f^{\leftarrow} \circ f^{\to})(A_1) \qquad \textit{and} \qquad (f^{\to} \circ f^{\leftarrow})(B_1) \subseteq B_1.$$

Moreover, equality holds in the first of these for all subsets A_1 of A if and only if f is injective; and equality holds in the second for all subsets B_1 of B if and only if f is surjective.

Proof. The inclusions obviously hold when each left hand side is $\varnothing$. So let $A_1 \neq \varnothing$ and $(f^{\to} \circ f^{\gets})(B_1) \neq \varnothing$. For every $x \in A_1$ we have clearly $(x) \in f^{\to}(A_1)$ so that $x \in f^{\gets}[f^{\to}(A_1)]$ and consequently $A_1 \subseteq f^{\gets}[f^{\to}(A_1)]$. As for the second inclusion, we note that if $y \in f^{\to}[f^{\gets}(B_1)]$ then we have $y = f(x)$ for some $x \in f^{\gets}(B_1)$ and so $y \in B_1$.

Suppose now that equality holds in the first for all subsets A_1 of A. Then in particular it holds for all singleton subsets of A. Consequently, given any $x, y \in A$ we have

$$\begin{aligned} f(x) = f(y) &\Rightarrow f^{\to}\{x\} = \{f(x)\} = \{f(y)\} = f^{\to}\{y\} \\ &\Rightarrow \{x\} = f^{\gets}[f^{\to}\{x\}] = f^{\gets}[f^{\to}\{y\}] = \{y\} \\ &\Rightarrow x = y \end{aligned}$$

and so f is injective. Conversely, suppose that f is injective. Then for every subset A_1 of A we have

$$f(x) \in f^{\to}(A_1) \Rightarrow (\exists y \in A_1) f(x) = f(y) \Rightarrow x = y \in A_1$$

whence we deduce that $f^{\gets}[f^{\to}(A_1)] \subseteq A_1$ and we have equality.

Suppose now that equality holds in the second for all subsets B_1 of B. Then in particular we have $f^{\to}[f^{\gets}(B)] = B$. Consequently, using Theorem 5.1 (j), $B = f^{\to}[f^{\gets}(B)] \subseteq f^{\to}(A) = \operatorname{Im} f$. But clearly $\operatorname{Im} f \subseteq B$; hence $\operatorname{Im} f = B$ and so f is surjective. Conversely, suppose that f is surjective. If B_1 is any subset of B then each $b \in B_1$ can be written in the form $b = f(a)$ for some $a \in A$; and clearly such an element a belongs to $f^{\gets}(B_1)$. Hence we have $B_1 \subseteq f^{\to}[f^{\gets}(B_1)]$ and the desired equality follows.

Definition. A mapping $f: A \to B$ is called **bijective** (or a **bijection**) if it is both injective and surjective.

We have already seen in Example 5.3 that identity maps are bijective. For another example, consider the mapping $f: \mathbf{Z} \to \mathbf{Z}$ given by $f(n) = n + 1$ for every $n \in \mathbf{Z}$. Clearly f is both injective and surjective and so is a bijection.

Bijections are characterised by the following result.

Theorem 5.9 *Let A, B be non-empty sets. Given any mapping $f: A \to B$, the following conditions are equivalent*:

(1) *f is a bijection*;
(2) *there exists $g: B \to A$ such that $f \circ g = id_B$ and $g \circ f = id_A$*;
(3) *f is both left and right cancellative.*

Proof. The proof is immediate from Theorems 5.4 and 5.6 except perhaps for (1) $\Rightarrow$ (2). To establish this, suppose that (1) holds. Then f is both injective and surjective so that, by Theorems 5.4 and 5.6,

(a) there exists $h_1: B \to A$ such that $h_1 \circ f = id_A$;

(b) there exists $h_2: B \to A$ such that $f \circ h_2 = id_B$.

We show that $h_1 = h_2$ whence the result follows with $g = h_1 = h_2$. We have, by Theorem 5.2,

$$h_2 = id_A \circ h_2 = (h_1 \circ f) \circ h_2 = h_1 \circ (f \circ h_2) = h_1 \circ id_B = h_1.$$

Remark. It is extremely important to note that when $f: A \to B$ is a bijection the mapping $g: B \to A$ such that $f \circ g = id_B$ and $g \circ f = id_A$ is *unique*. For, if $h: B \to A$ also satisfies these properties then the fact that f is both left and right cancellative gives $h = g$. This unique mapping g is called the **inverse of the bijection** f and will be denoted henceforth by f^{-1}. Note that this is not to be confused with $f^{\leftarrow}$ which is a mapping between power sets. When $f: A \to B$ is a bijection, f^{-1} and $f^{\leftarrow}$ are related by the property $f^{\leftarrow}\{y\} = \{f^{-1}(y)\}$ for all $y \in B$.

Theorem 5.10 *Suppose we have a diagram* $A \xrightarrow{f} B \xrightarrow{g} C$ *of non-empty sets and mappings. Then*

(1) *if f and g are injective so also is $g \circ f$;*

(2) *if f and g are surjective so also is $g \circ f$;*

(3) *if f and g are bijective so also is $g \circ f$ and in this case*

$$(g \circ f)^{-1} = f^{-1} \circ g^{-1}.$$

Proof. (1) If f and g are injective then for all $x, x^* \in A$ we have

$$\begin{aligned} (g \circ f)(x) = (g \circ f)(x^*) &\Rightarrow g[f(x)] = g[f(x^*)] \\ &\Rightarrow f(x) = f(x^*) \qquad [g \text{ injective}] \\ &\Rightarrow x = x^* \qquad [f \text{ injective}] \end{aligned}$$

and so $g \circ f$ is injective.

(2) If f and g are surjective then for each $z \in C$ there exists $y \in B$ such that $g(y) = z$, then an $x \in A$ such that $f(x) = y$. Thus for each $z \in C$ there exists $x \in A$ such that $z = g[f(x)]$. Consequently $g \circ f$ is surjective.

(3) If f and g are bijective then by (1) and (2) so also is $g \circ f$. Moreover, since

$$(g \circ f) \circ (f^{-1} \circ g^{-1}) = (g \circ f \circ f^{-1}) \circ g^{-1} = g \circ id_B \circ g^{-1} = g \circ g^{-1} = id_C$$

and similarly $(f^{-1} \circ g^{-1}) \circ (g \circ f) = id_A$ we see that the inverse of $g \circ f$ is $f^{-1} \circ g^{-1}$.

By way of applying the above results, we end the present section with a characterisation of the cartesian product of two sets. We shall call upon this characterisation in the next section when we shall seek to generalise this notion.

Definition. Let E_1 and E_2 be sets. By a **product** of E_1 and E_2 we shall mean a set P together with mappings $\pi_1: P \to E_1$ and $\pi_2: P \to E_2$ such that, for any set X and any mappings $f_1: X \to E_1$ and $f_2: X \to E_2$, there is a unique mapping $h: X \to P$ such that the diagram

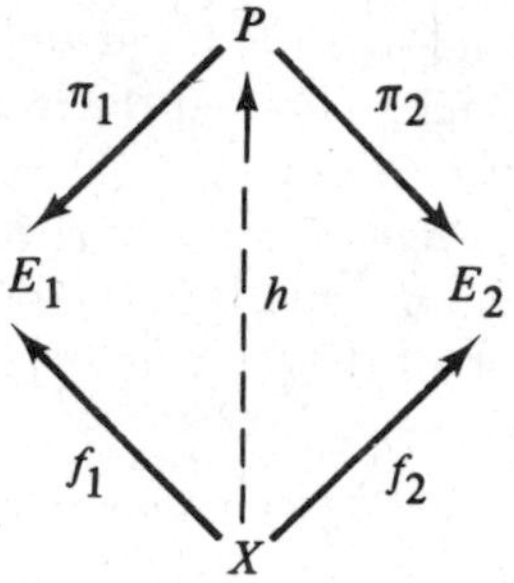

is commutative in that $\pi_1 \circ h = f_1$ and $\pi_2 \circ h = f_2$. We denote such a product of E_1 and E_2 by (P, π_1, π_2).

The reader will recognise here another "diagram-completing" problem. We shall meet with several others in the pages to follow.

Before discussing the existence of products, let us say something about the mappings π_1, π_2 in the above definition.

Theorem 5.11 *If (P, π_1, π_2) is a product of E_1 and E_2 then π_1 and π_2 are surjective.*

Proof. In the above diagram take $X = E_1$ and $f_1 = id_{E_1}$. Then we have $\pi_1 \circ h = id_{E_1}$ and so π_1 is surjective by Theorem 5.6. Similarly, by taking $X = E_2$ and $f_2 = id_{E_2}$ we see that π_2 is surjective.

As to the existence of products, we have:

Theorem 5.12 *If E_1 and E_2 are sets then $(E_1 \times E_2, pr_1, pr_2)$ is a product of E_1 and E_2.*

Proof. Consider any diagram of the form

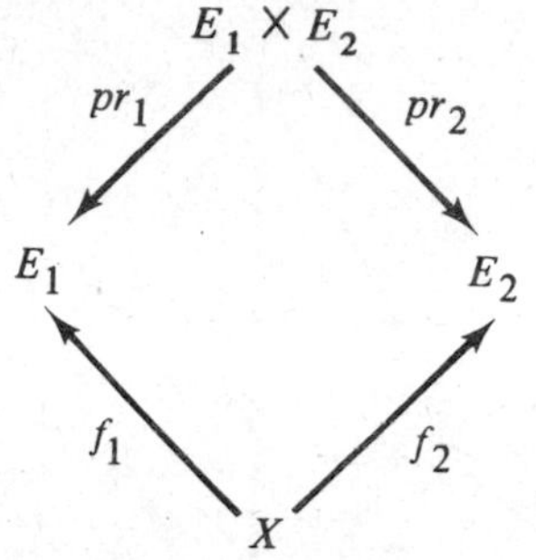

Define $h: X \to E_1 \times E_2$ by the prescription $(\forall x \in X)\ h(x) = (f_1(x), f_2(x))$. Then for every $x \in X$ we have $(pr_1 \circ h)(x) = pr_1(f_1(x), f_2(x)) = f_1(x)$, so that $pr_1 \circ h = f_1$; and similarly $pr_2 \circ h = f_2$. To establish the uniqueness of h, suppose that $k: X \to E_1 \times E_2$ is such that $pr_1 \circ k = f_1$ and $pr_2 \circ k = f_2$. Given any $x \in X$ let $k(x) = (e_1, e_2)$. Then we have $f_1(x) = (pr_1 \circ k)(x) = pr_1(e_1, e_2) = e_1$ and similarly $f_2(x) = e_2$ whence we have $k(x) = (e_1, e_2) = (f_1(x), f_2(x)) = h(x)$. Consequently $k = h$ and the result follows.

By saying that $E_1 \times E_2$ is characterised by the property of being a product, we mean precisely the following.

Theorem 5.13 *If (P, π_1, π_2) and (Q, σ_1, σ_2) are each products of the sets E_1, E_2 then there is a unique bijection $\theta: Q \to P$ such that $\pi_1 \circ \theta = \sigma_1$ and $\pi_2 \circ \theta = \sigma_2$.*

Proof. Consider the diagram

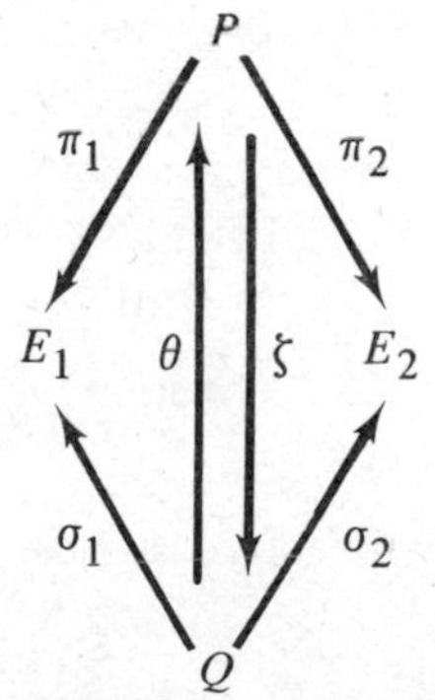

in which $\theta: Q \to P$ is the unique mapping such that $\pi_1 \circ \theta = \sigma_1$ and $\pi_2 \circ \theta = \sigma_2$ and $\zeta: P \to Q$ is the unique mapping such that $\sigma_1 \circ \zeta = \pi_1$ and

$\sigma_2 \circ \zeta = \pi_2$. Then it is clear that the mapping $h = \theta \circ \zeta : P \to P$ makes the diagram

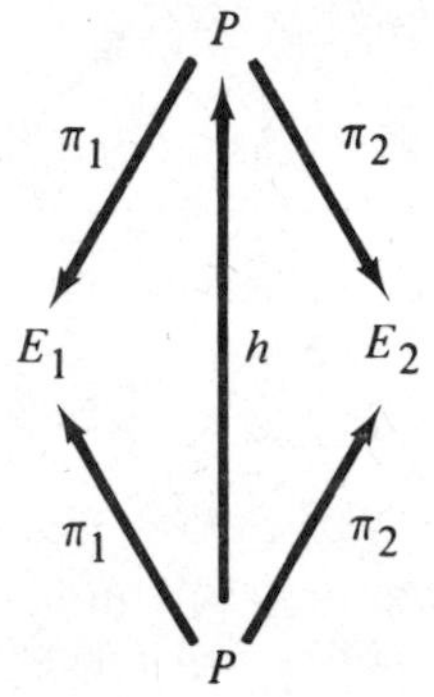

commutative. But clearly $h = id_P$ also makes this diagram commutative and from the definition of P as a product there can be only one such mapping. We conclude therefore that $\theta \circ \zeta = id_P$ whence we see that θ is surjective by Theorem 5.6. In a similar way we can argue that $\zeta \circ \theta = id_Q$ whence θ is injective by Theorem 5.4. Thus θ is a bijection (with $\theta^{-1} = \zeta$) and the result follows.

> *Remark.* Note that we cannot use the notion of a product of two sets as a *definition* of their cartesian product; for to formulate the notion of a product we require the notion of a mapping and we used cartesian products to define mappings! Later, we shall use the above characterisation as a guide in deciding how to define the cartesian product of an arbitrary collection of sets.

Exercises for §5

1. By considering the mapping $pr_1 : \mathbf{N} \times \mathbf{N} \to \mathbf{N}$ and the subsets $E_1 = \{(m, 1);\ m \in \mathbf{N}\}$ and $E_2 = \{(0, 2n);\ n \in \mathbf{N}\}$, show that equality does not hold in general in Theorem 5.1 (iii), (iv).

2. Prove that equality holds in Theorem 5.1 (iii) for all subsets E_1, E_2 of E if and only if f is injective.

[*Hint.* $\Rightarrow$: To show that f is injective suppose that $x \neq y$ and observe that $f^{\rightarrow}(\{x\} \cap \{y\}) = f^{\rightarrow}(\emptyset) = \emptyset$.]

3. Given $f : E \to F$, prove that the following are equivalent:
(1) f is a bijection;
(2) $f^{\rightarrow} \circ \complement_E = \complement_F \circ f^{\rightarrow}$.

[*Hint*. (2) $\Rightarrow$ (1): To show that f is surjective consider $f^{\rightarrow}(E)=f^{\rightarrow}(\complement_E(\emptyset))$. To show that f is injective use Exercise 5.2 and the de Morgan laws.]

Deduce that equality holds in Theorem 5.1 (iv) for all subsets E_1 of E if and only if f is injective.

[*Hint*. f is injective if and only if $f^+:E\rightarrow \mathrm{Im}\, f$ given by $f^+(x)=f(x)$ is bijective.]

4. Given $f:E\rightarrow F$ show that $f^{\rightarrow}\circ f^{\leftarrow}\circ f^{\rightarrow}=f^{\rightarrow}$ and that $f^{\leftarrow}\circ f^{\rightarrow}\circ f^{\leftarrow}=f^{\leftarrow}$.

[*Hint*. Use Theorem 5.8 and Theorem 5.1 (i), (j).]

5. Prove that if $f:E\rightarrow F$ and $A\subseteq E$, $B\subseteq F$ then

$$f^{\rightarrow}[A\cap f^{\leftarrow}(B)]=f^{\rightarrow}(A)\cap B.$$

[*Hint*. For $\subseteq$ use Theorems 5.1 and 5.8. For $\supseteq$ observe that if $x=f(a)\in B$ then $x=f(a)$ where $a\in f^{\leftarrow}(B)$.]

6. Given any mapping $f:\mathbf{R}\rightarrow\mathbf{R}$ show that
(a) f is surjective if and only if every line parallel to the x-axis meets the graph of f at least once;
(b) f is injective if and only if every line parallel to the x-axis meets the graph of f at most once;
(c) f is bijective if and only if every line parallel to the x-axis meets the graph of f precisely once.

Consider the mappings $f:\mathbf{R}\rightarrow\mathbf{R}$ given by

$$f(x)=x^2;\quad f(x)=x^3;\quad f(x)=x^3-x^2;\quad f(x)=x^2+2x+4.$$

Say whether each of these mappings is surjective, injective, neither or both.

7. Show from graphical considerations that if $a, b, c\in\mathbf{R}$ then the mapping $f:\mathbf{R}\rightarrow\mathbf{R}$ given by $f(x)=x^3+ax^2+bx+c$ is a bijection if and only if $a^2\leqslant 3b$.

[*Hint*. Use Exercise 6 and calculus.]

8. Consider the mappings $f:\mathbf{Z}\rightarrow\mathbf{Z}\times\mathbf{Z}$ and $g:\mathbf{Z}\times\mathbf{Z}\rightarrow\mathbf{Z}$ given by $f(m)=(m-1, 1)$ and $g(m, n)=m+n$. Show that $g\circ f=id_{\mathbf{Z}}$ whereas $f\circ g$ is neither surjective nor injective.

9. Consider the mapping $f:\mathbf{R}\rightarrow\mathbf{R}$ given by $f(x)=4x/(x^2+1)$. Determine $\mathrm{Im}\, f$ and find a subset $A=[-k, k]$ of $\mathbf{R}$ such that $f^{\rightarrow}(A)=\mathrm{Im}\, f$ and the mapping $g:A\rightarrow\mathrm{Im}\, f$, given by $g(a)=f(a)$ for every $a\in A$, is a bijection. Obtain a prescription giving g^{-1}.

10. If $\alpha:\mathbf{N}\to\mathbf{N}$ is given by $\alpha(n)=n+1$ show that there is no mapping $g:\mathbf{N}\to\mathbf{N}$ such that $\alpha\circ g=id_{\mathbf{N}}$ but that there are infinitely many maps $k:\mathbf{N}\to\mathbf{N}$ such that $k\circ\alpha=id_{\mathbf{N}}$.

[*Hint*. For every $p\in\mathbf{N}$ let $k_p:\mathbf{N}\to\mathbf{N}$ be given by

$$k_p(n)=\begin{cases} n-1 & \text{if } n\geqslant 1;\\ p & \text{if } n=0.]\end{cases}$$

If $\beta:\mathbf{N}\to\mathbf{N}$ is given by

$$\beta(n)=\begin{cases} n/2 & \text{if } n \text{ is even};\\ (n-1)/2 & \text{if } n \text{ is odd},\end{cases}$$

show that there is no mapping $f:\mathbf{N}\to\mathbf{N}$ such that $f\circ\beta=id_{\mathbf{N}}$ but that there are infinitely many maps $k:\mathbf{N}\to\mathbf{N}$ such that $\beta\circ k=id_{\mathbf{N}}$.

[*Hint*. For every $p\in\mathbf{N}$ define $k_p:\mathbf{N}\to\mathbf{N}$ by

$$k_p(n)=\begin{cases} 2n+1 & \text{if } n\neq p;\\ 2p & \text{if } n=p.]\end{cases}$$

11. Given a mapping $f:A\to A$ prove that the following are equivalent:
(1) f is a bijection with $f^{-1}=f$;
(2) $f\circ f=id_A$.

A mapping which satisfies these properties is called an **involution** on A. For which elements $a,b\in\mathbf{R}$ is the mapping $f_{a,b}:\mathbf{R}\to\mathbf{R}$ given by $f_{a,b}(x)=ax+b$ an involution?

12. Recall that for every $x\in\mathbf{R}$ we denote by $[x]$ the greatest integer less than or equal to x. Sketch the graphs of the functions $f,g:\mathbf{R}\to\mathbf{R}$ given by the prescriptions

$$f(x)=x[x];\quad g(x)=[x]+[-x].$$

What are the functions $f\circ g$ and $g\circ f$?

13. Let $f,g:\mathbf{R}\to\mathbf{R}$ be given by

$$f(x)=\begin{cases} x+2 & \text{if } x\leqslant 0;\\ 2\sqrt{(x+1)} & \text{if } x>0,\end{cases}\qquad g(x)=\begin{cases} -2x & \text{if } x\leqslant 0;\\ -x(x+2) & \text{if } x>0.\end{cases}$$

Sketch the graphs of f and g. Find prescriptions for the composite functions $f\circ g$ and $g\circ f$.

14. Given mappings $f:A\to B$ and $g:B\to C$ prove that
(a) $g\circ f$ injective $\Rightarrow f$ injective;
(b) $g\circ f$ surjective $\Rightarrow g$ surjective.

15. In the diagram $A \xrightarrow{f} B \xrightarrow{g} C \xrightarrow{h} D$ it is given that $g \circ f$ and $h \circ g$ are bijections. Prove that f, g, h are all bijections.

[*Hint*. Use Exercise 14.]

16. In the diagram $A \xrightarrow{f} B \xrightarrow{g} C \xrightarrow{h} A$ it is given that $h \circ g \circ f$ and $f \circ h \circ g$ are injective and $g \circ f \circ h$ is surjective. Prove that f, g, h are each bijective.

[*Hint*. Use Exercise 14.]

17. A diagram of the form

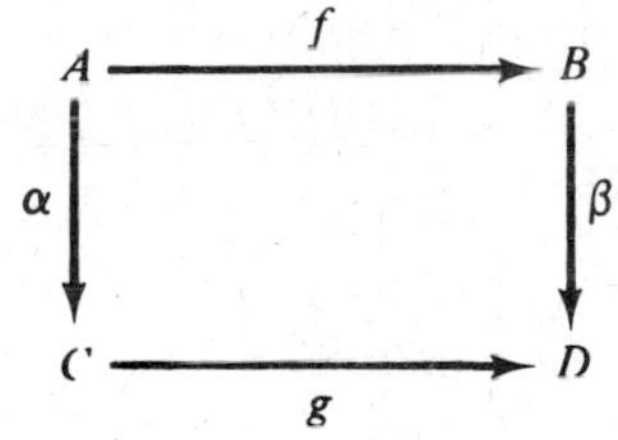

is said to be **commutative** if $g \circ \alpha = \beta \circ f$. Suppose that there is given the "cube" diagram

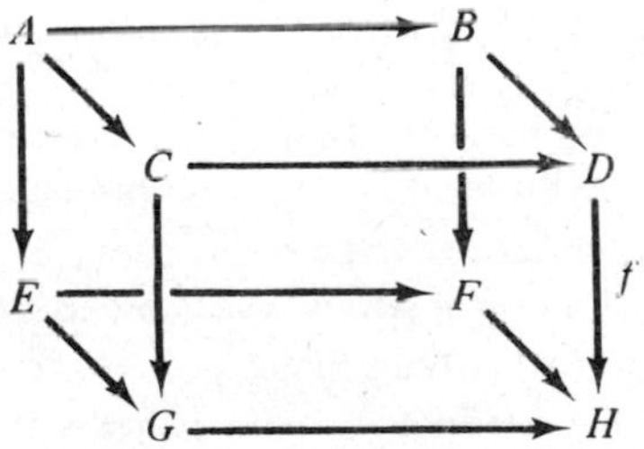

Suppose also that all the "faces" are commutative, except perhaps the top face. Prove that if f is injective then the top face is also commutative.

[*Hint*. f is left cancellative.]

18. Let E_1, E_2 be sets. Define a **coproduct** of E_1 and E_2 to be a set C together with mappings $i_1 : E_1 \to C$ and $i_2 : E_2 \to C$ such that, for any set X and any mappings $f_1 : E_1 \to X$ and $f_2 : E_2 \to X$, there is a unique

mapping $h: C \to X$ such that the diagram

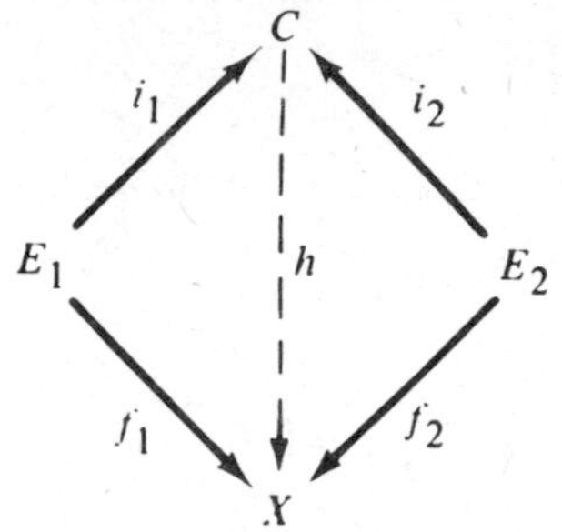

is commutative in that $h \circ i_1 = f_1$ and $h \circ i_2 = f_2$.

Prove that if (C, i_1, i_2) is a coproduct of E_1 and E_2 then i_1 and i_2 are injective.

Observing that the sets $E_1 \times \{1\}$ and $E_2 \times \{2\}$ are disjoint, define the **disjoint union** of E_1 and E_2 to be the set $E_1 \vee E_2 = (E_1 \times \{1\}) \cup (E_2 \times \{2\})$. Define also $i_1: E_1 \to E_1 \vee E_2$ and $i_2: E_2 \to E_1 \vee E_2$ by $i_1(e_1) = (e_1, 1)$ and $i_2(e_2) = (e_2, 2)$. Prove that $(E_1 \vee E_2, i_1, i_2)$ is a coproduct of E_1 and E_2.

If (C, i_1, i_2) and (D, j_1, j_2) are each coproducts of E_1 and E_2 prove that there is a unique bijection $\theta: D \to C$ such that $\theta \circ j_1 = i_1$ and $\theta \circ j_2 = i_2$.

§6. Indexed families; partitions; equivalence relations

We begin this section by introducing a convenient labelling device. Let I and E be sets and let $f: I \to E$ be a mapping, described by $i \mapsto f(i)$ for each $i \in I$. We often find it convenient to write x_i instead of $f(i)$ and write the mapping as $(x_i)_{i \in I}$ which we shall call a **family of elements of E indexed by I**. By abuse of language we refer to the x_i as the **elements of the family**. We shall assume in what follows that each indexing set is not empty.

Example 6.1. A very common index set in mathematics is the set $\mathbf{N}$ of natural numbers. A family of elements of E indexed by $\mathbf{N}$ (i.e., a mapping $n \mapsto x_n$ from $\mathbf{N}$ to E) is called a **sequence of elements of E** and is written $(x_n)_{n \in \mathbf{N}}$. To obtain a sequence of elements of E it is therefore equivalent to choose elements in E labelled $x_0, x_1, x_2, \ldots$.

> *Remark* 1. Note that there may well be repetitions among the elements of a family, in that it is possible to have $x_i = x_j$ when $i \neq j$. Of course, this will not happen when the mapping $i \mapsto x_i$ is

an injection, in which case we call $(x_i)_{i \in I}$ a **family of distinct elements of** E.

Remark 2. As we have already mentioned, many authors identify a mapping with its graph, thereby identifying the family $(x_i)_{i \in I}$ with the set $\{(i, x_i); i \in I\}$. In the case where the elements of the family are all distinct, some authors go even further and identify the mapping $(x_i)_{i \in I}$ with its image $\{x_i; i \in I\}$.

If, in the above definition, we replace E by $\mathbf{P}(E)$ then the mapping $f: I \to \mathbf{P}(E)$ may be described by $i \mapsto A_i$ where each A_i is a subset of E. This gives rise to an **indexed family** $(A_i)_{i \in I}$ **of subsets of** E.

If $(A_i)_{i \in I}$ is a family of subsets of a set E then we define the **intersection of the family** to be the subset of E given by

$$\bigcap_{i \in I} A_i = \{x \in E; (\forall i \in I)\ x \in A_i\}$$

and the **union of the family** to be the subset of E given by

$$\bigcup_{i \in I} A_i = \{x \in E; (\exists i \in I)\ x \subset A_i\}.$$

Thus we see that $\bigcap_{i \in I} A_i$ is the subset of E consisting of all the elements of E which belong to *every* element of the family; and $\bigcup_{i \in I} A_i$ is the subset of E consisting of all the elements of E which belong to *at least one* element of the family. In particular, it is clear that in the case where $I = \{1, 2\}$ we have $\bigcap_{i \in I} A_i = A_1 \cap A_2$ and $\bigcup_{i \in I} A_i = A_1 \cup A_2$.

Given any collection (= set) of sets, we shall assume the fundamental axiom that there exists a set which includes as a subset every set in the collection. [Note that this applies only to a *set* of sets, otherwise we lay ourselves open to the Russell paradox.] This axiom allows us to talk of an arbitrary **family of sets** (i.e., without specifying the set which includes each element of the family as a subset). The intersection and union of an arbitrary family of sets are then defined as above.

The following result is a generalisation of property (g) of §1 and shows that, given a family $(A_i)_{i \in I}$ of sets, the intersection of the family is the largest set included in every member of the family and, dually, the union of the family is the smallest set to include every member of the family.

Theorem 6.1 *If* $(A_i)_{i \in I}$ *is a family of sets then, for any set* X,

(1) $X \subseteq \bigcap_{i \in I} A_i \Leftrightarrow (\forall i \in I)\ X \subseteq A_i$;

(2) $\bigcup_{i\in I} A_i \subseteq X \Leftrightarrow (\forall i \in I)\ A_i \subseteq X.$

Proof. We prove (1), the proof of (2) being similar. Suppose that $X \subseteq A_i$ for every $i \in I$. Then given any $x \in X$ we have $x \in A_i$ for every $i \in I$ and so $x \in \bigcap_{i\in I} A_i$ whence $X \subseteq \bigcap_{i\in I} A_i$. Conversely, if $X \subseteq \bigcap_{i\in I} A_i$ then since clearly $\bigcap_{i\in I} A_i \subseteq A_j$ for every $j \in I$ we have $X \subseteq A_j$ for every $j \in I$.

Several other properties of sets given previously can be generalised to families of sets. Our next result illustrates one of these, namely the **general associativity** of intersection and union (see property (c) of §1). Roughly speaking, it says that in order to obtain the intersection (resp. union) of a family of sets we can write the family with an arbitrary grouping of terms and replace each group by its intersection (resp. union). Thus for example, as we would expect,

$A_1 \cap A_2 \cap A_3 \cap A_4 = (A_1 \cap A_2) \cap (A_3 \cap A_4) = (A_1 \cap A_2 \cap A_3) \cap A_4,$
etc.

Theorem 6.2 *Let* $(A_i)_{i\in I}$ *and* $(I_\lambda)_{\lambda\in\Lambda}$ *be indexed families of sets with* $I = \bigcup_{\lambda\in\Lambda} I_\lambda$. *Then*

$$\bigcap_{i\in I} A_i = \bigcap_{\lambda\in\Lambda}\Big(\bigcap_{i\in I_\lambda} A_i\Big) \quad \textit{and} \quad \bigcup_{i\in I} A_i = \bigcup_{\lambda\in\Lambda}\Big(\bigcup_{i\in I_\lambda} A_i\Big).$$

Proof. We establish the result for intersections, the proof for unions being similar. For every $\lambda \in \Lambda$ let $B_\lambda = \bigcap_{i\in I_\lambda} A_i$. Then, by formal application of the definitions, we have

$$\begin{aligned} x \in \bigcap_{i\in I} A_i &\Leftrightarrow (\forall i \in I)\quad x \in A_i \\ &\Leftrightarrow (\forall \lambda \in \Lambda)(\forall i \in I_\lambda)\quad x \in A_i \\ &\Leftrightarrow (\forall \lambda \in \Lambda)\quad x \in \bigcap_{i\in I_\lambda} A_i = B_\lambda \\ &\Leftrightarrow x \in \bigcap_{\lambda\in\Lambda} B_\lambda. \end{aligned}$$

Consequently,

$$\bigcap_{i\in I} A_i = \bigcap_{\lambda\in\Lambda} B_\lambda = \bigcap_{\lambda\in\Lambda}\Big(\bigcap_{i\in I_\lambda} A_i\Big).$$

We leave further general results of this nature to the exercises.

We now turn our attention to the problem of regarding a set as the union of "non-overlapping" subsets. To be more precise, by a **partitioning** of a non-empty set E we shall mean a family $(A_i)_{i\in I}$ of non-empty subsets of E such that

(1) $E = \bigcup_{i \in I} A_i$;

(2) the elements of $(A_i)_{i \in I}$ are **pairwise disjoint,** by which we mean that if $i, j \in I$ with $i \neq j$ then $A_i \cap A_j = \emptyset$.

If $(A_i)_{i \in I}$ is a partitioning of E then we note that repetitions cannot occur; for if $i \neq j$ and $A_i = A_j$ then $A_i = A_i \cap A_j = \emptyset$, a contradiction.

The image of the partitioning $(A_i)_{i \in I}$ is the set $\{A_i;\ i \in I\}$ and will be called a **partition** of E. [Note that the mapping does the partitioning and its effect (=image) is the partition.]

An intuitive picture of a partition of E is therefore that of a jig-saw puzzle: a collection of non-empty subsets which do not overlap and whose piecing together (=union) yields E.

Example 6.2. A set may be partitioned in several different ways. For example, if $E = \{a, b, c, d\}$ then $\{\{a\}, \{b, c\}, \{d\}\}$ and $\{\{a, b\}, \{c, d\}\}$ are each partitions of E.

Example 6.3. For each non-empty set E the family $(\{x\})_{x \in E}$ is a partitioning of E with associated partition $\{\{x\};\ x \in E\}$.

Example 6.4. If A is a subset of E with $\emptyset \subset A \subset E$ then $\{A, \complement_E(A)\}$ is a partition of E.

We shall now show that the notion of a partitioning is closely related to a particular type of relation which is an extremely powerful tool in algebra.

By a **binary relation** on a set E we shall mean a relation between E and itself. Of the properties which such a relation R may enjoy, the most commonly encountered in mathematics are the following: we say that R is

(a) **reflexive** if $(\forall x \in E)\quad xRx$;

(b) **symmetric** if $(\forall x, y \in E)\quad xRy \Rightarrow yRx$;

(c) **anti-symmetric** if $(\forall x, y \in E)\quad (xRy \text{ and } yRx) \Rightarrow x = y$;

(d) **transitive** if $(\forall x, y, z \in E)\quad (xRy \text{ and } yRz) \Rightarrow xRz$.

Note that to say that R is reflexive on E is equivalent to saying that the graph of R contains the diagonal of $E \times E$. In particular, for such a relation we have Dom $R = E$.

By an **equivalence relation** on a set E we shall mean a binary relation on E which is reflexive, symmetric and transitive. By an **order** on E we shall mean a binary relation on E which is reflexive, anti-symmetric and transitive. We shall look at orders in the next section; for the present, we shall concentrate on equivalence relations.

We often use the notation $x \equiv y\ (R)$ which is read "x is equivalent to y modulo R" instead of xRy. In this notation, then, R is an equivalence relation if and only if

(1) $(\forall x \in E)\quad x \equiv x(R)$;

(2) $(\forall x, y \in E)\quad x \equiv y(R) \Rightarrow y \equiv x(R)$;

(3) $(\forall x, y \in E)\quad (x \equiv y(R) \text{ and } y \equiv z(R)) \Rightarrow x \equiv z\ (R)$.

Example 6.5. Consider the binary relation "mod n" defined on $\mathbf{Z}$ by

$$x \equiv y \ (\text{mod } n) \Leftrightarrow (\exists k \in \mathbf{Z})\ x-y=kn.$$

In other words, x and y are equivalent modulo n if and only if they differ by an integer multiple of n. This is an equivalence relation on $\mathbf{Z}$. In fact,

(1) for every $x \in \mathbf{Z}$ we have $x-x=0=0n$, so $x \equiv x \ (\text{mod } n)$;
(2) for all $x, y \in \mathbf{Z}$ we have

$$\begin{aligned} x \equiv y \ (\text{mod } n) &\Rightarrow (\exists k \in \mathbf{Z})\ x-y=kn \\ &\Rightarrow (\exists k^* = -k \in \mathbf{Z})\ y-x=k^*n \\ &\Rightarrow y \equiv x \ (\text{mod } n); \end{aligned}$$

(3) for all $x, y, z \in \mathbf{Z}$,

$$\begin{aligned} (x \equiv y \ (\text{mod } n) \text{ and } y \equiv z \ (\text{mod } n)) &\Rightarrow (\exists k, t \in \mathbf{Z})\ x-y=kn,\ y-z=tn \\ &\Rightarrow (\exists p=k+t \in \mathbf{Z})\ x-z=pn \\ &\Rightarrow x \equiv z \ (\text{mod } n). \end{aligned}$$

Example 6.6. For any mapping $f: E \to F$ let R_f denote the binary relation on E given by

$$x \equiv y\,(R_f) \Leftrightarrow f(x)=f(y).$$

It is readily seen that R_f is an equivalence relation on E, called the **equivalence relation associated with the mapping** f.

If R is an equivalence relation on a set E then for every $x \in E$ we define the **equivalence class of** x **modulo** R to be the subset of E given by

$$x/R = \{y \in E;\ y \equiv x(R)\}.$$

Equivalent terminology in common use is the **R-class** of x and equivalent notation includes R_x, $[x]_R$ or simply $[x]$. The set of equivalence classes modulo R is (always) denoted by E/R so that our notation for the classes yields the visually appealing definition

$$E/R = \{x/R;\ x \in E\}.$$

The set E/R is called the **quotient set of** E **by** R.

Example 6.7. Consider the relation of equality on E defined by $x \equiv y(R) \Leftrightarrow x=y$. This is clearly an equivalence relation on E, the R-class of x being $x/R=\{x\}$. Thus $E/R=\{\{x\}; x \in E\}$ which should not be confused with $E=\{x; x \in E\}$ since, by definition, E/R is a set of sets.

Example 6.8. Consider the equivalence relation "mod n" of Example 6.5. Since, by definition, $y \in \mathbf{Z}$ belongs to the "mod n"-class of $x \in \mathbf{Z}$ if and only if y differs from x by an integer multiple of n, we see that the "mod n"-class of x is

$$x/\text{mod } n=\{\ldots, x-2n, x-n, x, x+n, x+2n, \ldots\}.$$

Theorem 6.3 *If R is an equivalence relation on E then*

(1) $(\forall x, y \in E)\quad x \in y/R \Leftrightarrow x \equiv y(R)$;

(2) $(\forall x, y \in E)\quad x/R=y/R \Leftrightarrow x \equiv y(R)$;

(3) $(\forall x, y \in E)\quad x/R \cap y/R \neq \emptyset \Rightarrow x/R=y/R$.

Proof. (1) is immediate from the definition of y/R. As for (2), suppose that $x/R=y/R$. Since R is reflexive on E we have, by (1), $t \in t/R$ for every $t \in E$. We therefore have $x \in y/R$ and so $x \equiv y(R)$. Conversely, if $x \equiv y(R)$ then for every $x^* \in x/R$ we have $x^* \equiv x(R)$ and so $x^* \equiv y(R)$ by transitivity, whence $x^* \in y/R$ and so $x/R \subseteq y/R$; the converse inclusion is obtained by interchanging x and y. Finally, to establish (3) it is equivalent, by virtue of (1) and (2), to show that if there exists $z \in E$ such that $z \equiv x(R)$ and $z \equiv y(R)$ then $x \equiv y(R)$. But if such an element z exists then by the symmetry of R we have $x \equiv z(R)$. Coupling this with $z \equiv y(R)$ and using the transitivity of R, we obtain $x \equiv y(R)$.

Corollary. *If R is an equivalence relation on E then any two equivalence classes modulo R are either disjoint or identical.*

Proof. If x/R and y/R are not disjoint then they have an element in common. The result is therefore immediate from (3) above.

The fundamental property which connects partitionings with equivalence relations is the following.

Theorem 6.4 *If R is an equivalence relation on a set E then the family* $(x/R)_{x \in E}$ *is a partitioning of E, the corresponding partition being E/R. Conversely, every partitioning of E yields an equivalence relation on E.*

Proof. Let R be an equivalence relation on E. By the reflexivity of R we have $x \in x/R$ for every $x \in E$. It follows immediately that $E=\bigcup_{x \in E} x/R$.

The previous Corollary now shows that $(x/R)_{x \in E}$ is a partitioning of E, the corresponding partition being $\{x/R;\ x \in E\}$.

Conversely, suppose that $P=(A_i)_{i\in I}$ is a partitioning of E. Define a binary relation R_P on E by setting

$$x \equiv y(R_P) \Leftrightarrow (\exists i \in I) \quad x, y \in A_i.$$

Since $E=\bigcup_{i\in I} A_i$ every element of E belongs to at least one of the sets A_i and so we have $x \equiv x(R_P)$ for every $x \in E$, whence R_P is reflexive on E. It is clear from the definition of R_P that R_P is symmetric. To show that R_P is also transitive, suppose that $x \equiv y(R_P)$ and $y \equiv z(R_P)$. Then there exist $i, j \in I$ with $x, y \in A_i$ and $y, z \in A_j$. Since then $A_i \cap A_j \neq \varnothing$ (because $y \in A_i$ and $y \in A_j$) we must have $A_i = A_j$ (for $(A_i)_{i\in I}$ is a partitioning) whence we deduce that $x \equiv z(R_P)$. Thus we see that R_P is an equivalence relation on E.

The notion of equivalence relation is a very powerful tool in mathematics. We are not in a position as yet to exhibit the full usefulness of this notion, for we simply do not have enough machinery at our disposal. However, we are in a position to show that, in a quite natural way, any mapping can be written as the composite of an injection, a bijection and a surjection.

For this purpose, we introduce the following important mapping. If R is an equivalence relation on a set E then we can define a mapping $\natural_R: E \to E/R$ by setting

$$(\forall x \in E) \qquad \natural_R(x) = x/R.$$

This mapping is clearly surjective. It is called the **canonical** (or **natural**) **surjection of** E **onto** E/R.

> *Remark.* Note that, from the definition of equality for mappings, we distinguish between the mapping $\natural_R$ from E to E/R described by $x \mapsto x/R$ and the partitioning $(x/R)_{x\in E}$ which is a mapping from E to $\mathbf{P}(E)$ also described by $x \mapsto x/R$.

We shall make use of the following result, which may be regarded as another special case of Theorem 5.3.

Theorem 6.5 *Let E be a set, let R be an equivalence relation on E, let $\natural_R: E \to E/R$ be the canonical surjection and let $f: E \to F$ be any mapping. The following conditions are then equivalent:*

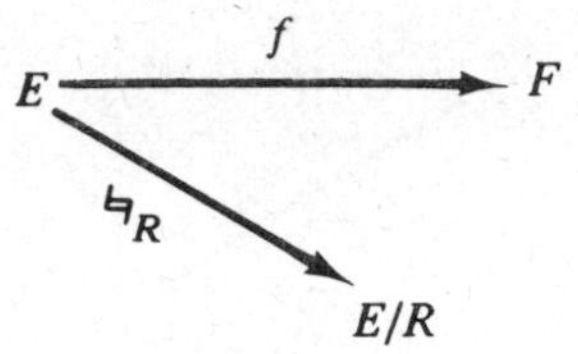

(1) *there exists* $f_*: E/R \to F$ *such that* $f_* \circ \natural_R = f$;

(2) $(\forall x, y \in E) \quad x \equiv y(R) \Rightarrow f(x) = f(y)$.

Moreover, if these conditions are satisfied, f_* *is unique and is*

(a) *injective if and only if* $(\forall x, y \in E)\ x \equiv y(R) \Leftrightarrow f(x) = f(y)$;

(b) *surjective if and only if* f *is surjective.*

Proof. The equivalence of (1) and (2) is immediate from Theorem 5.3 since clearly $x \equiv y(R) \Leftrightarrow \natural_R(x) = \natural_R(y)$.

Suppose then that (1) and (2) hold. The uniqueness of f_* follows from the fact that $\natural_R$ is surjective and hence right cancellative (see Theorem 5.6); for if g also satisfies $g \circ \natural_R = f$ then $g \circ \natural_R = f_* \circ \natural_R$ and so $g = f_*$.

To prove (a), we observe that f_* is injective if and only if

$$f_*(x/R) = f_*(y/R) \Rightarrow x/R = y/R;$$

i.e., since $f_*(x/R) = f(x)$ by (1), if and only if

$$f(x) = f(y) \Rightarrow x \equiv y(R).$$

Since (2) holds by hypothesis, the result follows.

As for (b), we note that since $f(x) = f_*(x/R)$ for all $x \in E$, every $y \in F$ is of the form $f(x)$ if and only if it is of the form $f_*(x/R)$. Hence f_* is surjective if and only if f is surjective.

As an application of the previous result, we now derive the following **canonical decomposition of a mapping.**

Theorem 6.6 *Every mapping can be written as the composite of an injection, a bijection and a surjection.*

Proof. Given any mapping $f: E \to F$, let $f^+: E \to \operatorname{Im} f$ be given by the prescription $f^+(x) = f(x)$ and let $\iota: \operatorname{Im} f \to F$ be the canonical injection. Then clearly the triangle

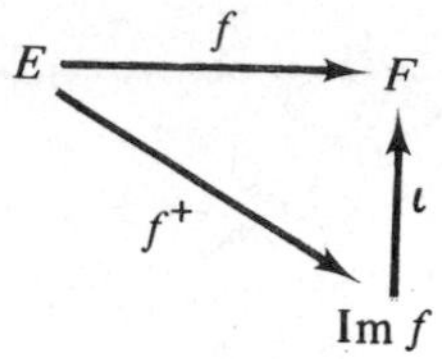

is commutative (in that $f=\iota\circ f^+$). Now let R_{f^+} be the equivalence relation associated with f^+ (Example 6.6) and consider the diagram

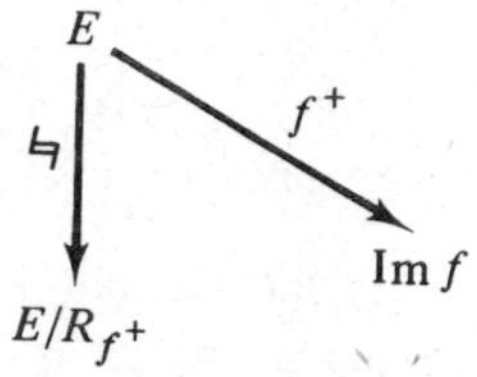

in which $\natural$ denotes the canonical surjection. We are in the situation of Theorem 6.5. Since condition (2) of that result is trivially satisfied (with R_{f^+} in place of R), so also is condition (1) and so there exists a mapping $f_*: E/R_{f^+} \to \mathrm{Im}\, f$ making the above triangle commutative. Moreover, since $(\forall x, y \in E)\ x \equiv y(R_{f^+}) \Leftrightarrow f^+(x)=f^+(y)$ we see from condition (a) of Theorem 6.5 that f_* is injective; and since f^+ is surjective so also is f_* by condition (b) of Theorem 6.5. Thus f_* is a bijection. We can now compound the above diagrams to form the diagram

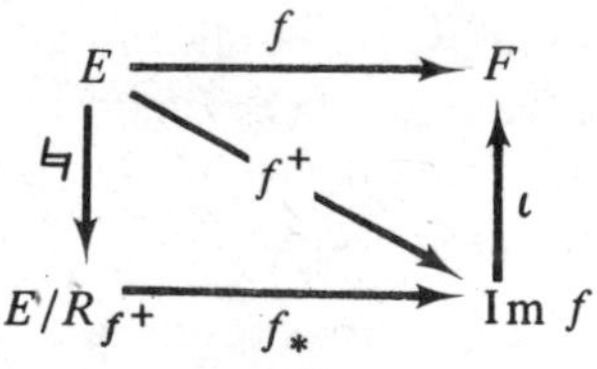

in which $f=\iota\circ f^+=\iota\circ f_*\circ\natural$. Thus f can be written as the composite of an injection, a bijection and a surjection.

We end the present section by returning to an arbitrary family of sets $(A_i)_{i\in I}$ and the problem of how to generalise the notion of cartesian product set. Guided by the characterisation of $E_1\times E_2$ as a product given at the end of the previous section, we introduce the following concept.

Definition. Let $(E_i)_{i\in I}$ be a family of sets. By a **product** of the family we shall mean a set P together with a family $(\pi_i)_{i\in I}$ of mappings with $\pi_i:P\to E_i$ for each $i\in I$ such that, for every set X and every family $(f_i)_{i\in I}$ of mappings with $f_i:X\to E_i$ for each $i\in I$, there is a unique mapping $h:X\to P$ such that, for every $i\in I$, the diagram

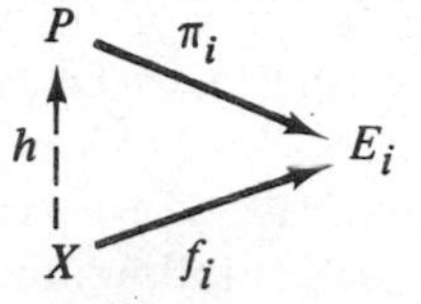

is commutative. We denote such a product by $(P,(\pi_i)_{i\in I})$.

Precisely as in Theorem 5.11, we see that if $(P, (\pi_i)_{i\in I})$ is a product of $(E_i)_{i\in I}$ then each of the mappings π_i is surjective.

We now discuss the existence of products. Given a family $(E_i)_{i\in I}$ of sets we denote by $\bigtimes_{i\in I} E_i$ the set consisting of all the mappings $f: I\to \bigcup_{i\in I} E_i$ such that $(\forall i\in I)\ f(i)\in E_i$. In other words, $\bigtimes_{i\in I} E_i$ is the set of all families $(x_i)_{i\in I}$ of elements of $\bigcup_{i\in I} E_i$ such that $(\forall i\in I)\ x_i\in E_i$. For every $j\in I$ we define $pr_j:\bigtimes_{i\in I} E_i\to E_j$ by the prescription $pr_j((x_i)_{i\in I})=x_j$. Then we have:

Theorem 6.7 *If* $(E_i)_{i\in I}$ *is a family of sets then* $\left(\bigtimes_{i\in I} E_i, (pr_i)_{i\in I}\right)$ *is a product of* $(E_i)_{i\in I}$.

Proof. Given any set X and any family $(f_i)_{i\in I}$ of mappings with $f_i:X\to E_i$ for each $i\in I$, consider the diagrams (one for each $i\in I$)

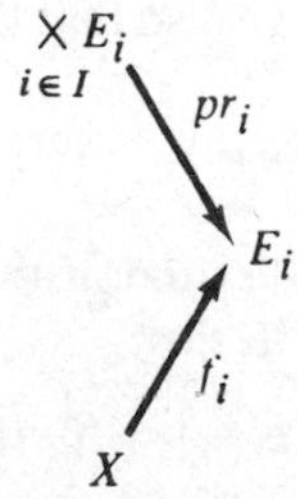

Define $h:X\to \bigtimes_{i\in I} E_i$ by $h(x)=(f_i(x))_{i\in I}$. Then the proof is essentially as in Theorem 5.12.

The reader will have no difficulty in extending the result of Theorem 5.13 to the case of products of arbitrary families of sets. These results therefore give a characterisation of the set $\bigtimes_{i\in I} E_i$ together with the family $(pr_i)_{i\in I}$ of (surjective) mappings as a product of the family $(E_i)_{i\in I}$, thereby generalising the situation described in §5. We shall therefore agree to call $\bigtimes_{i\in I} E_i$ the **cartesian product set** of the family $(E_i)_{i\in I}$ and, for each $j\in I$, the mapping $pr_j\colon \bigtimes_{i\in I} E_i \to E_j$ the **j-th projection.**

Note that in the particular case where $I=\{1, 2\}$ the family $(x_i)_{i\in\{1,2\}}$ has the graph $\{(1, x_1), (2, x_2)\}$. It follows that $(x_i)_{i\in\{1,2\}}$ and $(y_i)_{i\in\{1,2\}}$ are equal if and only if $x_1=y_1$ and $x_2=y_2$. Recalling the remarks made concerning the definition of ordered pairs, we can follow the common practice of identifying $(x_i)_{i\in\{1,2\}}$ with the ordered pair (x_1, x_2). More generally, when $I=\{1, \ldots, n\}$ we write the family $(x_i)_{i\in\{1,\ldots,n\}}$ as $(x_1, \ldots, x_n)$ and call this an **n-tuple**. When $I=\{1, \ldots, n\}$ we write $\bigtimes_{i\in I} E_i$ as $\bigtimes_{i=1}^{n} E_i$ or $E_1\times \ldots \times E_n$. In particular, when all the E_i are equal, say $E_i=E$ for $i=1, \ldots, n$, then we write $E_1\times \ldots \times E_n$ as E^n. Thus, for example,

$$E_1\times E_2\times E_3=\{(e_1, e_2, e_3);\ e_i\in E_i\};$$

$$E^3=\{(x, y, z);\ x, y, z\in E\}.$$

Exercises for §6

1. Let $(A_i)_{i\in I}$ be a family of subsets of a set E. Given any mapping $f\colon E\to F$, prove that

$$f^{\to}\Big(\bigcup_{i\in I} A_i\Big)=\bigcup_{i\in I} f^{\to}(A_i);\quad f^{\to}\Big(\bigcap_{i\in I} A_i\Big)\subseteq \bigcap_{i\in I} f^{\to}(A_i).$$

Prove that equality holds for all families $(A_i)_{i\in I}$ in the second of these if and only if f is injective. Prove also that if $(B_i)_{i\in I}$ is a family of subsets of F then

$$f^{\leftarrow}\Big(\bigcup_{i\in I} B_i\Big)=\bigcup_{i\in I} f^{\leftarrow}(B_i);\quad f^{\leftarrow}\Big(\bigcap_{i\in I} B_i\Big)=\bigcap_{i\in I} f^{\leftarrow}(B_i).$$

2. Prove the following generalisation of the de Morgan laws: if $(A_i)_{i\in I}$ is a family of subsets of a set E then

$$\complement_E\Big(\bigcup_{i\in I} A_i\Big)=\bigcap_{i\in I}\complement_E(A_i);\quad \complement_E\Big(\bigcap_{i\in I} A_i\Big)=\bigcup_{i\in I}\complement_E(A_i).$$

3. Determine $\bigcup_{r\in\mathbf{R}} X_r$ and $\bigcap_{r\in\mathbf{R}} X_r$ where, for every $r\in\mathbf{R}$, X_r is given by

(1) $\{(x, y)\in \mathbf{R}\times\mathbf{R};\ x^2+y^2=r^2\}$;

(2) $\{(x, y) \in \mathbf{R} \times \mathbf{R};\ x^2+y^2 \leqslant r^2\}$;
(3) $\{(x, y) \in \mathbf{R} \times \mathbf{R};\ x^2+y^2 \neq r^2\}$;
(4) $\{(x, y) \in \mathbf{R} \times \mathbf{R};\ x^2+y^2 > r^2\}$.

4. In the following list of sets and binary relations on them, say which of the relations are reflexive, which are symmetric and which are transitive:

(a) positive integers: $hcf\{x, y\}=1$;
(b) real numbers: $x<y$;
(c) $\mathbf{P}(E)$: $x \subseteq y$;
(d) lines in a plane: $x=y$ or x parallel to y;
(e) lines in a plane: x perpendicular to y;
(f) triangles in a plane: x congruent to y;
(g) females: x is a sister of y.

5. Let $E=\{1, 2\}$ and let R be the binary relation on E given by

$$x \equiv y(R) \Leftrightarrow x=y=2.$$

Show that R is not reflexive but is symmetric and transitive.

6. In the light of the previous exercise, criticise the following argument: if a binary relation R is symmetric and transitive then from $x \equiv y(R)$ we have $y \equiv x(R)$ and hence $x \equiv x(R)$; thus R is also reflexive.

7. Let E be the set of months in the year. Show that the binary relation R defined on E by $x \equiv y(R)$ if and only if x and y start on the same day of the week is an equivalence relation on E. Determine the equivalence classes for an ordinary year and for a leap year. How many Friday the thirteenths can there be in a year?

8. Show that the binary relation R defined on $\mathbf{N}\backslash\{0\}$ by

$$x \equiv y(R) \Leftrightarrow (\exists m \in \mathbf{Z})\ x=2^m y$$

is an equivalence relation. Show also that each equivalence class contains precisely one odd integer. If $c \equiv d(R)$ show that either c divides d or d divides c. Deduce that in any set of $n+1$ non-zero natural numbers, each of which is less than or equal to $2n$, there is at least one which is a factor of another.

9. Show that the binary relation R_n defined on $\mathbf{Z}$ by

$$x \equiv y(R_n) \Leftrightarrow x^2 \equiv y^2 \pmod{n}$$

is an equivalence relation. Show that there are four equivalence classes for R_6 and determine them. Show also that every R_6-class is a union of "mod 6"-classes.

10. If t is a given real number, consider the mapping $f_t: \mathbf{R} \to \mathbf{R}$ given by

$$f_t(x) = x^2 + tx + t^2.$$

Prove that

(a) $f_t^{\leftarrow}\{1\} = \emptyset$ if and only if $|t| > 2/\sqrt{3}$;
(b) $f_t^{\leftarrow}\{1\}$ is a singleton if and only if $|t| = 2/\sqrt{3}$;
(c) $f_t^{\leftarrow}\{1\}$ consists of two elements if and only if $|t| < 2/\sqrt{3}$.

Consider now the binary relation S defined on $\mathbf{R}$ by

$$x \equiv y(S) \Leftrightarrow x^3 - y^3 = x - y.$$

Show that S is an equivalence relation. Prove also that the S-class of $x \in \mathbf{R}$ consists of

(1) a single element if and only if $|x| > 2/\sqrt{3}$;
(2) two elements if and only if $|x| = 2/\sqrt{3}$ or $|x| = 1/\sqrt{3}$;
(3) three elements otherwise.

11. Show that the binary relation R defined on $\mathbf{R} \times \mathbf{R}$ by

$$(a, b) \equiv (c, d)(R) \Leftrightarrow a^2 + b^2 = c^2 + d^2$$

is an equivalence relation and determine the R-classes.

12. Show that the binary relation R defined on the set B of all bijections from $\mathbf{R}$ to itself by

$$f \equiv g(R) \Leftrightarrow (\exists h \in B)\; g = h^{-1} \circ f \circ h$$

is an equivalence relation. What is the R-class of $id_{\mathbf{R}}$?

13. Let $(E_i)_{i \in I}$ be a family of subsets of a set E such that $E = \bigcup_{i \in I} E_i$. Show that the binary relation R defined on E by

$$x \equiv y(R) \Leftrightarrow (\forall i \in I)\, \big(x, y \in E_i \text{ or } x, y \in \complement_E(E_i)\big)$$

is an equivalence relation. If X is a non-empty subset of E, prove that X is an R-class if and only if X is of the form

$$X_J = \bigcap_{i \in J} E_i \cap \bigcap_{i \in I \setminus J} \complement_E(E_i)$$

for some non-empty subset J of I.

[*Hint.* If X is an R-class let $x \in X$ and consider all the subsets E_i which contain x. Let J be the largest subset of I such that $x \in \bigcap_{i \in J} E_i$ and show that $x \in X_J$.

Now use the Corollary to Theorem 6.3.]

Hence show that each E_i is the union of a family of R-classes.

[*Hint.* For each $k \in I$ let $(X_J)_{J \in \Delta_k}$ be the family of all R-classes of elements of E_k. Show that $E_k = \bigcup_{J \in \Delta_k} X_J$.]

14. Let E, F be sets and $f: E \to F$ any mapping. If R, S are equivalence relations on E, F respectively, prove that the following are equivalent:
(1) there is a unique mapping $h: E/R \to F/S$ such that the diagram

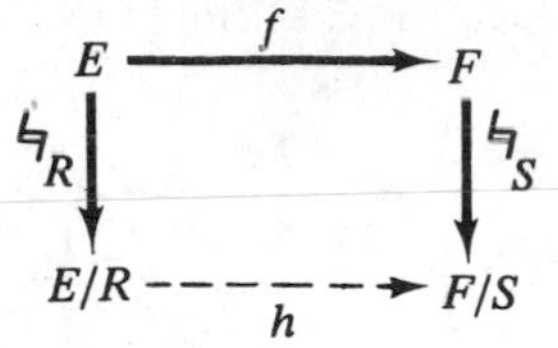

is commutative;
(2) $(\forall x, y \in E) \quad x \equiv y(R) \Rightarrow f(x) \equiv f(y)\,(S)$.

15. Let $(A_i)_{i \in I}$ be a family of sets. If $g: I \to I$ is a bijection, show that there is a bijection

$$g^*: \mathop{\times}_{i \in I} A_i \to \mathop{\times}_{i \in I} A_{g(i)}.$$

[*Hint.* Given any $f \in \mathop{\times}_{i \in I} A_i$ show that $f \circ g \in \mathop{\times}_{i \in I} A_{g(i)}$ and define g^* by $g^*(f) = f \circ g$.]

16. In this exercise we generalise the distributive laws of §1. Suppose that $((A_{i,\alpha})_{\alpha \in J_i})_{i \in I}$ is a family (indexed by I) of families of sets with $J = \mathop{\times}_{i \in I} J_i$. Prove that

$$\bigcup_{i \in I}\Big(\bigcap_{\alpha \in J_i} A_{i,\alpha}\Big) = \bigcap_{f \in J}\Big(\bigcup_{i \in I} A_{i,f(i)}\Big)$$

with a similar formula with $\cup$ and $\cap$ interchanged.

[*Hint.* Show that $x \in$ L.H.S. $\Rightarrow (\exists i \in I)(\forall f \in J)x \in A_{i,f(i)}$ and use the logical theorem $\exists\forall \Rightarrow \forall\exists$ to deduce that L.H.S. $\subseteq$ R.H.S. For the converse inclusion show that $x \notin$ L.H.S. $\Rightarrow (\forall i \in I)\{\alpha \in J_i;\ x \notin A_{i,\alpha}\} \neq \emptyset \Rightarrow (\exists f \in J)(\forall i \in I)x \notin A_{i,f(i)} \Rightarrow x \notin$ R.H.S.]

17. With the notation of the previous exercise, prove that

$$\mathop{\times}_{i \in I}\Big(\bigcup_{\alpha \in J_i} A_{i,\alpha}\Big) = \bigcup_{f \in J}\Big(\mathop{\times}_{i \in I} A_{i,f(i)}\Big)$$

with a similar formula with $\cup$ replaced by $\cap$.

[*Hint.* Show that $g \in$ L.H.S. $\Rightarrow (\forall i \in I)\{\alpha \in J_i;\, g(i) \in A_{i,\alpha}\} \neq \varnothing \Rightarrow (\exists f \in J)(\forall i \in I)g(i) \in A_{i,f(i)} \Rightarrow g \in$ R.H.S. For the converse inclusion, let $g \in$ R.H.S. and use the logical theorem $\exists\forall \Rightarrow \forall\exists$.]

18. **[Correspondence Theorem]** Let E be a set and let R be an equivalence relation on E. Let $\mathscr{A}$ be the set of partitionings of E associated with equivalence relations S on E such that $x \equiv y(R) \Rightarrow x \equiv y(S)$. Prove that there is a bijection θ from $\mathscr{A}$ onto the set of partitionings of E/R.

[*Hint.* Let f_S be the partitioning associated with S. Show that there is a unique mapping $h_S\colon E/R \to \mathbf{P}(E/R)$ making the diagram

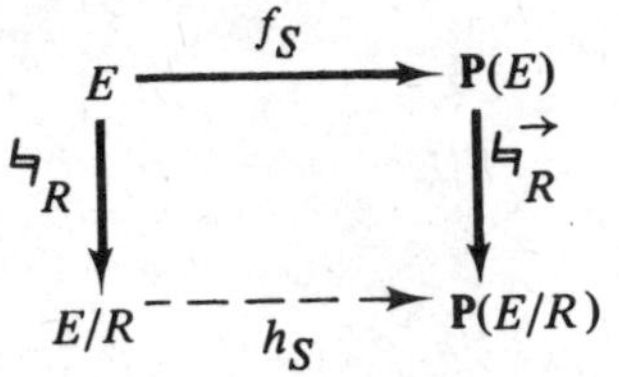

commutative. Show that h_S is a partitioning of E/R. Now define θ by $\theta(f_S) = h_S$. To show that θ is surjective let h be any partitioning of E/R and consider the map $\beta = \overleftarrow{\natural}_R \circ h \circ \natural_R$. Show that β is an element of $\mathscr{A}$, that $\beta = f_{R_\beta}$ and that $\theta(f_{R_\beta}) = h$.]

19. Given a mapping $f\colon E \to F$ let

$$E \xrightarrow{\natural} E/R_{f^+} \xrightarrow{f_*} \operatorname{Im} f \xrightarrow{\iota} F$$

be its canonical decomposition as described in Theorem 6.6. If A, B are sets such that in the diagram

$$E \xrightarrow{\alpha} A \xrightarrow{\beta} B \xrightarrow{\gamma} F$$

α is surjective, β is bijective, γ is injective and $f = \gamma \circ \beta \circ \alpha$, prove that there exist unique bijections $\theta\colon E/R_{f^+} \to A$ and $\zeta\colon \operatorname{Im} f \to B$ such that the following diagram is commutative:

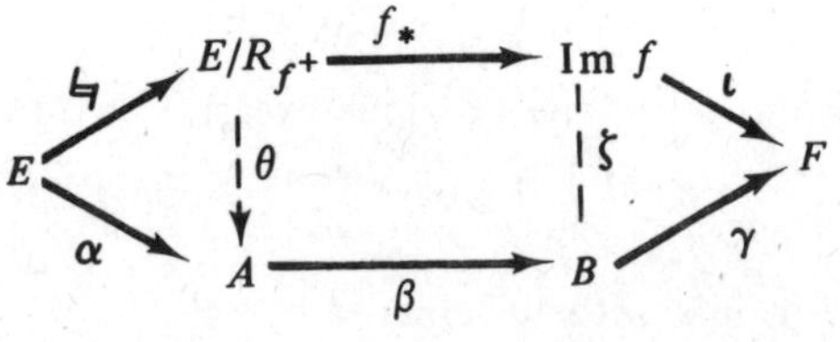

[*Hint.* Observe that $\operatorname{Im}\gamma = \operatorname{Im} f = \operatorname{Im}\iota$ and apply Theorem 5.5 (twice) to establish the existence of a unique bijection $\zeta : \operatorname{Im} f \to B$ such that $\gamma \circ \zeta = \iota$. Now let $\theta = \beta^{-1} \circ \zeta \circ f_*$ so that $\beta \circ \theta = \zeta \circ f_*$. Show that $\gamma \circ (\beta \circ \theta \circ \natural) = f = \gamma \circ \beta \circ \alpha$ and use Theorem 5.4(3) to obtain $\theta \circ \natural = \alpha$.]

20. Given the diagram of sets and mappings

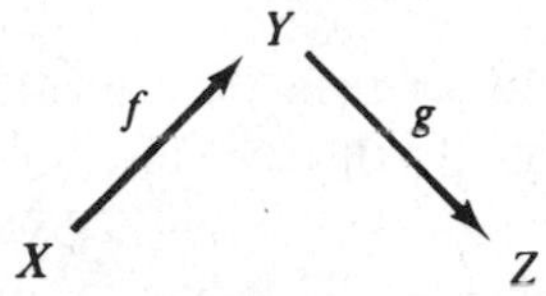

prove that the following conditions are equivalent:

(1) there exists $h : Z \to X$ such that $f \circ h \circ g = id_Y$;

(2) f is surjective and g is injective.

Deduce that if $\alpha : A \to B$ is any mapping then there exists a mapping $\beta : B \to A$ such that $\alpha \circ \beta \circ \alpha = \alpha$.

[*Hint.* Consider the canonical decomposition of α.]

§7. Order relations; ordered sets; order isomorphisms; lattices

Let E be a non-empty set. By an **order relation** (or simply an **order**) on E we shall mean a binary relation R on E which is reflexive, anti-symmetric and transitive. A set E together with an order on E is called an **ordered set**. If R is an order on E then we usually write R as $\leqslant$ which we read **is less than or equal to.** Thus $\leqslant$ is an order on E if and only if

(1) $(\forall x \in E)\quad x \leqslant x$;

(2) $(\forall x, y \in E)\quad (x \leqslant y \text{ and } y \leqslant x) \Rightarrow x = y$;

(3) $(\forall x, y, z \in E)\quad (x \leqslant y \text{ and } y \leqslant z) \Rightarrow x \leqslant z$.

We shall often denote an ordered set by $(E, \leqslant)$.

Example 7.1. Given any set E the relation $\subseteq$ of set inclusion is an order on $\mathbf{P}(E)$.

Example 7.2. Let $\mathbf{P} = \mathbf{N} \backslash \{0\}$ be the set of non-zero natural numbers. Let $|$ be the **divisibility relation** defined on $\mathbf{P}$ by

$$m \mid n \Leftrightarrow (\exists p \in \mathbf{P})\ n = mp.$$

Since $m = m1$ for every $m \in \mathbf{P}$ we see that $\mid$ is reflexive. If $m, n \in \mathbf{P}$ are such that $m \mid n$ and $n \mid m$ then there exist $a, b \in \mathbf{P}$ such that $n = ma$ and $m = nb$ from which we obtain $n = nba$. We must therefore have $b = a = 1$ whence $n = m$ and so $\mid$ is anti-symmetric. Finally, if $m, n, p \in \mathbf{P}$ are such that $m \mid n$ and $n \mid p$ then there exist $a, b \in \mathbf{P}$ such that $n = ma$ and $p = nb$ whence $p = mab$ and so $m \mid p$. Thus $\mid$ is also transitive and so is an order on $\mathbf{P}$.

If $(E, \leqslant)$ is an ordered set then we can often depict the order by means of a **Hasse diagram**, the interpretation of which is as follows: we represent the elements of E by points and represent $x \leqslant y$ by an ascending line segment, as in the diagram

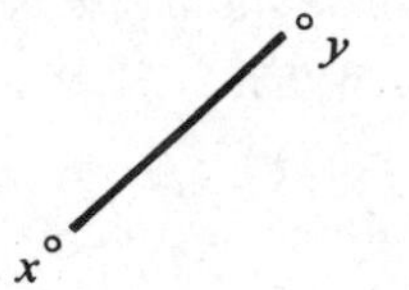

Example 7.3. Consider the set $E = \{1, 2, 3, 4, 6, 12\}$ consisting of all the positive factors of 12. If we order E by divisibility as in Example 7.2, the corresponding Hasse diagram is

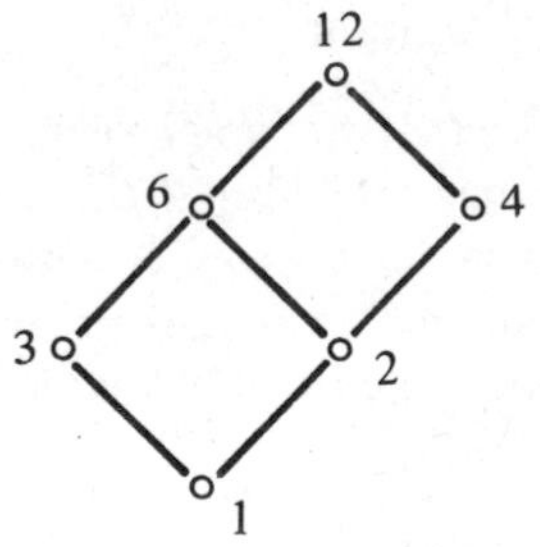

Note that in such a diagram we do not insert lines which are superfluous because of the transitivity of the order; for example, we do not join 2 and 12 directly by a line segment since $2 \mid 12$ is implied by $2 \mid 6$, $6 \mid 12$ and transitivity.

Example 7.4. If $E = \{1, 2, 3\}$ then

$$\mathbf{P}(E) = \{\emptyset, \{1\}, \{2\}, \{3\}, \{1, 2\}, \{1, 3\}, \{2, 3\}, E\}.$$

Ordering $\mathbf{P}(E)$ by set inclusion, we obtain the Hasse diagram

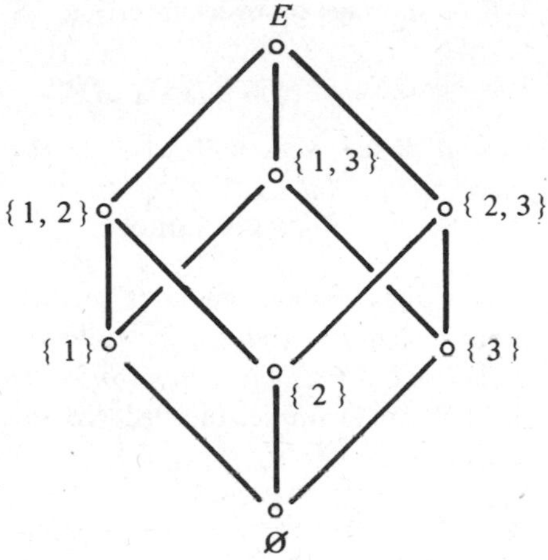

If $(E, \leqslant)$ is an ordered set and $x, y \in E$ are such that $x \leqslant y$ and $x \neq y$ then we shall write $x < y$ and say that x is **strictly less than** y. We note that the binary relation $<$ on E is transitive but is neither reflexive nor anti-symmetric. We shall also write $y \geqslant x$ as an alternative to $x \leqslant y$; the relation $\geqslant$ so defined is read **is greater than or equal to** and is also an order. Likewise, $y > x$ will signify $y \geqslant x$ and $y \neq x$. [In the terminology of Exercise 4.4, $\geqslant$ is the **converse** of $\leqslant$.]

A particularly important type of ordered set $(E, \leqslant)$ is that in which, for any two elements $x, y \in E$, either $x \leqslant y$ or $y \leqslant x$. Such an ordered set is said to be **totally ordered** or to form a **chain**. For example, $\mathbb{N}$ forms the chain $0 < 1 < 2 < 3 < \dots$. Note that in a totally ordered set $x \nleqslant y$ is equivalent to $y < x$.

> *Remark*. What we have defined as an **order** on a set many authors call a **partial order** and what we have called an **ordered set** is often called a **partially ordered set** or **poset**. As there is nothing which can even remotely be called "partial" in the definition of an order relation, we shall not use this terminology. We note also that totally ordered sets are often called **simply ordered** sets or **linearly ordered** sets.

We now consider mappings between ordered sets. Of especial importance are the following types of mapping. If $(E, \leqslant_1)$ and $(F, \leqslant_2)$ are ordered sets then a mapping $f: E \to F$ is called **isotone** or **order-preserving** (or **non-decreasing**) if and only if

$$(\forall x, y \in E) \quad x \leqslant_1 y \Rightarrow f(x) \leqslant_2 f(y).$$

We say that $f: E \to F$ is **antitone** or **order-inverting** (or **non-increasing**) if and only if

$$(\forall x, y \in E) \quad x \leqslant_1 y \Rightarrow f(x) \geqslant_2 f(y).$$

Example 7.5. If A and E are sets with $A \subseteq E$ then the mapping $t_A: \mathbf{P}(E) \to \mathbf{P}(E)$ given by $t_A(X) = A \cap X$ is isotone. The mapping $\complement_E: \mathbf{P}(E) \to \mathbf{P}(E)$ given by $X \mapsto \complement_E(X)$ is antitone.

As we shall be particularly interested later in bijections from one ordered set to another, we note carefully the following fact: *if E and F are ordered sets and $f: E \to F$ is an isotone bijection then f^{-1} need not be isotone.* Consider, for example, the ordered sets E, F described by the Hasse diagrams

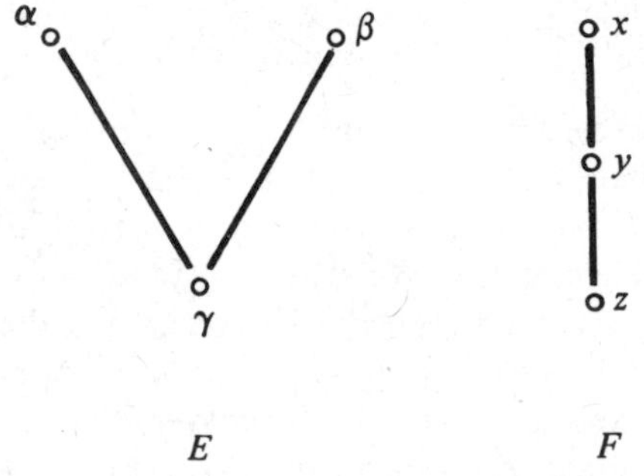

Let $f: E \to F$ be given by $f(\alpha) = x, f(\beta) = y, f(\gamma) = z$. It is readily seen that f is an isotone bijection. However, f^{-1} is not isotone; for example, we have $y < x$ but $f^{-1}(y) = \beta \nleqslant \alpha = f^{-1}(x)$.

Isotone bijections f for which f^{-1} is also isotone are of particular importance. If E, F are ordered sets then we say that $f: E \to F$ is an **order isomorphism** if it is an isotone bijection such that f^{-1} is also isotone. We say that ordered sets E, F are **order isomorphic** if there is an order isomorphism $f: E \to F$. We have the following characterisation of order isomorphisms.

Theorem 7.1 *If $(E, \leqslant_1)$ and $(F, \leqslant_2)$ are ordered sets then $f: E \to F$ is an order isomorphism if and only if*

(1) *f is surjective;*
(2) $(\forall x, y \in E) \quad x \leqslant_1 y \Leftrightarrow f(x) \leqslant_2 f(y)$.

Proof. If f is an order isomorphism then clearly f is surjective and

$(\forall x, y \in E)\ x \leqslant_1 y \Rightarrow f(x) \leqslant_2 f(y)$. Since f^{-1} is also isotone we have

$$f(x) \leqslant_2 f(y) \Rightarrow x = f^{-1}[f(x)] \leqslant_1 f^{-1}[f(y)] = y.$$

Thus we see that conditions (1) and (2) are necessary.

To prove that (1) and (2) are also sufficient, suppose that f is such that (1) and (2) hold. We deduce first from (2) that f is injective; for

$$\begin{aligned} f(x) = f(y) &\Rightarrow (f(x) \leqslant_2 f(y) \text{ and } f(y) \leqslant_2 f(x)) \\ &\Rightarrow (x \leqslant_1 y \text{ and } y \leqslant_1 x) \\ &\Rightarrow x = y. \end{aligned}$$

Thus we see that f is an isotone bijection. Now given any $\alpha, \beta \in F$ there exist unique elements $x, y \in E$ such that $\alpha = f(x)$ and $\beta = f(y)$. These unique elements are precisely $x = f^{-1}(\alpha)$ and $y = f^{-1}(\beta)$ and so, by (2),

$$\alpha \leqslant_2 \beta \Rightarrow f(x) \leqslant_2 f(y) \Rightarrow f^{-1}(\alpha) = x \leqslant_1 y = f^{-1}(\beta).$$

Thus f^{-1} is also isotone and so f is an order isomorphism.

In the case of totally ordered sets, a useful variant of Theorem 7.1 is the following.

Theorem 7.2 *If* $(E, \leqslant_1)$ *and* $(F, \leqslant_2)$ *are totally ordered sets then* $f: E \to F$ *is an order isomorphism if and only if*

(1) f *is surjective*;

(2) $(\forall x, y \in E)\ x <_1 y \Rightarrow f(x) <_2 f(y)$.

Proof. If f is an order isomorphism then clearly f is surjective and $x <_1 y \Rightarrow f(x) \leqslant_2 f(y)$, from which we have in fact that $f(x) <_2 f(y)$ since, f being injective, $f(x) = f(y)$ yields the contradiction $x = y$. The conditions are therefore necessary.

Conversely, suppose that f satisfies (1) and (2). Note that from (2) we have $x \leqslant_1 y \Rightarrow f(x) \leqslant_2 f(y)$. Suppose now that $f(x) \leqslant_2 f(y)$. Then we must have $x \leqslant_1 y$; for otherwise, by the fact that E is totally ordered, we would have $y <_1 x$ whence, by (2), the contradiction $f(y) <_2 f(x)$. Thus we have $x \leqslant_1 y \Leftrightarrow f(x) \leqslant_2 f(y)$ and this, with (1), shows that f is an order isomorphism by Theorem 7.1.

Remark. Note that if $(E, \leqslant_1)$ and $(F, \leqslant_2)$ are totally ordered sets and if $(\forall x, y \in E)\ x <_1 y \Rightarrow f(x) <_2 f(y)$ [i.e., f is **strictly isotone** (or **strictly increasing**)] then f is injective. For if $x \neq y$ then either $x <_1 y$ or $y <_1 x$ whence $f(x) <_2 f(y)$ or $f(y) <_2 f(x)$, so that $f(x) \neq f(y)$.

If E and F are ordered sets and $f: E \to F$ is an order isomorphism then we often say that F is an "isomorphic copy" of E. In an intuitive sense, this means that E, F differ only in notation for the elements and the orders.

Example 7.6. If $(E, \leqslant)$ is an ordered set then for every $x \in E$ let $[\leftarrow, x]=\{y \in E;\ y \leqslant x\}$. The mapping $f:(E, \leqslant)\rightarrow(\mathbf{P}(E), \subseteq)$ given by $f(x)=[\leftarrow, x]$ then satisfies $x \leqslant y \Leftrightarrow f(x) \subseteq f(y)$. If we let $\mathbf{P}^*(E)= \{[\leftarrow, x];\ x \in E\}$ and order $\mathbf{P}^*(E)$ by set inclusion then the mapping $g:E\rightarrow\mathbf{P}^*(E)$ given by $g(x)=[\leftarrow, x]$ is an order isomorphism. By way of illustration, let $E=\{1, 2, 3\}$; then we have the Hasse diagrams

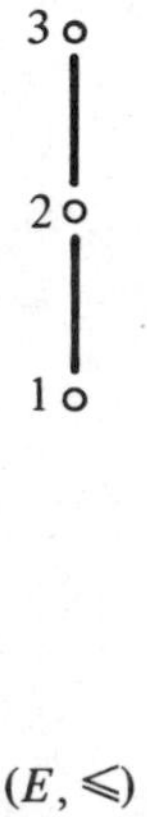

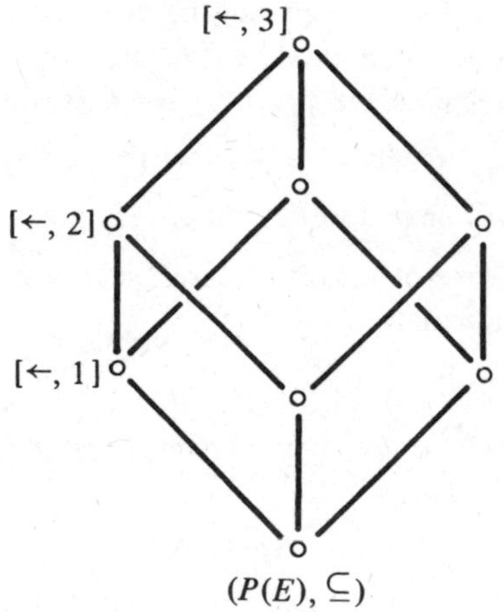

If $(E, \leqslant)$ is an ordered set then we say that $x \in E$ is a **least element** of E if $(\forall y \in E)\ x \leqslant y$. Equivalent terminology is **smallest element** or **minimum element.** We note that if such an element exists then it is necessarily unique; for if x and z are both least elements of E then for all $y \in E$ we have $x \leqslant y$ and $z \leqslant y$ whence in particular $x \leqslant z$ and $z \leqslant x$, so $z=x$. Dually, we say that $x \in E$ is a **greatest element** (or **largest element** or **maximum element**) of E if $(\forall y \in E)\ y \leqslant x$. If such an element exists then it is necessarily unique; for if x and z are each greatest elements of E then for all $y \in E$ we have $y \leqslant x$ and $y \leqslant z$ whence in particular $z \leqslant x$ and $x \leqslant z$, so $z=x$.

If X is a non-empty subset of the ordered set $(E, \leqslant)$ then we say that $x \in E$ is an **upper bound of** X if $(\forall y \in X)\ y \leqslant x$. Dually, we say that $x \in E$ is a **lower bound of** X if $(\forall y \in X)\ x \leqslant y$. We say that $x \in E$ is a **least upper bound of** X if x is the least element in the set of upper bounds of X; and dually that $x \in E$ is a **greatest lower bound of** X if x is the greatest element in the set of lower bounds of X. It follows from the above that least upper bounds and greatest lower bounds, when they exist, are unique.

> *Remark.* It should be noted that least upper bounds and greatest lower bounds for subsets X of E need not exist; and, even if they do exist, then although they are necessarily unique they need not be elements of X. For example, consider the set E consisting of the positive factors of 12 ordered by divisibility (Example 7.3). The subset $X=\{2, 3\}$ has as upper bounds the elements 6 and 12 and a least upper bound exists, namely 6, but $6 \notin \{2, 3\}$. On the other hand, in the set **P** of Example 7.2 we see that the subset $X=\{2n;\ n \in \mathbf{P}\}$ has no upper bounds in **P**.

If X is a non-empty subset of an ordered set E then the least upper bound of X, when it exists, is also called the **supremum** of X and is denoted by $\sup X$. Dually, the greatest lower bound of X, when it exists, is also called the **infimum** of X and is denoted by $\inf X$. A particularly important type of ordered set is one in which every two-element subset $X=\{x, y\}$ has both a supremum and an infimum: such an ordered set is called a **lattice**. In this case $\sup\{x, y\}$ is commonly written as $x \vee y$ or $x \cup y$ and is called the **union** of x and y; in a similar way $\inf\{x, y\}$ is commonly written as $x \wedge y$ or $x \cap y$ and is called the **intersection** of x and y.

Example 7.7. For every set E the power set $\mathbf{P}(E)$ is a lattice with respect to the order of set inclusion. In fact, for all $X, Y \in \mathbf{P}(E)$ we have, by property (g) of §1, $\inf\{X, Y\}=X \cap Y$ and $\sup\{X, Y\}=X \cup Y$.

Example 7.8. The ordered set of Example 7.2 is a lattice. In this case $\sup\{m, n\}$ is called the **least common multiple** of m and n and $\inf\{m, n\}$ is called the **greatest common divisor** of m and n.

Exercises for §7

1. Show that the only binary relation on a set E which is both an equivalence relation on E and an order on E is the relation of equality on E.

2. Let F be the set of all positive factors of 30. Draw a Hasse diagram for F when ordered by divisibility. Let E be the set of all prime factors of 30. Find an order isomorphism $f: \mathbf{P}(E) \to F$.

3. Show that the binary relations $\leqslant_1$ and $\leqslant_2$ defined on $\mathbf{R} \times \mathbf{R}$ by

$$(x, y) \leqslant_1 (a, b) \Leftrightarrow x \leqslant a \text{ and } y \leqslant b;$$

$$(x, y) \leqslant_2 (a, b) \Leftrightarrow x < a \text{ or } (x=a \text{ and } y \leqslant b)$$

are order relations. Show that $(\mathbf{R}\times\mathbf{R}, \leqslant_2)$ is a chain whereas $(\mathbf{R}\times\mathbf{R}, \leqslant_1)$ is not. Show also that $(\mathbf{R}\times\mathbf{R}, \leqslant_1)$ is a lattice and find expressions for the supremum and infimum of any two-element subset.

4. An ordered set $(E, \leqslant)$ is said to be **well-ordered** if every non-empty subset of E has a smallest element. Prove that every well-ordered set is a chain with a smallest element.

[*Hint.* Given any $x, y \in E$ consider $\inf\{x, y\}$.]

5. Show that for every set E the power set $\mathbf{P}(E)$ is a chain under $\subseteq$ if and only if E is either $\emptyset$ or a singleton.

6. By a **convex region** of $\mathbf{R}\times\mathbf{R}$ we mean a non-empty subset X of $\mathbf{R}\times\mathbf{R}$ with the property that if $a, b \in X$ then all the points on the line segment joining a to b also belong to X. Show that, when ordered by set inclusion, the set C of convex regions of $\mathbf{R}\times\mathbf{R}$ is a lattice in which $\inf\{X, Y\}= X\cap Y$ but $\sup\{X, Y\}\neq X\cup Y$ in general.

7. A binary relation R on a set E is called a **pre-order** on E if it is reflexive and transitive. Suppose that R is a pre-order on E and let R^* be the binary relation defined on E by

$$aR^*b \Leftrightarrow (aRb \text{ and } bRa).$$

Show that R^* is an equivalence relation on E. Show also that the relation $\leqslant$ defined on E/R^* by

$$a/R^* \leqslant b/R^* \Leftrightarrow aRb$$

is an order.

8. Let L and M be lattices. If $f:L\to M$ is a bijection prove that the following conditions are equivalent:

(1) f is an order isomorphism;
(2) $(\forall x, y \in L)\quad f(x \wedge y)=f(x) \wedge f(y)$;
(3) $(\forall x, y \in L)\quad f(x \vee y)=f(x) \vee f(y)$.

9. Let E, F be sets and suppose that F is ordered under $\leqslant$. Show that the set Map (E, F) of all mappings $f: E\to F$ is ordered under the relation $\leqslant$ given by

$$f\leqslant g \Leftrightarrow (\forall x \in E)\quad f(x)\leqslant g(x).$$

Show that if $(F, \leqslant)$ is a lattice then so also is (Map $(E, F), \leqslant$).

10. Let $f: A\to B$ be a mapping from an ordered set A to an ordered set B. Then f is said to be **residuated** if it is isotone and there exists an isotone mapping $g: B\to A$ such that, in the notation of the previous

exercise, $f \circ g \leqslant id_B$ and $g \circ f \geqslant id_A$. Prove that if f is residuated then at most one such g exists and, when it does, then for every $y \in B$ the element $g(y)$ is the greatest element in the set $\{x \in A; f(x) \leqslant y\}$. If X is any subset of a set E prove that the mapping $f: \mathbf{P}(E) \to \mathbf{P}(E)$ given by $f(Y) = X \cap Y$ is residuated and find the corresponding mapping g.

§8. Equipotent sets; cardinal arithmetic; N

Definition. We say that sets E and F are **equipotent** if there exists a bijection $f: E \to F$.

The following are two trivial examples of sets which are equipotent.

Example 8.1. If A, B are sets then $A \times B$ and $B \times A$ are equipotent. In fact the mapping $f: A \times B \to B \times A$ given by $(a, b) \mapsto (b, a)$ is clearly a bijection.

Example 8.2. If $f: E \to F$ is injective then E and $\operatorname{Im} f$ are equipotent. In fact the mapping $f^+: E \to \operatorname{Im} f$ given by $f^+(x) = f(x)$ for every $x \in E$ is clearly a bijection.

A more sophisticated example, which turns out to be very important, is given in the following result, in which we show that for every set A the power set $\mathbf{P}(A)$ is equipotent to the set of all mappings from A to a two-element set $\{0, 1\}$. For this purpose we define, for every subset B of A, the **characteristic function of** B to be the mapping $\chi_B: A \to \{0, 1\}$ given by

$$\chi_B(x) = \begin{cases} 1 & \text{if } x \in B; \\ 0 & \text{if } x \in \complement_A(B). \end{cases}$$

In what follows we let $\operatorname{Map}(E, F)$ denote the set of all mappings $f: E \to F$.

Theorem 8.1 *For every set A the sets $\mathbf{P}(A)$ and $\operatorname{Map}(A, \{0, 1\})$ are equipotent.*

Proof. Define $f: \mathbf{P}(A) \to \operatorname{Map}(A, \{0, 1\})$ and $g: \operatorname{Map}(A, \{0, 1\}) \to \mathbf{P}(A)$ by the prescriptions $f(B) = \chi_B$ and $g(\theta) = \theta^{\leftarrow}\{1\}$. Since, for every $x \in A$,

$$\begin{aligned} \chi_{\theta^{\leftarrow}\{1\}}(x) &= \begin{cases} 1 & \text{if } x \in \theta^{\leftarrow}\{1\}; \\ 0 & \text{if } x \notin \theta^{\leftarrow}\{1\}, \end{cases} \\ &= \begin{cases} 1 & \text{if } \theta(x) = 1; \\ 0 & \text{if } \theta(x) = 0, \end{cases} \\ &= \theta(x), \end{aligned}$$

we see that, for every $\theta \in \text{Map}(A, \{0, 1\})$,

$$(f \circ g)(\theta) = f[\theta^{\leftarrow}\{1\}] = \chi_{\theta^{\leftarrow}\{1\}} = \theta$$

and so $f \circ g$ is the identity map on $\text{Map}(A, \{0, 1\})$. Also, from the definition of χ_B, we have $\chi_B^{\leftarrow}\{1\} = B$ and so, for every subset B of A, we have

$$(g \circ f)(B) = g[f(B)] = g(\chi_B) = \chi_B^{\leftarrow}\{1\} = B$$

and so $g \circ f$ is the identity map on $\mathbf{P}(A)$. It follows that f and g are mutually inverse bijections whence $\mathbf{P}(A)$ and $\text{Map}(A, \{0, 1\})$ are equipotent.

We now make the hypothesis that associated with every set E there is a set, called the **cardinal** of E and written Card E, such that

(1) Card E is equipotent to E;

(2) a set F is equipotent to E if and only if Card F = Card E.

[Note that, since Card E is a set, (1) and (2) imply that Card (Card E) = Card E.]

In what follows we shall use bold face lower case letters to denote cardinals.

Suppose now that we are given a set S of cardinals. Define a relation $\leqslant$ on S by

$$\mathbf{a} \leqslant \mathbf{b} \Leftrightarrow \text{there is an injection of } \mathbf{a} \text{ into } \mathbf{b}.$$

It is clear that $\leqslant$ is reflexive on S and, since the composite of two injections is an injection, $\leqslant$ is also transitive. That $\leqslant$ is anti-symmetric, and hence an order on S, follows from the following important result.

Theorem 8.2 [**Schröder–Bernstein**] *If A, B are sets and if there exist injections $f: A \to B$, $g: B \to A$ then A and B are equipotent.*

Proof. Let $\complement_A : \mathbf{P}(A) \to \mathbf{P}(A)$ be given by $X \mapsto \complement_A(X)$ and let $\complement_B$ be defined similarly on $\mathbf{P}(B)$. Now define $\zeta : \mathbf{P}(A) \to \mathbf{P}(A)$ by

$$\zeta = \complement_A \circ (g^{\rightarrow} \circ \complement_B \circ f^{\rightarrow}).$$

Since $\complement_A$, $\complement_B$ are inclusion-inverting and $f^{\rightarrow}$, $g^{\rightarrow}$ are inclusion-preserving, it is readily seen that ζ is inclusion-preserving. Consider now the collection $F = \{X \in \mathbf{P}(A);\ X \subseteq \zeta(X)\}$, noting that F is not empty since $\emptyset \in F$. Let $G = \bigcup \{X;\ X \in F\}$ be the union of all the sets in the collection F. Then for every $X \in F$ we have $X \subseteq G$ and so $X \subseteq \zeta(X) \subseteq \zeta(G)$. It follows from Theorem 6.1 (2) that $G \subseteq \zeta(G)$ and so we have $\zeta(G) \subseteq \zeta[\zeta(G)]$ whence $\zeta(G) \in F$ and consequently $\zeta(G) \subseteq G$. We therefore have $\zeta(G) = G$. Since $\complement_A \circ \complement_A$ is the identity map on $\mathbf{P}(A)$, we obtain from this

$$\complement_A(G) = (\complement_A \circ \zeta)(G) = (g^{\rightarrow} \circ \complement_B \circ f^{\rightarrow})(G),$$

which we can write in the form

$$\complement_A(G) = g^{\rightarrow}[\complement_B(f^{\rightarrow}(G))].$$

The situation is thus summarised by the following diagram:

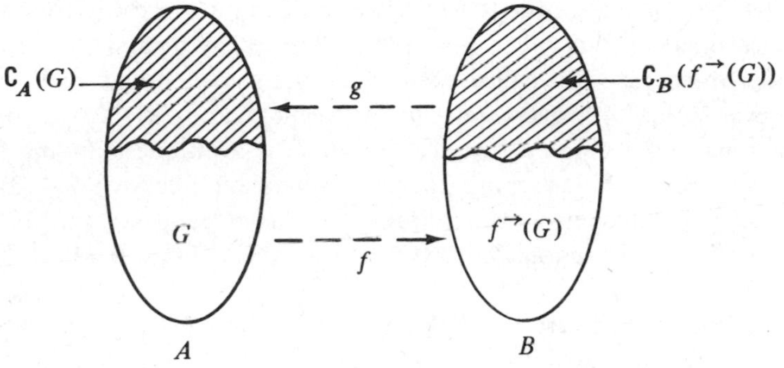

Since f and g are by hypothesis injective, this configuration shows that we can define a bijection $h: A \to B$ by the prescription

$$h(x) = \begin{cases} f(x) & \text{if } x \in G; \\ \text{the unique element of } g^{\leftarrow}\{x\} & \text{if } x \in \complement_A(G). \end{cases}$$

Hence A and B are equipotent.

Corollary. *If* $\mathbf{a}$ *and* $\mathbf{b}$ *are cardinals such that* $\mathbf{a} \leqslant \mathbf{b}$ *and* $\mathbf{b} \leqslant \mathbf{a}$ *then* $\mathbf{a} = \mathbf{b}$.

Proof. Since $\mathbf{a} \leqslant \mathbf{b}$ there is an injection $f: \mathbf{a} \to \mathbf{b}$; and since $\mathbf{b} \leqslant \mathbf{a}$ there is an injection $g: \mathbf{b} \to \mathbf{a}$. By the theorem, $\mathbf{a}$ and $\mathbf{b}$ are equipotent, whence $\mathbf{a} = \text{Card } \mathbf{a} = \text{Card } \mathbf{b} = \mathbf{b}$.

The previous result allows us to assert that, on any *set* of cardinals, $\leqslant$ is an order. Much more can be said about this order. We say that a non-empty ordered set E is **well-ordered** if every non-empty subset X of E has a smallest element. Note that this is equivalent to saying that inf X exists and belongs to X. In particular, if a non-empty set E is well-ordered, inf E exists and so E itself has a smallest element. Concerning the order $\leqslant$ on cardinals, we have the following important result, the proof of which we omit since it is too hard to give here.

Theorem 8.3 **[Zermelo]** *Every set of cardinals is well-ordered with respect to* $\leqslant$.

Corollary. *Every set of cardinals is totally ordered under* $\leqslant$.

Proof. Let E be a set of cardinals and let $\mathbf{a}, \mathbf{b} \in E$. Consider the subset $X = \{\mathbf{a}, \mathbf{b}\}$ of E. Since E is well-ordered, inf X exists and belongs to X. Thus either inf $X = \mathbf{a}$ or inf $X = \mathbf{b}$. Since, by definition, inf $X \leqslant \mathbf{a}$ and inf $X \leqslant \mathbf{b}$ we deduce that either $\mathbf{a} \leqslant \mathbf{b}$ or $\mathbf{b} \leqslant \mathbf{a}$. Thus E is totally ordered under $\leqslant$.

Two important cardinals which we shall require are the following. The cardinal of $\emptyset$ is called **zero** and is denoted by $\mathbf{0}$; and the cardinal of the singleton $\{\emptyset\}$ is called **one** (or **unity**) and is denoted by $\mathbf{1}$. Since, for every object x, every relation between $\emptyset$ and $\{x\}$ is (vacuously) functional, with graph $\emptyset$, we see that there is a unique mapping from $\emptyset$ to $\{x\}$. Since this mapping satisfies (vacuously) Theorem 5.4 (2), we shall extend the courtesy of regarding $\emptyset$ as a set by also regarding this mapping as an injection. Since $\emptyset \subset \{x\}$ we then obtain $\mathbf{0} < \mathbf{1}$.

Definition. Given cardinals $\mathbf{a}, \mathbf{b}$ we define the **product** of $\mathbf{a}$ and $\mathbf{b}$ to be the cardinal $\mathbf{ab} = \text{Card } \mathbf{a} \times \mathbf{b}$.

We have the following immediate properties of this product:

Theorem 8.4 *If* $\mathbf{a}, \mathbf{b}, \mathbf{c}$ *are cardinals then*

(1) $\mathbf{ab} = \mathbf{ba}$;

(2) $\mathbf{a(bc)} = \mathbf{(ab)c}$;

(3) $\mathbf{1a} = \mathbf{a}$;

(4) $\mathbf{0a} = \mathbf{0}$.

Proof. (1) is immediate from the equipotence of $\mathbf{a} \times \mathbf{b}$ and $\mathbf{b} \times \mathbf{a}$ under the bijection $(x, y) \mapsto (y, x)$; (2) follows from the equipotence of $\mathbf{a} \times (\mathbf{b} \times \mathbf{c})$ and $(\mathbf{a} \times \mathbf{b}) \times \mathbf{c}$ under the bijection $(x, (y, z)) \mapsto ((x, y), z)$; (3) follows from the equipotence of $\{\emptyset\} \times \mathbf{a}$ and $\mathbf{a}$ under the bijection $(\emptyset, x) \mapsto x$; and (4) follows from $\emptyset \times \mathbf{a} = \emptyset$.

There is another important method of combining two cardinals to form another cardinal. Given cardinals $\mathbf{a}, \mathbf{b}$ let A, B be *disjoint* sets such that $\mathbf{a} = \text{Card } A$ and $\mathbf{b} = \text{Card } B$. [Note that we have no difficulty in choosing A and B disjoint; we need only take $A = \mathbf{a} \times \{\mathbf{0}\}$ and $B = \mathbf{b} \times \{\mathbf{1}\}$. Then Card $A = \mathbf{a1} = \mathbf{a}$ and Card $B = \mathbf{b1} = \mathbf{b}$ with $A \cap B = \emptyset$.] Then we define the **sum** of $\mathbf{a}$ and $\mathbf{b}$ to be the cardinal $\mathbf{a} + \mathbf{b} = \text{Card } A \cup B$.

Remark. In the terminology of Exercise 5.18, $\mathbf{a} + \mathbf{b}$ is then the cardinal of any coproduct of $\mathbf{a}$ and $\mathbf{b}$.

Theorem 8.5 *If* $\mathbf{a}, \mathbf{b}, \mathbf{c}$ *are cardinals then*

(1) $\mathbf{a}+\mathbf{b}=\mathbf{b}+\mathbf{a}$;
(2) $\mathbf{a}+(\mathbf{b}+\mathbf{c})=(\mathbf{a}+\mathbf{b})+\mathbf{c}$;
(3) $\mathbf{a}+\mathbf{0}=\mathbf{a}$.

Proof. (1) is immediate from the fact that $A \cup B=B \cup A$. To prove (2), let A, B, C be sets such that Card $A=\mathbf{a}$, Card $B=\mathbf{b}$, Card $C=\mathbf{c}$ and $A \cap B=B \cap C=C \cap A=\emptyset$. [To see that such a choice is possible, define $\mathbf{2}=\text{Card}\,\{\emptyset, \{\emptyset\}\}$ and choose $A=\mathbf{a}\times\{\mathbf{0}\}$, $B=\mathbf{b}\times\{\mathbf{1}\}$, $C=\mathbf{c}\times\{\mathbf{2}\}$.] Then we have $\mathbf{a}+\mathbf{b}=\text{Card}\ A \cup B$ and $\mathbf{b}+\mathbf{c}=\text{Card}\ B \cup C$. Since $(A \cup B) \cap C=(A \cap C) \cup (B \cap C)=\emptyset \cup \emptyset=\emptyset$ we deduce that Card $[(A \cup B) \cup C]=\text{Card}\ A \cup B+\text{Card}\ C=(\mathbf{a}+\mathbf{b})+\mathbf{c}$; and since $A \cap (B \cup C)=(A \cap B) \cup (A \cap C)=\emptyset \cup \emptyset=\emptyset$ we deduce that Card $[A \cup (B \cup C)]=\text{Card}\ A+\text{Card}\ B \cup C=\mathbf{a}+(\mathbf{b}+\mathbf{c})$. The result now follows from the fact that $(A \cup B) \cup C=A \cup (B \cup C)$. As for (3), we observe that for any set A we have $A=A \cup \emptyset$ with $A \cap \emptyset=\emptyset$ and so Card $A=\text{Card}\ A+\text{Card}\ \emptyset$.

Addition and multiplication of cardinals as defined above are connected in the following way:

Theorem 8.6 *If* $\mathbf{a}, \mathbf{b}, \mathbf{c}$ *are cardinals then* $\mathbf{a}(\mathbf{b}+\mathbf{c})=\mathbf{ab}+\mathbf{ac}$.

Proof. Let A, B, C be such that Card $A=\mathbf{a}$, Card $B=\mathbf{b}$, Card $C=\mathbf{c}$ and choose B, C such that $B \cap C=\emptyset$. Then since $(A\times B) \cap (A\times C)=A\times(B \cap C)=A\times\emptyset=\emptyset$ we have Card $[(A\times B) \cap (A\times C)]=\text{Card}\ A\times B+\text{Card}\ A\times C=\mathbf{ab}+\mathbf{ac}$; and since $(A\times B) \cup (A\times C)=A\times(B\cup C)$ we have Card $[(A\times B) \cup (A\times C)]=\text{Card}\ A\times(B \cup C)=\mathbf{a}(\mathbf{b}+\mathbf{c})$.

The next two results exhibit the relationship of the ordering of cardinals with addition and multiplication.

Theorem 8.7 *If* $\mathbf{a}, \mathbf{b}, \mathbf{c}$ *are cardinals then*

$$\mathbf{a}\leqslant\mathbf{b} \Rightarrow (\mathbf{ac}\leqslant\mathbf{bc} \textit{ and } \mathbf{a}+\mathbf{c}\leqslant\mathbf{b}+\mathbf{c}).$$

Proof. Suppose that $\mathbf{a}\leqslant\mathbf{b}$; then there is an injection $f:\mathbf{a}\to\mathbf{b}$. The mapping $g:\mathbf{a}\times\mathbf{c}\to\mathbf{b}\times\mathbf{c}$ given by $g(a, c)=(f(a), c)$ is then also an injection, whence $\mathbf{ac}\leqslant\mathbf{bc}$.

Suppose now that $\mathbf{c}$ is chosen such that $\mathbf{a} \cap \mathbf{c}=\emptyset=\mathbf{b} \cap \mathbf{c}$ (such a choice always being possible, of course). Then the mapping $h:\mathbf{a} \cup \mathbf{c}\to \mathbf{b} \cup \mathbf{c}$ given by

$$h(x)=\begin{cases} f(x) & \text{if } x \in \mathbf{a};\\ x & \text{if } x \in \mathbf{c},\end{cases}$$

is also an injection and so $\mathbf{a}+\mathbf{c}\leqslant\mathbf{b}+\mathbf{c}$.

Theorem 8.8 *If* $\mathbf{a}$ *and* $\mathbf{b}$ *are cardinals then* $\mathbf{a} \leqslant \mathbf{b}$ *if and only if there is a cardinal* $\mathbf{c}$ *such that* $\mathbf{b} = \mathbf{a} + \mathbf{c}$.

Proof. Since $\mathbf{a} \leqslant \mathbf{b}$ there is a bijection from $\mathbf{a}$ onto a subset E of $\mathbf{b}$ (see Example 8.2). Let $F = \complement_{\mathbf{b}}(E)$ so that $\mathbf{b} = E \cup F$ and $E \cap F = \emptyset$. Then $\mathbf{b} = \text{Card } E + \text{Card } F = \mathbf{a} + \text{Card } F$. Conversely, suppose that there is a cardinal $\mathbf{c}$ with $\mathbf{b} = \mathbf{a} + \mathbf{c}$. Then $\mathbf{b}$ is the union of two disjoint sets A, C with Card $A = \mathbf{a}$ and Card $C = \mathbf{c}$. The canonical injection of A into $\mathbf{b}$ then gives $\mathbf{a} \leqslant \mathbf{b}$.

The following property of the cardinal $\mathbf{1}$ will play an important rôle in what follows.

Theorem 8.9 *If* $\mathbf{a}$ *and* $\mathbf{b}$ *are cardinals then*

$$\mathbf{a} + \mathbf{1} = \mathbf{b} + \mathbf{1} \Rightarrow \mathbf{a} = \mathbf{b}.$$

Proof. Suppose that $\mathbf{a} + \mathbf{1} = \mathbf{b} + \mathbf{1} = \mathbf{c}$, say. Then there is a set C with Card $C = \mathbf{c}$ which has subsets A, B and elements α, β such that Card $A = \mathbf{a}$, Card $B = \mathbf{b}$ and $\complement_C(A) = \{\alpha\}$, $\complement_C(B) = \{\beta\}$. Now if $\alpha = \beta$ we have $A = B$ and so $\mathbf{a} = \mathbf{b}$. On the other hand, if $\alpha \neq \beta$ then we must have $\beta \in A$ and $\alpha \in B$ whence, by the distributive laws, $\{\beta\} \cup (A \cap B) = A$ and $\{\alpha\} \cup (A \cap B) = B$. Since $\{\beta\} \cap A \cap B = \emptyset = \{\alpha\} \cap A \cap B$ we deduce that $\mathbf{a} = \text{Card } A = \mathbf{1} + \text{Card } A \cap B = \text{Card } B = \mathbf{b}$.

We now make the observation that for every cardinal $\mathbf{a}$ we have $\mathbf{a} \leqslant \mathbf{a} + \mathbf{1}$. In fact, if $\mathbf{a} \neq \mathbf{0}$ and $\mathbf{a} = \text{Card } A$ then $A \neq \emptyset$ and the set $(A \times \{\emptyset\}) \cup \{\emptyset\}$ has cardinal $\mathbf{a} + \mathbf{1}$. The mapping given by $x \mapsto (x, \emptyset)$ is then an injection of A into $(A \times \{\emptyset\}) \cup \{\emptyset\}$, so that $\mathbf{a} \leqslant \mathbf{a} + \mathbf{1}$. If, on the other hand, $\mathbf{a} = \mathbf{0}$ then as we have seen above $\mathbf{0} < \mathbf{1} = \mathbf{0} + \mathbf{1}$.

This observation gives rise to the following definition.

Definition. Let $\mathbf{a}$ be a cardinal. Then we say that $\mathbf{a}$ is **finite** if $\mathbf{a} < \mathbf{a} + \mathbf{1}$ and **infinite** if $\mathbf{a} = \mathbf{a} + \mathbf{1}$. If E is any set then we say that E is **finite** (resp. **infinite**) if Card E is finite (resp. infinite).

The following result gives a characterisation of finite sets and is often used as a definition of finite sets.

Theorem 8.10 *If X is a set, the following are equivalent:*

(1) *X is finite;*

(2) *the only subset of X which is equipotent to X is X itself.*

Proof. (1) $\Rightarrow$ (2): Suppose that X has a proper subset Y such that Card $Y = \text{Card } X$. Then there exists $\alpha \in \complement_X(Y)$ and $Y \subset Y \cup \{\alpha\} \subseteq X$.

The canonical injections now give Card X = Card $Y \leqslant$ Card $Y + \mathbf{1} \leqslant$ Card X whence Card X = Card $Y + \mathbf{1}$ = Card $X + \mathbf{1}$ and so X is not finite.

(2) $\Rightarrow$ (1): Suppose that X is not finite, so that Card X = Card $X + \mathbf{1}$. If α is any object not belonging to X there is therefore a bijection $f: X \cup \{\alpha\} \to X$. If f_X denotes the restriction of f to X then we have $\operatorname{Im} f_X = X \backslash \{f(\alpha)\}$ which is a proper subset of X equipotent to X (see Example 8.2).

As an immediate application of this result, we have:

Theorem 8.11 *Let X be a finite set and $f: X \to X$ any mapping. Then f is injective if and only if f is surjective.*

Proof. If f is injective then X is equipotent to the subset $\operatorname{Im} f$. Since X is finite it follows from Theorem 8.10 that $\operatorname{Im} f = X$, whence f is surjective.

Conversely, if f is surjective then by Theorem 5.6 there is a mapping $g: X \to X$ such that $f \circ g = id_X$. By Theorem 5.4, g is injective and so, by the previous paragraph, g is surjective. Hence g is a bijection whence so also is f since then $f = g^{-1}$. Consequently f is injective.

> *Remark.* Note that this result shows that for mappings from a *finite* set to itself the injections, surjections and bijections coincide.

A fundamental axiom in the theory of sets is that there exists at least one infinite set. It can be shown that this axiom is equivalent to saying that the finite cardinals form a *set* (which is, moreover, an infinite set). We denote this set by **N**. To see that **N** is indeed infinite, consider the mapping $\zeta: \mathbf{N} \to \mathbf{N}$ given by $\zeta(\mathbf{a}) = \mathbf{a} + \mathbf{1}$. By Theorem 8.9 we see that ζ is injective; but it is not surjective since $(\forall \mathbf{a} \in \mathbf{N})\ \zeta(\mathbf{a}) = \mathbf{a} + \mathbf{1} \geqslant \mathbf{0} + \mathbf{1} = \mathbf{1} > \mathbf{0}$ and so there is no $\mathbf{a} \in \mathbf{N}$ such that $\zeta(\mathbf{a}) = \mathbf{0}$. We conclude from Theorem 8.11 that **N** cannot be finite.

> *Remark.* The reader will not fail to observe that we have reserved the notation **N** for the set of natural numbers. We beg his patience; the explanation for the present usage of **N** will follow in due course.

For the remainder of this section we shall concentrate on further important properties of the set **N** of finite cardinals.

Theorem 8.12 *Let* $\mathbf{n} \in \mathbf{N}$; *then*

(1) $\mathbf{0} \leqslant \mathbf{n}$;

(2) *if* $\mathbf{0} < \mathbf{n} \leqslant \mathbf{1}$ *then* $\mathbf{n} = \mathbf{1}$;

(3) *if* $\mathbf{a}$ *is a cardinal such that* $\mathbf{a} \leqslant \mathbf{n}$ *then* $\mathbf{a} \in \mathbf{N}$.

Proof. (1) Since $\mathbf{n} = \mathbf{n} + \mathbf{0}$ we deduce from Theorem 8.8 that $\mathbf{0} \leqslant \mathbf{n}$.

(2) Let $\mathbf{0} < \mathbf{n} \leqslant \mathbf{1}$ and let X be a set with Card $X = \mathbf{n}$. Since $\mathbf{0} < \mathbf{n}$ we see that X is not empty. Let $x \in X$; then the canonical injection $\{x\} \to X$ gives $\mathbf{1} = \text{Card}\,\{x\} \leqslant \text{Card}\,X = \mathbf{n}$. The Schröder–Bernstein Theorem now gives $\mathbf{n} = \mathbf{1}$.

(3) If $\mathbf{a} \leqslant \mathbf{n}$ then by Theorem 8.8 there is a cardinal $\mathbf{b}$ such that $\mathbf{n} = \mathbf{a} + \mathbf{b}$. Then by Theorem 8.5 we have

$$(\mathbf{a}+\mathbf{1})+\mathbf{b} = \mathbf{a}+(\mathbf{1}+\mathbf{b}) = \mathbf{a}+(\mathbf{b}+\mathbf{1}) = (\mathbf{a}+\mathbf{b})+\mathbf{1} = \mathbf{n}+\mathbf{1} \neq \mathbf{n} = \mathbf{a}+\mathbf{b}$$

from which we deduce that $\mathbf{a} + \mathbf{1} \neq \mathbf{a}$ and so $\mathbf{a}$ is finite.

Corollary. *If* $\mathbf{x}$ *is a finite cardinal and* $\mathbf{y}$ *is an infinite cardinal then* $\mathbf{x} < \mathbf{y}$.

Proof. Suppose that $\mathbf{x} \not< \mathbf{y}$. Then by the Corollary to Theorem 8.3 we have $\mathbf{y} \leqslant \mathbf{x}$ and so, by (3) of Theorem 8.12, we obtain the contradiction that $\mathbf{y}$ is finite. Hence we must have $\mathbf{x} < \mathbf{y}$.

We note in passing that an equivalent formulation of the above corollary is: *if* Y *is an infinite set then every finite set* X *is equipotent to a subset of* Y.

Theorem 8.13 *Suppose that* $\mathbf{n} \in \mathbf{N}$ *with* $\mathbf{n} \neq \mathbf{0}$. *Then*

(1) *there is a unique* $\mathbf{m} \in \mathbf{N}$ *such that* $\mathbf{m} + \mathbf{1} = \mathbf{n}$;

(2) $\mathbf{m}$ *is such that* $(\forall \mathbf{p} \in \mathbf{N}) \quad \mathbf{p} \leqslant \mathbf{m} \Leftrightarrow \mathbf{p} < \mathbf{n}$.

Proof. (1) Since $\mathbf{N}$ is totally ordered we deduce from parts (1) and (2) of Theorem 8.12 that $\mathbf{1} \leqslant \mathbf{n}$. By Theorem 8.8 there is a cardinal $\mathbf{m}$ such that $\mathbf{n} = \mathbf{m} + \mathbf{1}$; and since $\mathbf{m} \leqslant \mathbf{n}$ we have $\mathbf{m} \in \mathbf{N}$ by part (3) of Theorem 8.12. To show the uniqueness of $\mathbf{m}$, we observe that if $\mathbf{p} \in \mathbf{N}$ is such that $\mathbf{n} = \mathbf{p} + \mathbf{1}$ then $\mathbf{p} = \mathbf{m}$ by Theorem 8.9.

(2) Suppose that $\mathbf{p} \in \mathbf{N}$ is such that $\mathbf{p} < \mathbf{n}$ where $\mathbf{n} = \mathbf{m} + \mathbf{1}$. By Theorem 8.8 there exists $\mathbf{a} \in \mathbf{N}$ such that $\mathbf{n} = \mathbf{p} + \mathbf{a}$ with $\mathbf{a} > \mathbf{0}$. By (1) there exists $\mathbf{b} \in \mathbf{N}$ such that $\mathbf{a} = \mathbf{b} + \mathbf{1}$. Thus we have

$$\mathbf{m}+\mathbf{1} = \mathbf{n} = \mathbf{p}+\mathbf{a} = \mathbf{p}+(\mathbf{b}+\mathbf{1}) = (\mathbf{p}+\mathbf{b})+\mathbf{1}$$

and so, by Theorem 8.9, $\mathbf{m} = \mathbf{p} + \mathbf{b} \geqslant \mathbf{p} + \mathbf{0} = \mathbf{p}$. Conversely, it is clear that if $\mathbf{p} \leqslant \mathbf{m} \in \mathbf{N}$ then $\mathbf{p} < \mathbf{m} + \mathbf{1} = \mathbf{n}$.

Remark. The unique cardinal $\mathbf{m}$ such that $\mathbf{m}+\mathbf{1}=\mathbf{n}$ is often written as $\mathbf{n}-\mathbf{1}$.

Theorem 8.14 **[Induction Principle]** *If* E *is a subset of* $\mathbf{N}$ *such that*

(1) $\mathbf{0} \in E$;

(2) $(\forall \mathbf{n} \in E) \quad \mathbf{n}+\mathbf{1} \in E$,

then E *coincides with* $\mathbf{N}$.

Proof. To show that $E=\mathbf{N}$ we show that $\complement_{\mathbf{N}}(E)=\emptyset$. Suppose, by way of obtaining a contradiction, that $\complement_{\mathbf{N}}(E)\neq\emptyset$. Since $\mathbf{N}$ is well-ordered (Theorem 8.3) the subset $\complement_{\mathbf{N}}(E)$ has a smallest element, $\mathbf{n}^*$ say. Now $\mathbf{n}^*$ cannot be $\mathbf{0}$ by condition (1) so, applying Theorem 8.13 (1), there exists $\mathbf{m} \in \mathbf{N}$ such that $\mathbf{n}^*=\mathbf{m}+\mathbf{1}$. This implies that $\mathbf{m}<\mathbf{n}^*$ and so $\mathbf{m} \notin \complement_{\mathbf{N}}(E)$ since $\mathbf{n}^*$ is the smallest element of this set. We therefore have $\mathbf{m} \in E$. By condition (2) we deduce that $\mathbf{n}^*=\mathbf{m}+\mathbf{1} \in E$ which is a contradiction. We conclude, therefore, that $\complement_{\mathbf{N}}(E)=\emptyset$ whence $E=\mathbf{N}$.

A variation of the above result which is sometimes useful is the following.

Theorem 8.15 **[Second Principle of Induction]** *If* E *is a subset of* $\mathbf{N}$ *such that*

(1) $\mathbf{0} \in E$;

(2) $(\forall \mathbf{n} \in \mathbf{N}) \quad \mathbf{n} \in E$ *whenever* $(\forall \mathbf{r}<\mathbf{n}) \quad \mathbf{r} \in E$,

then E *coincides with* $\mathbf{N}$.

Proof. By way of obtaining a contradiction, suppose that $\complement_{\mathbf{N}}(E)\neq\emptyset$. Let $\mathbf{n}^*$ be the smallest element of $\complement_{\mathbf{N}}(E)$. Then we have $\mathbf{n}^*\neq\mathbf{0}$ by (1) and $(\forall \mathbf{r}<\mathbf{n}^*)\ \mathbf{r} \in E$. By (2) we obtain the contradiction $\mathbf{n}^* \in E$. We conclude that $\complement_{\mathbf{N}}(E)=\emptyset$ and hence that $E=\mathbf{N}$.

To illustrate the uses of induction, we prove the following two important results.

Theorem 8.16 *If* $\mathbf{m}, \mathbf{n} \in \mathbf{N}$ *then* $\mathbf{m}+\mathbf{n} \in \mathbf{N}$ *and* $\mathbf{mn} \in \mathbf{N}$.

Proof. We observe first that if $\mathbf{k} \in \mathbf{N}$ then $\mathbf{k}+\mathbf{1} \in \mathbf{N}$; for

$$\begin{aligned} \mathbf{k}+\mathbf{1} \notin \mathbf{N} &\Rightarrow \mathbf{k}+\mathbf{1}=(\mathbf{k}+\mathbf{1})+\mathbf{1} \\ &\Rightarrow \mathbf{k}=\mathbf{k}+\mathbf{1} \text{ by Theorem 8.9} \\ &\Rightarrow \mathbf{k} \notin \mathbf{N}. \end{aligned}$$

Now let $\mathbf{m}$ be any element of $\mathbf{N}$ and define

$$E_{\mathbf{m}}=\{\mathbf{n} \in \mathbf{N};\ \mathbf{m}+\mathbf{n} \in \mathbf{N}\}.$$

Since $\mathbf{m}+\mathbf{0}=\mathbf{m} \in \mathbf{N}$ we have $\mathbf{0} \in E_{\mathbf{m}}$. Suppose by way of induction that $\mathbf{n} \in E_{\mathbf{m}}$. Then by the above observation we have

$$\mathbf{m}+(\mathbf{n}+\mathbf{1})=(\mathbf{m}+\mathbf{n})+\mathbf{1} \in \mathbf{N}$$

so that $\mathbf{n}+\mathbf{1} \in E_{\mathbf{m}}$. It follows by the induction principle that $E_{\mathbf{m}}=\mathbf{N}$; i.e., for any given $\mathbf{m} \in \mathbf{N}$ we have $\mathbf{m}+\mathbf{n} \in \mathbf{N}$ for every $\mathbf{n} \in \mathbf{N}$.

As for products, let $\mathbf{m}$ be any element of $\mathbf{N}$ and define

$$F_{\mathbf{m}}=\{\mathbf{n} \in \mathbf{N};\ \mathbf{mn} \in \mathbf{N}\}.$$

Since $\mathbf{m0}=\mathbf{0} \in \mathbf{N}$ we see that $\mathbf{0} \in F_{\mathbf{m}}$. Suppose now that $\mathbf{n} \in F_{\mathbf{m}}$. Then by the result just proved for addition we have

$$\mathbf{m}(\mathbf{n}+\mathbf{1})=\mathbf{mn}+\mathbf{m1}=\mathbf{mn}+\mathbf{m} \in \mathbf{N},$$

so that $\mathbf{n}+\mathbf{1} \in F_{\mathbf{m}}$. It follows by the induction principle that $F_{\mathbf{m}}=\mathbf{N}$; i.e., for any given $\mathbf{m} \in \mathbf{N}$ we have $\mathbf{mn} \in \mathbf{N}$ for every $\mathbf{n} \in \mathbf{N}$.

Theorem 8.17 *Let* $\mathbf{m}, \mathbf{n}, \mathbf{p} \in \mathbf{N}$. *Then*

$$\mathbf{m}+\mathbf{p}=\mathbf{n}+\mathbf{p} \Rightarrow \mathbf{m}=\mathbf{n};$$

and, if $\mathbf{p} \neq \mathbf{0}$,

$$\mathbf{mp}=\mathbf{np} \Rightarrow \mathbf{m}=\mathbf{n}.$$

Proof. Let $E_{\mathbf{m},\mathbf{n}}=\{\mathbf{p} \in \mathbf{N};\ \mathbf{m}+\mathbf{p}=\mathbf{n}+\mathbf{p} \Rightarrow \mathbf{m}=\mathbf{n}\}$. It is clear that $\mathbf{0} \in E_{\mathbf{m},\mathbf{n}}$. Moreover, given any $\mathbf{p} \in E_{\mathbf{m},\mathbf{n}}$ we have

$$\begin{aligned}\mathbf{m}+(\mathbf{p}+\mathbf{1})=\mathbf{n}+(\mathbf{p}+\mathbf{1}) &\Rightarrow (\mathbf{m}+\mathbf{p})+\mathbf{1}=(\mathbf{n}+\mathbf{p})+\mathbf{1}\\ &\Rightarrow \mathbf{m}+\mathbf{p}=\mathbf{n}+\mathbf{p}\\ &\Rightarrow \mathbf{m}=\mathbf{n}\end{aligned}$$

and so $\mathbf{p}+\mathbf{1} \in E_{\mathbf{m},\mathbf{n}}$. We deduce by the induction principle that $E_{\mathbf{m},\mathbf{n}}=\mathbf{N}$.

As for the second statement, we note the restriction $\mathbf{p} \neq \mathbf{0}$; this is necessary since the property is false if $\mathbf{p}=\mathbf{0}$: for $\mathbf{m0}=\mathbf{0}$ for every $\mathbf{m} \in \mathbf{N}$. Given $\mathbf{m}, \mathbf{n} \in \mathbf{N}$ suppose then that $\mathbf{p} \neq \mathbf{0}$ is such that $\mathbf{mp}=\mathbf{np}$. Since $\mathbf{N}$ is totally ordered, there is no loss in generality in supposing that $\mathbf{m} \leqslant \mathbf{n}$, in which case there exists $\mathbf{q} \in \mathbf{N}$ with $\mathbf{m}+\mathbf{q}=\mathbf{n}$ by Theorem 8.8. Then we have

$$\mathbf{mp}+\mathbf{0}=\mathbf{mp}=\mathbf{np}=(\mathbf{m}+\mathbf{q})\mathbf{p}=\mathbf{mp}+\mathbf{qp}$$

from which we deduce, by the first part of the theorem, that $\mathbf{qp}=\mathbf{0}$. Now if Q, P are sets such that Card $Q=\mathbf{q}$ and Card $P=\mathbf{p} \neq \mathbf{0}$ then clearly $P \neq \varnothing$ and $Q \times P$ is equipotent to $\varnothing$. Thus Q must be empty and so $\mathbf{q}=\mathbf{0}$. Consequently $\mathbf{n}=\mathbf{m}+\mathbf{q}=\mathbf{m}+\mathbf{0}=\mathbf{m}$.

> *Remark.* The proof of the second statement of Theorem 8.17 may also be given in an inductive manner. For this purpose, we require a more general form of Theorem 8.14. This is as follows (the easy details are left to the reader). *For every* $\mathbf{m} \in \mathbf{N}$ *let* $[\mathbf{m}, \rightarrow] = \{\mathbf{p} \in \mathbf{N}; \mathbf{m} \leqslant \mathbf{p}\}$. *Then if E is a subset of* $[\mathbf{m}, \rightarrow]$ *such that* (1) $\mathbf{m} \in E$ *and* (2) $(\forall \mathbf{n} \in E)\ \mathbf{n}+\mathbf{1} \in E$, *we have* $E = [\mathbf{m}, \rightarrow]$. This is often referred to as *induction from the anchor point* **m**. In Theorem 8.14, the anchor point is $\mathbf{m}=\mathbf{0}$. Theorem 8.17(2) can now be proved by induction from the anchor point **1**.

Two further important properties of **N** are given in the following results.

Theorem 8.18 *Let* $\mathbf{a}, \mathbf{b} \in \mathbf{N}$ *with* $\mathbf{b} \neq \mathbf{0}$. *Then* $(\exists \mathbf{n} \in \mathbf{N})\ \mathbf{a} < \mathbf{nb}$.

Proof. Since $\mathbf{1} \leqslant \mathbf{b}$ we have

$$(\mathbf{a}+\mathbf{1})\mathbf{b} = \mathbf{ab} + \mathbf{1b} \geqslant \mathbf{a1} + \mathbf{1b} = \mathbf{a} + \mathbf{b} \geqslant \mathbf{a} + \mathbf{1} > \mathbf{a}$$

and since $\mathbf{a} \in \mathbf{N}$ we have $\mathbf{a}+\mathbf{1} \in \mathbf{N}$ by Theorem 8.16. Hence we have the result with $\mathbf{n} = \mathbf{a}+\mathbf{1}$.

> *Remark.* Theorem 8.18 is often referred to by saying that the order of **N** is **archimedean.**

Theorem 8.19 *Let* $\mathbf{a}, \mathbf{b} \in \mathbf{N}$ *with* $\mathbf{b} \neq \mathbf{0}$. *Then there exist* $\mathbf{q}, \mathbf{r} \in \mathbf{N}$ *with* $\mathbf{0} \leqslant \mathbf{r} < \mathbf{b}$ *and* $\mathbf{a} = \mathbf{bq} + \mathbf{r}$.

Proof. Let $P_{\mathbf{a},\mathbf{b}} = \{\mathbf{p} \in \mathbf{N}; \mathbf{a} < \mathbf{pb}\}$, noting that $P_{\mathbf{a},\mathbf{b}} \neq \varnothing$ by Theorem 8.18. Since **N** is well-ordered, $P_{\mathbf{a},\mathbf{b}}$ has a smallest element, $\mathbf{p}^*$ say. Note that $\mathbf{p}^* \neq \mathbf{0}$ (for $\mathbf{0b} = \mathbf{0} \leqslant \mathbf{a}$ and so $\mathbf{0} \notin P_{\mathbf{a},\mathbf{b}}$) and so there exists $\mathbf{q} \in \mathbf{N}$ such that $\mathbf{p}^* = \mathbf{q}+\mathbf{1}$ (Theorem 8.13). This element **q** satisfies $\mathbf{bq} \leqslant \mathbf{a}$; for $\mathbf{q} < \mathbf{q}+\mathbf{1} = \mathbf{p}^*$ and $\mathbf{p}^*$ is the smallest element **p** of **N** satisfying $\mathbf{bp} > \mathbf{a}$. We thus have $\mathbf{bq} \leqslant \mathbf{a} < \mathbf{b}(\mathbf{q}+\mathbf{1})$. It follows by Theorem 8.8 that there exists $\mathbf{r} \in \mathbf{N}$ such that $\mathbf{a} = \mathbf{bq} + \mathbf{r}$; moreover, we must have $\mathbf{0} \leqslant \mathbf{r} < \mathbf{b}$ since otherwise $\mathbf{r} \geqslant \mathbf{b}$ and this would give $\mathbf{a} = \mathbf{bq} + \mathbf{r} \geqslant \mathbf{bq} + \mathbf{b} = \mathbf{b}(\mathbf{q}+\mathbf{1})$ in contradiction to $\mathbf{a} < \mathbf{bp}^* = \mathbf{b}(\mathbf{q}+\mathbf{1})$.

> *Remark.* The result of Theorem 8.19 is often referred to as the **euclidean division of a by b.** Note from the above proof that **q** and **r** are *unique*.

Example 8.3. As an application of Theorem 8.19, let us define $\mathbf{a} \in \mathbf{N}$ to be **even** if it is of the form $\mathbf{2n} = \mathbf{n}+\mathbf{n}$ for some $\mathbf{n} \in \mathbf{N}$; and **odd** if it is of the form $\mathbf{2n}+\mathbf{1}$ for some $\mathbf{n} \in \mathbf{N}$. Then *every* $\mathbf{a} \in \mathbf{N}$ *is either even or odd.* In fact, given $\mathbf{a} \in \mathbf{N}$ we have, by Theorem 8.19, elements $\mathbf{q}, \mathbf{r} \in \mathbf{N}$

such that $\mathbf{a}=\mathbf{2q}+\mathbf{r}$ where $\mathbf{0}\leqslant\mathbf{r}<\mathbf{2}=\mathbf{1}+\mathbf{1}$. Since necessarily $\mathbf{0}\leqslant\mathbf{r}\leqslant\mathbf{1}$ we have either $\mathbf{r}=\mathbf{0}$ or $\mathbf{r}=\mathbf{1}$ whence the result follows.

We end the present section by discussing another method of combining two cardinals to obtain a new cardinal. If A, B are sets, write the set of all mappings $f: B\to A$ as A^B. If C, D are sets with C equipotent to A and D equipotent to B then C^D is equipotent to A^B. In fact let $\alpha: C\to A$ and $\beta: D\to B$ be bijections; then for every $f\in A^B$ we have $\alpha^{-1}\circ f\circ\beta\in C^D$ and since α, β are bijections the assignment $f\mapsto\alpha^{-1}\circ f\circ\beta$ is injective. Consequently Card $A^B\leqslant$ Card C^D and a similar argument reveals the opposite inequality. We may therefore make the following definition.

Definition. Given cardinals **a, b** let A, B be sets with Card $A=\mathbf{a}$, Card $B=\mathbf{b}$. Then we define the **exponential of a by b** to be the cardinal $\mathbf{a}^{\mathbf{b}}=$ Card A^B.

It is clear from this definition that for every cardinal **a** we have $\mathbf{a}^{\mathbf{0}}=\mathbf{1}$; for the only mapping from $\emptyset$ to A is that whose graph is $\emptyset$.

The principal properties of exponentiation of cardinals are given in the following result.

Theorem 8.20 *If* **a, b, c** *are cardinals then*

(1) $\mathbf{a}^{\mathbf{b}+\mathbf{c}}=\mathbf{a}^{\mathbf{b}}\mathbf{a}^{\mathbf{c}}$;

(2) $(\mathbf{ab})^{\mathbf{c}}=\mathbf{a}^{\mathbf{c}}\mathbf{b}^{\mathbf{c}}$;

(3) $(\mathbf{a}^{\mathbf{b}})^{\mathbf{c}}=\mathbf{a}^{\mathbf{bc}}$.

Proof. (1) is immediate from the fact that if B, C are disjoint sets then the mapping $A^{B\cup C}\to A^B\times A^C$ given by $f\mapsto(f_B, f_C)$ is a bijection. To establish (2), we note that for every $f\in(A\times B)^C$ we have $(\forall x\in C)$ $f(x)=(g(x), h(x))$ for some $g\in A^C$, $h\in B^C$. It is clear that the assignment $f\mapsto(g, h)$ is a bijection. Finally, to establish (3), given any $g\in A^{B\times C}$ consider the mappings $g_c\in A^B$ given by $g_c(b)=g(b, c)$. We observe that the family $(g_c)_{c\in C}$ belongs to $(A^B)^C$ and that $g\mapsto(g_c)_{c\in C}$ is a bijection.

We leave to the reader the task of showing that we can generalise sums and products of cardinals as follows. If $(E_\alpha)_{\alpha\in A}$ is a family of sets then we define the **product of the family** $(\text{Card } E_\alpha)_{\alpha\in A}$ to be the cardinal $\prod_{\alpha\in A}\text{Card } E_\alpha=\text{Card } \bigtimes_{\alpha\in A} E_\alpha$; and we define the **sum of the family** $(\text{Card } E_\alpha)_{\alpha\in A}$ to be the cardinal $\sum_{\alpha\in A}\text{Card } E_\alpha=\text{Card }\bigcup_{\alpha\in A} F_\alpha$ where $(F_\alpha)_{\alpha\in A}$ is a family of *disjoint* sets with F_α equipotent to E_α for every $\alpha\in A$. The definitions of **ab** and **a + b** are effectively produced by taking $A=\{0, 1\}$. In the case where all the E_α are the same, say each $E_\alpha=E$, we have $\prod_{\alpha \in A} \text{Card } E_\alpha = \text{Card } E^A$; for in this case $\bigtimes_{\alpha \in A} E_\alpha$ is the set of all mappings $f: A \to \bigcup_{\alpha \in A} E_\alpha$ such that $(\forall \alpha \in A)\ f(\alpha) \in E_\alpha = E$ so that, in this case, $\bigtimes_{\alpha \in A} E_\alpha$ coincides with E^A. If, furthermore, $A = \{1, \ldots, n\}$ then we obtain a generalisation of the situation described at the end of §6.

Exercises for §8

1. [Another proof of the Schröder–Bernstein theorem.]
Let $f: A_1 \to E_1$ and $g: E_1 \to A_1$ be given injective mappings and let $B_1 = \text{Im } f$ and $F_1 = \text{Im } g$. For $n \geqslant 2$ define

$$A_n = g^{\to}(B_{n-1}),\ B_n = f^{\to}(A_n),\ E_n = f^{\to}(F_{n-1}),\ F_n = g^{\to}(E_n).$$

Establish the chains

$$A_1 \supset F_1 \supset \ldots \supset A_n \supset F_n \supset A_{n+1} \supset F_{n+1} \supset \ldots;$$

$$E_1 \supset B_1 \supset \ldots \supset E_n \supset B_n \supset E_{n+1} \supset B_{n+1} \supset \ldots.$$

Show that if $A = \bigcap_{n \geqslant 1} A_n$ and $E = \bigcap_{n \geqslant 1} E_n$ then the set $\{A_n \backslash F_n,\ F_n \backslash A_{n+1},\ A;\ n \geqslant 1\}$ is a partition of A_1 and the set $\{E_n \backslash B_n,\ B_n \backslash E_{n+1},\ E;\ n \geqslant 1\}$ is a partition of E_1. Show also that $f^{\to}(A) = E$ and deduce that $f^*: A \to E$ given by $f^*(x) = f(x)$ for all $x \in A$ is a bijection. Show further, by considering the restrictions of f to $A_n \backslash F_n$ and $E_n \backslash B_n$, that for every n there are bijections $f_n: A_n \backslash F_n \to B_n \backslash E_{n+1}$ and $g_n: E_n \backslash B_n \to F_n \backslash A_{n+1}$. Finally, construct a bijection from A_1 to E_1 by means of the mappings f_n, g_n^{-1} and f^*.

2. If $(E_\alpha)_{\alpha \in A}$ is a family of sets recall that $\sum_{\alpha \in A} \text{Card } E_\alpha$ is, by definition, $\text{Card} \bigcup_{\alpha \in A} F_\alpha$ where $(F_\alpha)_{\alpha \in A}$ is a family of *disjoint* sets with F_α equipotent to E_α for every $\alpha \in A$. Prove that $\text{Card} \bigcup_{\alpha \in A} E_\alpha \leqslant \sum_{\alpha \in A} \text{Card } E_\alpha$.

[*Hint.* For every $\alpha \in A$ let $F_\alpha = E_\alpha \times \{\alpha\}$.]

3. Let $(E, \leqslant)$ be an ordered set. Prove that E is well-ordered if and only if

(1) $\leqslant$ is a total order;

(2) E has a smallest element α;

(3) if F is a subset of E such that (*a*) $\alpha \in F$ and (*b*) $(\forall r \in E)\ r \in F$ whenever $\{x \in E;\ x < r\} \subseteq F$, then $F = E$.

[*Hint.* $\Rightarrow$: Observe that in the case where $E = \mathbf{N}$ condition (3) is simply the second principle of induction; so to establish (3) assume that $F \neq E$ and consider inf $\complement_E(F)$.

$\Leftarrow$: It is equivalent to show that if F is any subset of E having no smallest element then $F = \emptyset$; so suppose that F is such a subset and apply (3) to $\complement_E(F)$ to show that $\complement_E(F) = E$.]

§9. Recursion; characterisation of N

Our principal objective now is to give a characterisation of the set $\mathbf{N}$ of finite cardinals and thereby justify the use of the term "natural number" for "finite cardinal". This characterisation is a consequence of the so-called **Principle of Recursive Definition** (see Theorem 9.2, below). As we shall see, several other important facts are consequences of this theorem. By way of preparation, we introduce the following notation.

If $\mathbf{m}, \mathbf{n} \in \mathbf{N}$ are such that $\mathbf{m} \leqslant \mathbf{n}$ then we shall write $[\mathbf{m}, \mathbf{n}] = \{\mathbf{x} \in \mathbf{N}; \mathbf{m} \leqslant \mathbf{x} \leqslant \mathbf{n}\}$ and $[\mathbf{m}, \mathbf{n}[= \{\mathbf{x} \in \mathbf{N}; \mathbf{m} \leqslant \mathbf{x} < \mathbf{n}\}$. It follows readily from Theorem 8.13 that $[\mathbf{m}, \mathbf{n}] = [\mathbf{m}, \mathbf{n}[\cup \{\mathbf{n}\}$; for $[\mathbf{m}, \mathbf{n}[= [\mathbf{m}, \mathbf{n} - \mathbf{1}]$ where $\mathbf{n} - \mathbf{1}$ is the unique element of $\mathbf{N}$ such that $(\mathbf{n} - \mathbf{1}) + \mathbf{1} = \mathbf{n}$. We shall also write $[\mathbf{m}, \rightarrow] = \{\mathbf{x} \in \mathbf{N}; \mathbf{x} \geqslant \mathbf{m}\}$.

Since the proof of the Principle of Recursive Definition is somewhat complicated, we shall first explain what this result says and give a simple illustrative application of it. For this purpose, we define the mapping $(+): \mathbf{N} \rightarrow \mathbf{N}$ by the prescription $(+)(\mathbf{n}) = \mathbf{n} + \mathbf{1}$. This mapping is called the **successor function** on $\mathbf{N}$ with $\mathbf{n} + \mathbf{1}$ being called the **successor** of $\mathbf{n}$ for every $\mathbf{n} \in \mathbf{N}$. [The reason for this terminology is that the set $\{\mathbf{x} \in \mathbf{N}; \mathbf{n} < \mathbf{x} < \mathbf{n} + \mathbf{1}\}$ is empty by Theorem 8.13.] The Principle of Recursive Definition then says that if E is a non-empty set, if $a \in E$ and if $f: E \rightarrow E$ is any mapping then for every $\mathbf{m} \in \mathbf{N}$ there is a unique mapping $g: [\mathbf{m}, \rightarrow] \rightarrow E$ such that $g(\mathbf{m}) = a$ and the diagram

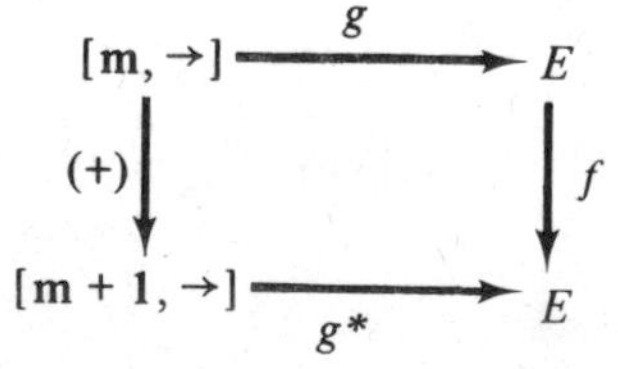

is commutative [in that $g^* \circ (+) = f \circ g$] where g^* is the restriction of g to $[\mathbf{m} + \mathbf{1}, \rightarrow]$.

Example 9.1. By way of illustrating an application of the above, suppose that $\mathbf{m} \in \mathbf{N}$ and let $f_{\mathbf{m}}: \mathbf{N} \to \mathbf{N}$ be given by $f_{\mathbf{m}}(\mathbf{n}) = \mathbf{n} + \mathbf{m}$. By the above, there is a unique mapping $g: \mathbf{N} \to \mathbf{N}$ such that $g(\mathbf{0}) = \mathbf{0}$ and $(\forall \mathbf{r} \in \mathbf{N})\ g(\mathbf{r}+\mathbf{1}) = g(\mathbf{r}) + \mathbf{m}$. It is here that we see the recursive nature of g in that $g(\mathbf{r}+\mathbf{1})$ is given in terms of $g(\mathbf{r})$. Now it is clear that the mapping from **N** to **N** described by $\mathbf{r} \mapsto \mathbf{mr}$ also satisfies each of the above properties and so must coincide with g. Thus we see that the product **mr** can be expressed as a sum $\mathbf{r}+\mathbf{r}+\ldots+\mathbf{r}$; in other words, multiplication of finite cardinals may be defined recursively in terms of addition.

In establishing the Principle of Recursive Definition we shall use the following result which is of a similar flavour.

Theorem 9.1 *Let E be a non-empty set, let $a \in E$ and let $f: E \to E$ be any mapping. Then for all $\mathbf{m}, \mathbf{n} \in \mathbf{N}$ with $\mathbf{m} \leqslant \mathbf{n}$ there is one and only one mapping $g_{\mathbf{n}}: [\mathbf{m}, \mathbf{n}] \to E$ such that*

$$(*)\quad \begin{cases} g_{\mathbf{n}}(\mathbf{m}) = a; \\ (\forall \mathbf{r} \in [\mathbf{m}, \mathbf{n}[)\quad g_{\mathbf{n}}(\mathbf{r}+\mathbf{1}) = f[g_{\mathbf{n}}(\mathbf{r})]. \end{cases}$$

Proof. Let $S_{\mathbf{m}}$ denote the set of all $\mathbf{n} \in \mathbf{N}$ such that $\mathbf{n} \geqslant \mathbf{m}$ and there exists a unique mapping $g_{\mathbf{n}}: [\mathbf{m}, \mathbf{n}] \to E$ satisfying (*). We note that $S_{\mathbf{m}} \neq \emptyset$ since it contains **m**; for taking $\mathbf{n} = \mathbf{m}$ we have $[\mathbf{m}, \mathbf{m}] = \{\mathbf{m}\}$ and $[\mathbf{m}, \mathbf{m}[= \emptyset$ so only one such mapping is possible, namely that given by $g_{\mathbf{m}}(\mathbf{m}) = a$. Suppose now, by way of using induction from the anchor point **m** (see the remark following Theorem 8.17), that $\mathbf{t} \in S_{\mathbf{m}}$. Let the unique associated mapping be $g_{\mathbf{t}}$ and define $h: [\mathbf{m}, \mathbf{t}+\mathbf{1}] \to E$ by

$$h(\mathbf{x}) = \begin{cases} g_{\mathbf{t}}(\mathbf{x}) & \text{if } \mathbf{x} \in [\mathbf{m}, \mathbf{t}]; \\ f[g_{\mathbf{t}}(\mathbf{t})] & \text{if } \mathbf{x} = \mathbf{t}+\mathbf{1}. \end{cases}$$

Since $g_{\mathbf{t}}$ satisfies (*) on $[\mathbf{m}, \mathbf{t}]$ it is clear that h satisfies (*) on $[\mathbf{m}, \mathbf{t}+\mathbf{1}]$. Now if $k: [\mathbf{m}, \mathbf{t}+\mathbf{1}] \to E$ also satisfies (*) then clearly the restriction k' of k to $[\mathbf{m}, \mathbf{t}]$ satisfies (*) on $[\mathbf{m}, \mathbf{t}]$ and so we must have $k' = g_{\mathbf{t}}$ by the uniqueness of $g_{\mathbf{t}}$. We thus have

$$\begin{cases} (\forall \mathbf{x} \in [\mathbf{m}, \mathbf{t}])\quad k'(\mathbf{x}) = g_{\mathbf{t}}(\mathbf{x}) = h(\mathbf{x}); \\ k(\mathbf{t}+\mathbf{1}) = f[k'(\mathbf{t})] = f[g_{\mathbf{t}}(\mathbf{t})] = h(\mathbf{t}+\mathbf{1}). \end{cases}$$

This then shows that $k = h$ and so h is unique. We deduce from this that $\mathbf{t}+\mathbf{1} \in S_{\mathbf{m}}$ and so, by induction from the anchor point **m**, we have $S_{\mathbf{m}} = [\mathbf{m}, \to]$, thereby establishing the result.

Theorem 9.2 **[Principle of Recursive Definition]** *Let E be a non-empty set, let $a \in E$ and let $f: E \to E$ be any mapping. Then for every $\mathbf{m} \in \mathbf{N}$ there* *is one and only one mapping* $g:[\mathbf{m}, \rightarrow] \rightarrow E$ *such that*

$$(**)\begin{cases} g(\mathbf{m})=a; \\ (\forall \mathbf{r}>\mathbf{m})\ g(\mathbf{r}+\mathbf{1})=f[g(\mathbf{r})]. \end{cases}$$

Proof. Given $\mathbf{n} \geqslant \mathbf{m}$, let $g_{\mathbf{n}}$ be the unique mapping satisfying (*) of Theorem 9.1. Then it is readily seen that the restriction of $g_{\mathbf{n+1}}$ to $[\mathbf{m}, \mathbf{n}]$ satisfies (*) on $[\mathbf{m}, \mathbf{n}]$ and so must coincide with $g_{\mathbf{n}}$. Consider now the mapping $g:[\mathbf{m}, \rightarrow] \rightarrow E$ given by

$$(\forall \mathbf{n} \geqslant \mathbf{m}) \quad g(\mathbf{n})=g_{\mathbf{n}}(\mathbf{n}).$$

We have $g(\mathbf{m})=g_{\mathbf{m}}(\mathbf{m})=a$ and, for every $\mathbf{n} \geqslant \mathbf{m}$,

$$g(\mathbf{n}+\mathbf{1})=g_{\mathbf{n+1}}(\mathbf{n}+\mathbf{1})=f[g_{\mathbf{n+1}}(\mathbf{n})]=f[g_{\mathbf{n}}(\mathbf{n})]=f[g(\mathbf{n})].$$

Thus we have constructed a mapping g satisfying (**). Suppose now that $h:[\mathbf{m}, \rightarrow] \rightarrow E$ also satisfies (**). Then clearly the restriction of h to $[\mathbf{m}, \mathbf{n}]$ satisfies (**) on $[\mathbf{m}, \mathbf{n}]$ and so coincides with $g_{\mathbf{n}}$. It follows that, for every $\mathbf{n} \geqslant \mathbf{m}$, $h(\mathbf{n})=g_{\mathbf{n}}(\mathbf{n})=g(\mathbf{n})$ and so $h=g$. Thus g is the only mapping satisfying (**).

Theorem 9.2 is instrumental in establishing many important results. We shall now proceed to establish one of these, namely that (in a sense to be made precise) the ordered set $(\mathbf{N}, \leqslant)$ is essentially the only one of its type; other consequences of Theorem 9.2 are developed in the exercises.

Definition. By a **naturally ordered set** (or a **Peano set**) we shall mean a set E on which there is defined a total order $\leqslant$ such that

(1) E has a smallest element α;
(2) every $x \in E$ has a successor $x^+ \in E$ (in that, for every $x \in E$, there is an $x^+ \in E$ such that $x<x^+$ and $x<y \leqslant x^+ \Rightarrow y=x^+$);
(3) If X is a subset of E such that

$$\text{(i) } \alpha \in X \qquad \text{and} \qquad \text{(ii) } (\forall x \in X) \quad x^+ \in X,$$

then X coincides with E.

Example 9.2. $(\mathbf{N}, \leqslant)$ is a naturally ordered set; for $\mathbf{N}$ has a smallest element $\mathbf{0}$, every $\mathbf{n} \in \mathbf{N}$ has $\mathbf{n}^+=\mathbf{n}+\mathbf{1}$ as successor and the induction principle holds in $\mathbf{N}$.

Theorem 9.3 *There is, to within order isomorphism, a unique naturally ordered set.*

Proof. We achieve the result by supposing that E is a naturally ordered set and showing that there is an order isomorphism $f:\mathbf{N}\to E$. This will then show that all naturally ordered sets are order isomorphic to $\mathbf{N}$; and hence any two such structures are order isomorphic to each other (since the composite of two order isomorphisms is also an order isomorphism).

Suppose then that E is a naturally ordered set. Let α be the smallest element of E and let $(+):E\to E$ be the successor function $x \mapsto x^+$. By Theorem 9.2 there is a unique mapping $g:\mathbf{N}\to E$ such that

$$\begin{cases} g(\mathbf{0})=\alpha; \\ (\forall \mathbf{r} \in \mathbf{N}) \quad g(\mathbf{r}+\mathbf{1})=[g(\mathbf{r})]^+. \end{cases}$$

Now on the one hand these equalities show that $\alpha \in \operatorname{Im} g$ and that for every $x=g(\mathbf{r}) \in \operatorname{Im} g$ we have

$$x^+=[g(\mathbf{r})]^+=g(\mathbf{r}+\mathbf{1}) \in \operatorname{Im} g.$$

Since E is naturally ordered, we deduce that $\operatorname{Im} g=E$ and so g is surjective. On the other hand, if we let

$$F=\{\mathbf{x} \in \mathbf{N};\ (\forall \mathbf{y} \in \mathbf{N}) \quad \mathbf{y}<\mathbf{x} \Rightarrow g(\mathbf{y})<g(\mathbf{x})\}$$

then clearly $\mathbf{0} \in F$ and, moreover, $(\forall \mathbf{x} \in F)\ \mathbf{x}+\mathbf{1} \in F$; for if $\mathbf{x} \in F$ then

$$\mathbf{y}<\mathbf{x}+\mathbf{1} \Rightarrow \mathbf{y}\leqslant\mathbf{x} \Rightarrow g(\mathbf{y})\leqslant g(\mathbf{x})<[g(\mathbf{x})]^+=g(\mathbf{x}+\mathbf{1}).$$

It therefore follows by the induction principle that $F=\mathbf{N}$. Combining these facts with Theorem 7.2, we conclude that g is an order isomorphism.

Because of this important result, we shall agree to take the hitherto undefined term "natural number" to be synonymous with the term "finite cardinal". For, whatever our intuitive idea of the set of natural numbers may have been up until now, we at least agree that it forms a naturally ordered set. By the above result, we may therefore select any naturally ordered set and decide to call this the set of natural numbers; and the obvious such set we have to hand is the set $\mathbf{N}$ of finite cardinals.

This identification having been made, we shall henceforth drop the convention that finite cardinals be written in bold face type and write in the usual way

$$\mathbf{N}=\{0,\ 1=0^+,\ 2=1^+=1+1,\ 3=2^+=2+1=(1+1)+1, \ldots\}$$

where $0=\operatorname{Card}\varnothing$, $1=\operatorname{Card}\{\varnothing\}$, $2=\operatorname{Card}\{\varnothing, \{\varnothing\}\}$, $3=\operatorname{Card}\{\varnothing, \{\varnothing\},$

$\{\emptyset, \{\emptyset\}\}\}$, . . . , and all the theorems concerning **N** may now be read with "natural number" replacing "finite cardinal".

Since, by induction, we have Card $\{1, \ldots, n\} = n$ and since $\{1, \ldots, n\}$ consists of n elements we often refer to Card X, when X is a *finite* set, as the **number of elements** in X.

Exercises for §9

1. Prove by induction that, for every $n \in \mathbf{N}$,

(a) $3^{2n+4} \equiv 2^{2n} \pmod 5$;
(b) $3^{6n} \equiv 1 \pmod 7$.

2. If t is any odd integer prove that $t^2 \equiv 1 \pmod 8$.
Hence prove by induction that, for $n \geqslant 1$,

$$t^{2^n} \equiv 1 \pmod{2^{n+2}}.$$

[*Hint.* $t^{2^{n+1}} - 1 = (t^{2^n} - 1)(t^{2^n} + 1)$ and $t^{2^n} + 1$ is even.]

3. Determine $\{n \in \mathbf{N};\ 2^n \geqslant (n+1)^2\}$.

4. Let $n \mapsto x_n$ be an injection from **N** to **N** such that

$$(1)\ x_0 = a;\ (2)\ n < m \Rightarrow x_n < x_m.$$

(α) Show that $\{x_n;\ n \in \mathbf{N}\}$ has no upper bounds in **N**.

[*Hint.* If k is an upper bound, $[0, k]$ is finite.]

(β) Deduce that for every $m \geqslant a$ there is a unique $k \in \mathbf{N}$ such that $x_k \leqslant m < x_{k+1}$.

[*Hint.* Given m, consider $E_m = \{x_t;\ m < x_t\}$. Observe that $E_m \neq \emptyset$ and use the fact that **N** is well-ordered.]

Suppose now that $p \neq 0$ is fixed in **N**.

(γ) Prove that for every $m \neq 0$ in **N** there is a unique $k \in \mathbf{N}$ such that $p^k \leqslant m < p^{k+1}$.

[*Hint.* Show that $k \mapsto p^k$ is injective and use (β).]

(δ) Prove that for every $m \neq 0$ in **N** there is a unique $k \in \mathbf{N}$ and a unique $m_k < p$ such that

$$m_k p^k \leqslant m < (m_k + 1)p^k.$$

[*Hint.* Let k be as in (γ) and apply euclidean division to the pair p^k, m.]

Deduce from the above that for every $m \neq 0$ in **N** there is a unique $k \in \mathbf{N}$ and unique elements $m_k, \ldots, m_1, m_0$ with each $m_i < p$ such that

$$(*) \qquad m = m_k p^k + \ldots + m_1 p + m_0 = \sum_{t=0}^{k} m_t p^t.$$

We call (*) the **p-adic expansion** of m and represent this by writing $m =_p m_k m_{k-1} \dots m_1 m_0$ (where this is *not* a product but simply the m_i written in juxtaposition in the order shown). For example, in the everyday case when $p=10$ we write

$$2.10^5+0.10^4+0.10^3+9.10^2+2.10^1+5.10^0$$

as 200925. In the **binary** ($p=2$) case we have, for example,

$$1.2^4+1.2^3+0.2^2+1.2^1+1.2^0 =_2 11011.$$

Determine

(a) the 3-adic expansion of 81;

(b) the 5-adic expansion of 134;

(c) the 12-adic expansion of 1860.

[*Hint.* Use euclidean division; in (*c*) a symbol for the number eleven will be required.]

5. Use the Principle of Recursive Definition to show that addition on **N** is unique in that it is the only mapping $f:\mathbf{N}\times\mathbf{N}\to\mathbf{N}$ such that

(1) $(\forall m \in \mathbf{N})\quad f(m, 0)=m$;

(2) $(\forall m, n \in \mathbf{N})\quad f(m, n+1)=[f(m, n)]^+$.

[*Hint.* Given any $m \in \mathbf{N}$, show that there is a unique mapping $f_m:\mathbf{N}\to\mathbf{N}$ such that $f_m(0)=m$ and $(\forall r \in \mathbf{N})\ f_m(r+1)=[f_m(r)]^+$. Deduce that $f:\mathbf{N}\times\mathbf{N}\to\mathbf{N}$ given by $f(m, n)=f_m(n)$ satisfies (1) and (2). To show the uniqueness of f let f^* also satisfy (1) and (2). For every $m \in \mathbf{N}$ let $S_m=\{n \in \mathbf{N}; f(m, n)=f^*(m, n)\}$ and show by induction that $S_m=\mathbf{N}$.]

6. Show that multiplication on **N** is the only mapping $f:\mathbf{N}\times\mathbf{N}\to\mathbf{N}$ such that

(1) $(\forall m \in \mathbf{N})\quad f(m, 0)=0$;

(2) $(\forall m, n \in \mathbf{N})\quad f(m, n+1)=f(m, n)+m$.

[*Hint.* Proceed as in the previous exercise with the successor function $n \mapsto n^+$ replaced by the translation $t_m:n \mapsto n+m$.]

7. Prove the following generalisation of the Principle of Recursive Definition. Let E be a non-empty set, let $a \in E$ and let $(f_n)_{n\geqslant 1}$ be a family of mappings $f_n:E\to E$. Prove that, for every $m \in \mathbf{N}$, there is one and only one mapping $g:[m, \to]\to E$ such that

$$\begin{cases} g(m)=a; \\ (\forall r\geqslant m)\quad g(r+1)=f_{r+1}[g(r)]. \end{cases}$$

[*Hint.* Modify the proofs of Theorems 9.1 and 9.2; in Theorem 9.1 define $h(t+1)=f_{t+1}[g_t(t)]$ and in Theorem 9.2 observe that $g_{n+1}(n+1)=f_{n+1}[g_{n+1}(n)]$.]

Deduce that there is a unique mapping $f:\mathbf{N}\to\mathbf{N}$ such that

$$\begin{cases} f(0)=1; \\ (\forall n \in \mathbf{N}) \quad f(n+1)=(n+1)f(n). \end{cases}$$

This unique mapping is usually written $n \mapsto n!$ where $f(n)=n!$ is called **factorial** n. Thus we have $0!=1$, $1!=1.0!=1$, $2!=2.1!=2.1$, $3!=3.2!=3.2.1, \ldots, n!=n(n-1)\ldots 3.2.1$.

8. Consider the relation $\preccurlyeq$ defined on $\mathbf{N}$ by

$$m \preccurlyeq n \Leftrightarrow \begin{cases} \textit{either } m \text{ is even and } n \text{ is odd} \\ \textit{or } m, n \text{ are of the same parity and } m \leqslant n. \end{cases}$$

Show that $(\mathbf{N}, \preccurlyeq)$ is a well-ordered set which is not naturally ordered.

9. Let E, F be sets consisting of n, m elements respectively with $m \leqslant n$. Show that the number of injections from F into E is $n!/(n-m)!$.

[*Hint.* Observe that if $m=0$ then there is $1=n!/(n-0)!$ injection of F into E (namely that whose graph is $\varnothing$). Now proceed by induction: if Card $F=m+1$ and Card $E=n \geqslant m+1$ observe that there is no loss in generality in supposing that $F=[0, m]$ and $E=[0, n-1]$. Let x be a fixed element of F, let Inj (F, E) be the set of injections of F into E and let $g:\text{Inj}(F, E)\to E$ be given by $g(\theta)=\theta(x)$. Note that for every $y \in E$ there exists $\theta \in \text{Inj}(F, E)$ such that $\theta(x)=y$ so that g is surjective. Let R_g be the equivalence relation associated with g and for every $y \in E$ consider the equivalence class $y/R_g=\{\theta; g(\theta)=y\}$. Show that Card $y/R_g=\text{Card Inj}(\complement_F\{x\}, \complement_E\{y\})$ by considering the assignment $\theta\to\theta^*$ where θ^* is the restriction of θ to $\complement_F\{x\}$. Now apply the induction hypothesis and use the fact that $\{y/R_g; y \in E\}$ is a partition of Inj (F, E).]

Deduce from the above that if X is a set consisting of n elements then the number of bijections from X to itself is $n!$.

10. Let X be a set consisting of n elements. Show that the number of subsets of X containing precisely r elements $(0 \leqslant r \leqslant n)$ is $n!/r!(n-r)!$.

[*Hint.* Let P be the collection of all subsets of X having r elements and let E be a fixed element of P. Define $h:\text{Inj}(E, X)\to P$ by $h(\theta)=\text{Im }\theta$. Note that h is surjective. Let R_h be the associated equivalence relation on Inj (E, X) and for every $Y \in P$ consider the equivalence class $Y/R_h=\{\theta; h(\theta)=Y\}$. Show that if $f \in \text{Bij}(E, E)$ and $\theta \in Y/R_h$ then $\theta\circ f \in Y/R_h$. Deduce that the mapping $f \mapsto \theta\circ f$ from Bij(E, E) to Y/R_h

is a bijection. Since $\{Y/R_h;\ Y \in P\}$ is a partition of Inj (E, X) we then have Card Inj $(E, X)=$ Card P Card Bij(E, E). Now apply both parts of Exercise 9 to determine Card P.]

Note that the above shows that $n!/r!(n-r)!$ belongs to **N**. We define

$$\binom{n}{m}=\begin{cases} n!/m!\,(n-m)! & \text{if } m \leqslant n; \\ 0 & \text{otherwise} \end{cases}$$

and call $\binom{n}{m}$ a **binomial coefficient**. [Because of the above result $\binom{n}{m}$ is often read "*n choose m*".] Show that, for every $n \in \mathbf{N}$,

(1) $\sum_{r=0}^{n} \binom{n}{r}=2^n$;

[*Hint*. Use the above result and Theorem 8.1.]

(2) $\binom{n}{0}=\binom{n}{n}=1$;

(3) $\binom{n+1}{m+1}=\binom{n}{m}+\binom{n}{m+1}$;

(4) if $m \leqslant n$ then $\binom{n}{n-m}=\binom{n}{m}$.

11. Deduce from Exercises 9 and 10 that $(\forall n \in \mathbf{N})\ n! \leqslant 2^n$. What then is wrong with the following argument? Suppose that $2^n \leqslant n!$; then for all $n \geqslant 1$ we have $2^{n+1}=2.2^n \leqslant 2.n! \leqslant (n+1).n!=(n+1)!$ and so, by induction, $2^n \leqslant n!$ for all $n \geqslant 1$.

§10. Infinite cardinals

In this section we shall concentrate on infinite cardinals. These do not behave like finite cardinals; for example, as the following shows, the cancellation laws of Theorem 8.17 do not hold for infinite cardinals.

Example 10.1. If $\mathbf{a}$ is an infinite cardinal then by definition we have $\mathbf{a}=\mathbf{a}+1$. Thus $\mathbf{a}+0=\mathbf{a}+1$ and so, if $\mathbf{a}$ were cancellative for addition, we would have the contradiction $0=1$. The same holds for multiplication: consider the mapping $f:\{0, 1\} \times \mathbf{N} \to \mathbf{N}$ given by

$$f(x, n)=\begin{cases} 2n & \text{if } x=0; \\ 2n+1 & \text{if } x=1. \end{cases}$$

Using Example 8.3, it is readily verified that f is a bijection. Thus we have 2 Card $\mathbf{N}=$ Card $\mathbf{N}$ so if the (infinite) cardinal Card $\mathbf{N}$ were cancellative for multiplication we would have the contradiction $2=1$.

Definition. A set E is called **denumerable** if Card E=Card $\mathbf{N}$; i.e., if there is a bijection $f:\mathbf{N}\to E$.

The following two results give examples of denumerable sets.

Theorem 10.1 *Every infinite subset of* $\mathbf{N}$ *is denumerable.*

Proof. Let A be an infinite subset of $\mathbf{N}$ and for every $m \in \mathbf{N}$ let $A_m = A \cap [m+1, \to]$ and $B_m = A \cap [\leftarrow, m]$. Then we have $A_m \cup B_m = A \cap \mathbf{N} = A$ and $A_m \cap B_m = \emptyset$, so that Card A=Card A_m+Card B_m. We deduce from this that A_m is infinite; for the contrary would imply, since B_m is finite, that A is finite by Theorem 8.16. Since $\mathbf{N}$ is well-ordered, A_m has a smallest element which we shall denote by α_m. Now define $f:A\to A$ by setting $(\forall m \in A)\, f(m)=\alpha_m$. By Theorem 9.2 there is a unique mapping $g:\mathbf{N}\to A$ such that, if α denotes the smallest element of A,

$$\begin{cases} g(0)=\alpha; \\ (\forall r \in \mathbf{N}) \quad g(r+1)=f[g(r)]=\alpha_{g(r)}. \end{cases}$$

Since for every $m \in A$ we have $\alpha_m \geqslant m+1 > m$ it follows that

$$(\forall r \in \mathbf{N}) \quad g(r) < \alpha_{g(r)} = g(r+1).$$

Consequently g is injective and so Card $\mathbf{N} \leqslant$ Card A. The canonical injection of A into $\mathbf{N}$ and the Schröder–Bernstein Theorem now show that Card A=Card $\mathbf{N}$, whence A is denumerable.

Theorem 10.2 $\mathbf{N}\times\mathbf{N}$ *is denumerable.*

Remark. This result, first proved by Cantor, is somewhat of a surprise in that, roughly speaking, it says that $\mathbf{N}\times\mathbf{N}$ has the "same number of elements" as $\mathbf{N}$; and phrased in this imprecise manner it certainly contradicts intuition. For the proof which follows, it will be helpful to refer to the following picture:

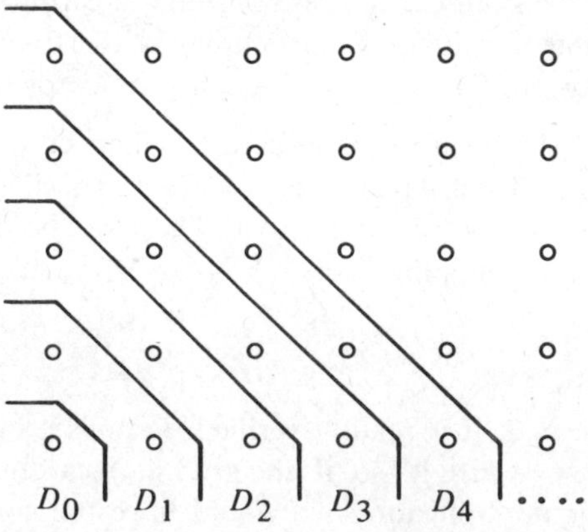

Proof. For every $n \in \mathbf{N}$ let

$$D_n = \{(i, j);\ i+j=n\} = \{(0, n), (1, n-1), \ldots, (n, 0)\}.$$

It is clear that $\{D_n;\ n \in \mathbf{N}\}$ is a partition of $\mathbf{N} \times \mathbf{N}$. For every $n \in \mathbf{N}$ let

$$J_n = \left[\sum_{r=0}^{n} r, \left(\sum_{r=0}^{n+1} r\right) - 1\right].$$

It is clear that $\{J_n;\ n \in \mathbf{N}\}$ is a partition of $\mathbf{N}$. Now it is readily seen that the mapping $f_n : [\leftarrow, n] \to D_n$ given by the prescription

$$f_n(p) = (n-p, p)$$

is a bijection. Likewise, so is the mapping $g_n : J_n \to [\leftarrow, n]$ given by

$$g_n\left(\sum_{r=0}^{n} r + q\right) = q.$$

Consequently $f_n \circ g_n : J_n \to D_n$ is a bijection and so Card J_n = Card D_n for every $n \in \mathbf{N}$. Thus we have

$$\begin{aligned}\text{Card } \mathbf{N} \times \mathbf{N} = \text{Card} \bigcup_{n \in \mathbf{N}} D_n &= \sum_{n \in \mathbf{N}} \text{Card } D_n \\ &= \sum_{n \in \mathbf{N}} \text{Card } J_n = \text{Card} \bigcup_{n \in \mathbf{N}} J_n = \text{Card } \mathbf{N}\end{aligned}$$

and so $\mathbf{N} \times \mathbf{N}$ is denumerable.

Corollary. *If the sets E, F are denumerable then so also is $E \times F$.*

Proof. From bijections $f : E \to \mathbf{N}$ and $g : F \to \mathbf{N}$ we can construct a bijection $(x, y) \mapsto (f(x), g(y))$ of $E \times F$ to $\mathbf{N} \times \mathbf{N}$.

Definition. A set E is said to be **countable** if it is finite or denumerable.

Theorem 10.3 *If E is a non-empty set then the following conditions are equivalent*:

(1) *E is countable*;

(2) *there is an injection $g : E \to \mathbf{N}$*;

(3) *there is a surjection $f : \mathbf{N} \to E$.*

Proof. (1) $\Rightarrow$ (2): Suppose that E is countable. Then (2) is clear if E is denumerable. If, on the other hand, E is finite then by the Corollary to Theorem 8.12 we have Card $E \leqslant$ Card $\mathbf{N}$ and so there is an injection $g : E \to \mathbf{N}$.

(2) ⇒ (3): If (2) holds then by Theorem 5.4 there is a mapping $f:\mathbf{N}\to E$ such that $f\circ g=id_{\mathbf{N}}$; and, by Theorem 5.6, f is surjective.

(3) ⇒ (2): This is analogous to (2) ⇒ (3).

(2) ⇒ (1): Let $g:E\to\mathbf{N}$ be injective. If E is finite then (1) holds. If E is infinite then Im g, being equipotent to E, is an infinite subset of $\mathbf{N}$ and so, by Theorem 10.1, is denumerable. Thus we have Card $E=$ Card Im $g=$ Card $\mathbf{N}$ and so E is denumerable.

Theorem 10.4 *Let $(E_i)_{i\in I}$ be a family of sets each of which is countable. Then if the index set I is countable so also is $\bigcup_{i\in I} E_i$.*

Proof. If I is countable then by Theorem 10.3 there is an injection $g:I\to\mathbf{N}$. There is therefore no loss in generality if we make the assumption that I is a subset of $\mathbf{N}$. The obvious injections

$$\mathbf{N}\to\{0\}\times\mathbf{N}\to I\times\mathbf{N}\to\mathbf{N}\times\mathbf{N}$$

together with the Schröder–Bernstein Theorem and Theorem 10.3 now show that $I\times\mathbf{N}$ is denumerable. Since each E_i is countable, there exists, by Theorem 10.3, a surjection $f_i:\mathbf{N}\to E_i$ for each $i\in\mathbf{N}$. Consider therefore the mapping $f:I\times\mathbf{N}\to\bigcup_{i\in I} E_i$ given by $f(i,n)=f_i(n)$. Since each f_i is surjective, so also is f. The diagram

$$\mathbf{N}\xrightarrow{\text{bij}} I\times\mathbf{N}\xrightarrow{f}\bigcup_{i\in I} E_i$$

now yields a surjection of $\mathbf{N}$ onto $\bigcup_{i\in I} E_i$ and the result follows from Theorem 10.3.

Our aim now is to prove that every infinite set contains a denumerable subset. For this purpose, we require an axiom of set theory which we have not mentioned as yet (though we have tacitly assumed it in a previous result!). This is Zermelo's **Axiom of Choice** and is as follows:

for every family $(E_\alpha)_{\alpha\in A}$ of non-empty sets there is a mapping $f: A\to\bigcup_{\alpha\in A} E_\alpha$ with the property that $(\forall\alpha\in A)\ f(\alpha)\in E_\alpha$.

Roughly speaking, the axiom of choice says that, for any family $(E_\alpha)_{\alpha\in A}$ of non-empty sets, there is a "choice function" which allows us to select at will an element from each of the E_α. We note that we have effectively used the axiom of choice in the proof of Theorem 5.5: there we considered a family $(f^{\leftarrow}[g(x)])_{x\in C}$ of subsets of B and we selected an element y_x from each of the sets in the family. We also used this axiom in defining the cartesian product of an arbitrary family of sets.

The following important result illustrates situations where the axiom of choice is used (often without explicit mention).

Theorem 10.5 *Every infinite set contains a denumerable subset.*

Proof. Let E be an infinite set. We show first for every $n \in \mathbf{N}$ there is a subset E_n of E having n elements. This is obvious for $n=0$, the subset in question being $E_0=\emptyset$. Suppose now that E_n is a subset of E with Card $E_n=n$. Since E is infinite we have $\complement_E(E_n)\neq\emptyset$. Let $x \in \complement_E(E_n)$ and define $E_{n+1}=E_n \cup \{x\}$. Then clearly $E_{n+1} \subset E$ with Card $E_{n+1}=n+1$. This then shows that the property holds for all n by induction. Thus, associated with every $n \in \mathbf{N}$, there is a set of subsets of E each of which has cardinal n. We now use the axiom of choice to select a family $(E_n)_{n\in\mathbf{N}}$ of such sets and define a family $(F_n)_{n\in\mathbf{N}}$ of non-empty disjoint subsets of E by setting

$$F_n = E_{2^n} \cap \bigcap_{t=0}^{n-1} \complement_E(E_{2^t}).$$

[To see that every $F_n\neq\emptyset$, suppose that $F_n=\emptyset$ for some n; then we would have

$$E_{2^n} \subseteq \complement_E \bigcap_{t=0}^{n-1} \complement_E(E_{2^t}) = \bigcup_{t=0}^{n-1} E_{2^t}$$

which would give

$$2^n = \text{Card } E_{2^n} \leqslant \sum_{t=0}^{n-1} \text{Card } E_{2^t} = \sum_{t=0}^{n-1} 2^t = 2^n - 1,$$

a contradiction. To see that the F_n are disjoint, we note that if $n\neq m$ then either $n<m$ or $n>m$, and in either case we have $F_n \cap F_m=\emptyset$ from the definition of F_n.]

We can now apply the axiom of choice to select an element from each F_n, in so doing forming a family $(x_n)_{n\in\mathbf{N}}$ of distinct elements of E. Then $\{x_n;\, n \in \mathbf{N}\}$ is a denumerable subset of E.

Corollary. Card $\mathbf{N}$ *is the smallest infinite cardinal.*

Proof. If E is an infinite set then, as was shown above, there is an injection $n \mapsto x_n$ of $\mathbf{N}$ into E whence Card $\mathbf{N} \leqslant$ Card E.

We shall now use Theorem 10.5 to show that every infinite set admits a partition into a set of denumerable subsets. For this purpose, we require another axiom of set theory. The axiom in question is **Zorn's axiom** (often called **Zorn's lemma**) and is stated in terms of the following concepts.

Definitions. An ordered set E is called **inductively ordered** if every non-empty totally ordered subset of E has an upper bound. An element m of an ordered set E is called a **maximal element** of E if, for any $x \in E$, $m \leqslant x \Rightarrow m = x$; in other words, if $\{x \in E;\, x > m\} = \emptyset$.

Zorn's axiom is then as follows:

every inductively ordered set contains a maximal element.

Although we shall not require the result, we mention that, surprisingly enough, Zorn's axiom is equivalent to the axiom of choice.

Theorem 10.6 *Every infinite set admits a partition into a set of denumerable subsets.*

Proof. Let E be an infinite set and let S denote the set of pairs $(X, \mathscr{X})$ where X is a non-empty subset of E and $\mathscr{X}$ is a partition of X into a set of denumerable subsets. That $S \neq \emptyset$ follows from the observation that, by Theorem 10.5, E contains a denumerable subset F and so $(F, \{F\})$ belongs to S. We now order S by writing

$$(X, \mathscr{X}) \leqslant (Y, \mathscr{Y}) \Leftrightarrow X \subseteq Y, \mathscr{X} \preccurlyeq \mathscr{Y}$$

where by $\mathscr{X} \preccurlyeq \mathscr{Y}$ we mean that every $\mathscr{X}$-class is a $\mathscr{Y}$-class [or, more precisely, if P is an equivalence relation on X associated with $\mathscr{X}$ and Q is an equivalence relation on Y associated with $\mathscr{Y}$ then $(\forall x \in X)$ $x/P = x/Q$]. We show first that S is inductively ordered.

Suppose that T is a non-empty totally ordered subset of S, say $T = \{(X_i, \mathscr{X}_i);\, i \in I\}$. Define $X = \bigcup_{i \in I} X_i$ and $\mathscr{X} = \bigvee_{i \in I} \mathscr{X}_i$ where $\bigvee$ denotes supremum relative to the order $\preccurlyeq$. If A, B are elements of $\mathscr{X}$ with $A \neq B$ then we have $A \in \mathscr{X}_i$, $B \in \mathscr{X}_j$ for some indices i, j. Since T is totally ordered we can assume that $(X_i, \mathscr{X}_i) \leqslant (X_j, \mathscr{X}_j)$ so that $\mathscr{X}_i \preccurlyeq \mathscr{X}_j$ and A, B both belong to $\mathscr{X}_j$ whence we have $A \cap B = \emptyset$. Now given any $x \in X$ we have $x \in X_i$ for some i and so $x \in C$ for some $C \in \mathscr{X}_i$. This then shows that $\mathscr{X}$ is a partition of X. Since the elements of each $\mathscr{X}_i$ are denumerable subsets of E it follows that $(X, \mathscr{X}) \in S$ and is clearly an upper bound for T. Thus S is inductively ordered.

Applying Zorn's axiom to S, let $(Z, \mathscr{Z})$ be a maximal element of S. If $Z = E$ then there is nothing more to prove. Suppose then that $Z \neq E$. If $\complement_E(Z)$ is infinite then, by Theorem 10.5, it contains a denumerable subset Z^* and so $(Z \cup Z^*, \mathscr{Z} \cup \{Z^*\})$ belongs to S, contradicting the maximality of $(Z, \mathscr{Z})$. Hence $\complement_E(Z)$ is finite. Given any $Z_0 \in \mathscr{Z}$ the set $Z_0 \cup \complement_E(Z)$ is then denumerable by Theorem 10.4, so we see that

$$\mathscr{Z}\backslash\{Z_0\} \cup \{Z_0 \cup \complement_E(Z)\}$$

is a partition of E into a set of denumerable subsets.

We shall now use the above result to establish the following property of infinite cardinals.

Theorem 10.7 *Let* $\mathbf{a}$ *be an infinite cardinal. Then for every cardinal* $\mathbf{b}$ *with* $0<\mathbf{b}\leqslant \text{Card } \mathbf{N}$ *we have* $\mathbf{ab}=\mathbf{a}$.

Proof. Let A, B be sets with $\text{Card } A=\mathbf{a}$ and $\text{Card } B=\mathbf{b}$. Suppose first that $\mathbf{b}=\text{Card } \mathbf{N}$ so that B is denumerable. By Theorem 10.6 there is a partition $\{A_i; i \in I\}$ of A into denumerable subsets and

$$A \times B=\Big(\bigcup_{i\in I} A_i\Big) \times B=\bigcup_{i\in I} (A_i \times B).$$

Since the sets $A_i \times B$ are disjoint and since every $A_i \times B$ is denumerable (by the Corollary to Theorem 10.2) we deduce that

$$\mathbf{ab}=\text{Card } A \times B=\text{Card} \bigcup_{i\in I} A_i=\text{Card } A=\mathbf{a}.$$

Suppose now that $0<\mathbf{b}<\text{Card } \mathbf{N}$, in which case B is finite. By the first part we have

$$\text{Card } A \leqslant \text{Card } A \times B \leqslant \text{Card } A \times \mathbf{N}=\text{Card } A$$

and so, by the Schröder–Bernstein Theorem, $\mathbf{a}=\mathbf{ab}$.

We shall soon extend Theorem 10.7 to cover the case where $\mathbf{b}$ is any cardinal. For the present, we require Theorem 10.7 in order to establish the following result.

Theorem 10.8 *If* $\mathbf{a}$ *is an infinite cardinal then* $\mathbf{a}^2=\mathbf{a}$.

Proof. Let E be a set with $\text{Card } E=\mathbf{a}$. Let S be the set consisting of pairs (F,f) where F is an infinite subset of E and $f: F \to F \times F$ is a bijection. We note that $S \neq \emptyset$ since, by Theorem 10.5, E has a denumerable subset D and there is a bijection $f: D \to D \times D$ by the Corollary to Theorem 10.2. Now order S by the rule

$$(F,f) \leqslant (G,g) \Leftrightarrow F \subseteq G \text{ and } g \text{ extends } f.$$

[By saying that g extends f we mean that $(\forall x \in F)\ g(x)=f(x)$.]

To show that S is inductively ordered, let $T=\{(F_i,f_i); i \in I\}$ be a totally ordered subset of S. Define $F=\bigcup_{i\in I} F_i$ and define a map

$f: F \to F \times F$ by $f(x) = f_i(x)$ whenever $x \in F_i$. [That f is well-defined follows from the fact that if also $x \in F_j$ with $j \neq i$ then, T being totally ordered, either $F_i \subseteq F_j$ and f_j extends f_i or vice-versa, so that in either case $f_i(x) = f_j(x)$.] To see that f is injective, let $x \in F_i$ and $y \in F_j$ be such that $f(x) = f(y)$. Since T is totally ordered, either $x, y \in F_i$ in which case $f_i(x) = f_i(y)$ and $x = y$, or $x, y \in F_j$ in which case again $x = y$. To see that f is surjective, let $(x, y) \in F \times F$; then $x \in F_i$, $y \in F_j$ for some i, j. Since T is totally ordered, either $x, y \in F_i$, in which case there exists $b \in F_i$ such that $f(b) = f_i(b) = (x, y)$, or $x, y \in F_j$, in which case there exists $b \in F_j$ such that $f(b) = f_j(b) = (x, y)$. Thus we see that (F, f) is an element of S and is clearly an upper bound for T. We can therefore apply Zorn's axiom to obtain a maximal element (G, g) of S. We shall now show that Card $G = \mathbf{a}$, which will establish the theorem.

Suppose in fact that Card $G = \mathbf{b} \neq \mathbf{a}$. Then since $(G, g) \in S$ we have $\mathbf{b}^2 = \mathbf{b}$ and, by Theorem 10.7, $\mathbf{b} = 2\mathbf{b} = 3\mathbf{b}$. Now since $\mathbf{b} < \mathbf{a}$ we must have Card $\complement_E(G) > \mathbf{b}$; for otherwise Card $\complement_E(G) \leqslant \mathbf{b}$ and

$$\mathbf{a} = \text{Card } E = \text{Card } (G \cup \complement_E(G)) = \text{Card } G + \text{Card } \complement_E(G) \leqslant 2\mathbf{b} = \mathbf{b},$$

a contradiction. There is therefore a subset H of $\complement_E(G)$ which is equipotent to G. Now we note that

$$(H \cup G) \times (H \cup G) = (H \times H) \cup (H \times G) \cup (G \times H) \cup (G \times G),$$

the four sets on the right being mutually disjoint. Since H and G are equipotent, we have

$$\text{Card } H \times G = \text{Card } G \times H = \text{Card } G \times G = \mathbf{b}^2 = \mathbf{b}$$

and so

$$\text{Card } [(H \times G) \cup (G \times H) \cup (G \times G)] = 3\mathbf{b} = \mathbf{b}.$$

Consequently, there is a bijection $k: H \to (H \times G) \cup (G \times H) \cup (G \times G)$. The mapping $h: H \cup G \to (H \cup G) \times (H \cup G)$ given by

$$h(x) = \begin{cases} g(x) & \text{if } x \in G; \\ k(x) & \text{if } x \in H, \end{cases}$$

is then a bijection which clearly extends g. We thus see that $(H \cup G, h) \in S$ and this contradicts the maximality of (G, g). We therefore deduce that $\mathbf{b} = \mathbf{a}$ and the theorem is proved.

A simple inductive argument yields the following.

Corollary. *If* $\mathbf{a}$ *is an infinite cardinal then* $(\forall n \geqslant 1)\ \mathbf{a}^n = \mathbf{a}$.

Sums and products of infinite cardinals can now be summarised in the following general result, from which we see that the arithmetic of infinite cardinals is much simpler than that of finite cardinals.

Theorem 10.9 *If* $\mathbf{a}, \mathbf{b}$ *are non-zero cardinals at least one of which is infinite then*

$$\mathbf{ab} = \mathbf{a} + \mathbf{b} = \sup\{\mathbf{a}, \mathbf{b}\}.$$

Proof. Suppose that $\mathbf{b}$ is infinite. Then if $\mathbf{a} \leqslant \mathbf{b}$ we have, by Theorems 8.7 and 10.8, $\mathbf{ab} \leqslant \mathbf{b}^2 = \mathbf{b}$. Since $1 \leqslant \mathbf{a}$ we also have $\mathbf{b} \leqslant \mathbf{ab}$ and so, by the Schröder–Bernstein Theorem, $\mathbf{ab} = \mathbf{b} = \sup\{\mathbf{a}, \mathbf{b}\}$. If, on the other hand, $\mathbf{b} \leqslant \mathbf{a}$ then $\mathbf{a}$ is also infinite and a similar proof yields $\mathbf{ab} = \mathbf{a} = \sup\{\mathbf{a}, \mathbf{b}\}$. As for sums, let $\mathbf{b}$ again be infinite. If $\mathbf{a} \leqslant \mathbf{b}$ then since $0 < \mathbf{a}$ we have $\mathbf{b} \leqslant \mathbf{a} + \mathbf{b} \leqslant \mathbf{b} + \mathbf{b}$. But if $\mathbf{b} = \text{Card } B$ then

$$\mathbf{b} + \mathbf{b} = \text{Card } [(B \times \{1\}) \cup (B \times \{2\})] = \text{Card } B \times \{1, 2\} = 2\mathbf{b} = \mathbf{b}$$

by Theorem 10.7. Thus we have $\mathbf{a} + \mathbf{b} = \mathbf{b} = \sup\{\mathbf{a}, \mathbf{b}\}$. If, on the other hand, $\mathbf{b} \leqslant \mathbf{a}$ then $\mathbf{a}$ is also infinite and a similar proof yields $\mathbf{a} + \mathbf{b} = \mathbf{a} = \sup\{\mathbf{a}, \mathbf{b}\}$.

We end this section by showing that for any given cardinal $\mathbf{a}$ there is a cardinal strictly greater than $\mathbf{a}$.

Theorem 10.10 *For every cardinal* $\mathbf{a}$ *we have* $\mathbf{a} < 2^{\mathbf{a}}$.

Proof. Let A be such that Card $A = \mathbf{a}$. Then by Theorem 8.1 we have Card $\mathbf{P}(A) = 2^{\mathbf{a}}$ and the injection $x \mapsto \{x\}$ of A into $\mathbf{P}(A)$ shows that $\mathbf{a} \leqslant 2^{\mathbf{a}}$. Now if we had $\mathbf{a} = 2^{\mathbf{a}}$ then there would exist an injection $g: \mathbf{P}(A) \to A$. By Theorem 5.4 we would then have a mapping $h: A \to \mathbf{P}(A)$ such that $h \circ g$ is the identity map on $\mathbf{P}(A)$; and by Theorem 5.6 we see that h is surjective. Consider now the element of $\mathbf{P}(A)$ given by

$$B = \{x \in A;\ x \notin h(x)\}.$$

Since h is surjective there exists $y \in A$ such that $h(y) = B$. Now if $y \in B$ we have the contradiction $y \notin h(y) = B$; and if $y \notin B$ we have the contradiction $y \in h(y) = B$. We conclude, therefore, that no such injection g can exist and that, consequently, $\mathbf{a} < 2^{\mathbf{a}}$.

Remark. If we write Card $\mathbf{N}$ as $\aleph_0$ (**aleph zero**) then by the above results we have the infinite chain of cardinals

$$0 < 1 < 2 < \ldots < \aleph_0 < 2^{\aleph_0} < 2^{2^{\aleph_0}} < \ldots.$$

Now it can be shown (see the exercises for §19) that if we write Card $\mathbf{R}$ as $\mathbf{c}$ then $\mathbf{c} = 2^{\aleph_0}$; in other words, Card $\mathbf{R}$ = Card $\mathbf{P}(\mathbf{N})$. A celebrated problem, which was first proposed by Cantor and which has since intrigued mathematicians, is the following: *do there exist cardinals lying strictly between* $\aleph_0$ *and* $\mathbf{c}$? The assertion that there are none, which is equivalent to the assertion that every non-countable set contains a subset equipotent to **R**, is known as the **continuum hypothesis**; and the more general assertion that there are no cardinals between other successive pairs in the above chain is known as the **generalised continuum hypothesis**. It has been shown as recently as 1963 by the American mathematician P. J. Cohen that these questions are essentially unanswerable in that these hypotheses are independent of the usual axioms of set theory. Thus the reader may choose to accept or reject these hypotheses depending on the needs of the mathematics he wishes to develop!

Exercises for §10

1. Prove that every subset of a countable set is countable.

[*Hint*. Use Theorem 10.3.]

2. If E is a countable set and R is an equivalence relation on E prove that E/R is countable.

[*Hint*. Use Theorem 10.3.]

3. Let E and F be countable sets. Show that $E \times F$ is countable.

4. Establish the denumerability of $\mathbf{N} \times \mathbf{N}$ by considering the mapping $f: \mathbf{N} \times \mathbf{N} \to \mathbf{N}$ given by

$$f(n, m) = \begin{cases} 0 & \text{if } m = n = 0; \\ 2^{n-1}(2m-1) & \text{otherwise.} \end{cases}$$

[*Hint*. Observe that every $p \in \mathbf{N}$ is of the form $2^a b$ where b is odd.]

5. Using calculus and Exercise 5.6, show that the mapping $f:]0, 1[\to \mathbf{R}$ given by

$$f(x) = \frac{1-2x}{x(1-x)}$$

is a bijection. Deduce, using the Schröder–Bernstein Theorem, that

$$\text{Card }]0, 1[= \text{Card } [0, 1] = \text{Card } \mathbf{R}.$$

6. Prove that if $S = \{(x, y) \in \mathbf{R} \times \mathbf{R};\ x^2 + y^2 \leqslant 1\}$ then Card $S =$ Card $\mathbf{R}$.

[*Hint*. Use Exercise 5.]

7. **[Hilbert's Hotel]** The rooms (all single) of a certain infinite hotel are indexed by **N**. Moreover, the hotel is full. A countable number of unexpected guests arrive. Show how the manager can accommodate them all, including his original guests, one person to a room!

[*Hint.* By Theorem 9.1 the set of odd-numbered rooms is denumerable; consider the injection $n \mapsto 2n$.]

8. Let $\mathbf{a}$ be a cardinal and let I be a set equipotent to $\mathbf{a}$. For every $i \in I$ let $a_i = 1$. Show that $\mathbf{a} = \sum_{i \in I} a_i$.

[*Hint.* Use the fact that every non-empty set admits a partition into singleton sets.]

Deduce that if $\mathbf{b}$ is a cardinal and $\mathbf{b}_i = \mathbf{b}$ for every $i \in I$ then $\mathbf{ab} = \sum_{i \in I} \mathbf{b}_i$.

9. Let E be an infinite set and for every $n \in \mathbf{N}$ let $\mathscr{F}_n$ denote the set whose elements are the subsets of E having n elements. Show that Card $\mathscr{F}_n \leqslant$ Card E.

[*Hint.* Recall that E^n is the set of all mappings $f: [1, n] \to E$ and use the fact that for every $X \in \mathscr{F}_n$ there is a bijection $g: [1, n] \to X$.]

Hence show that if $\mathbf{F}(E)$ denotes the set of all finite subsets of E then Card $\mathbf{F}(E) =$ Card E.

[*Hint.* To show that Card $\mathbf{F}(E) \leqslant$ Card E observe that $\mathbf{F}(E) = \bigcup\{\mathscr{F}_n; n \in \mathbf{N}\}$ and use Exercise 8 to show that Card $\mathbf{F}(E) \leqslant$ Card E Card $\mathbf{N}$. For the converse inequality use the injection $x \mapsto \{x\}$.]

10. Let E be an infinite set. Show that

$$\text{Card } \mathbf{P}(E \times E) = \text{Card } \mathbf{P}(E).$$

Show also that the mapping from the set of partitions on E to $\mathbf{P}(E \times E)$ given by

$$\{A_i;\, i \in I\} \mapsto \bigcup_{i \in I} A_i \times A_i$$

is an injection and deduce that the set of partitions on E is equipotent to $\mathbf{P}(E)$.

2

Algebraic structures and the number system

§11. Laws of composition; semigroups; morphisms

If E is a non-empty set then by a **law of composition** on E we shall mean a mapping $f: E \times E \to E$ described by writing $(x, y) \mapsto f(x, y)$. It is very common to write $f(x, y)$ simply as $x+y$ (**additive** notation) or as xy (**multiplicative** notation). However, since we shall attach a particular significance to these notations later, we shall use the notation $x \top y$. [The symbol $\top$ is called **truc** ("trook") and is French for "thingummy-jig"! The idea it conveys is that what we call our law of composition does not matter, for what we are really interested in are sets of objects and mappings between them.] We shall refer to $\top$ as the law of composition defined on E by $(x, y) \mapsto x \top y$.

Example 11.1. The mappings from $\mathbf{N} \times \mathbf{N}$ to $\mathbf{N}$ described by $(m, n) \mapsto m+n$ and $(m, n) \mapsto mn$ are laws of composition on $\mathbf{N}$, by Theorem 8.15. They are called respectively **addition** and **multiplication**.

Example 11.2. The mappings from $\mathbf{P}(E) \times \mathbf{P}(E)$ to $\mathbf{P}(E)$ described by $(X, Y) \mapsto X \cap Y$ and $(X, Y) \mapsto X \cup Y$ are laws of composition on $\mathbf{P}(E)$.

Example 11.3. The mapping from $\mathrm{Map}(E, E) \times \mathrm{Map}(E, E)$ to $\mathrm{Map}(E, E)$ given by $(f, g) \mapsto f \circ g$ is a law of composition on $\mathrm{Map}(E, E)$.

By a **groupoid** $(E, \top)$ we shall mean a pair consisting of a set E together with a law of composition $\top$ on E. When the law of composition is clear we shall often refer to E itself as a groupoid.

Let $(E, \top)$ be a groupoid. By a **neutral element** (or **identity element**) with respect to $\top$ we shall mean an element $e \in E$ such that

$$(\forall x \in E) \quad e \top x = x = x \top e.$$

It should be observed that a groupoid need not have a neutral element; for example, $(\mathbf{N}\backslash\{0\}, +)$ is a groupoid in which no neutral element exists since, by virtue of the cancellation laws, the only possible candidate is 0 which does not belong to the set. However, as the following result shows, if a groupoid does have a neutral element then it has precisely one.

Theorem 11.1 *A groupoid has at most one neutral element.*

Proof. Suppose that e, e^* are neutral elements of the groupoid $(E, \top)$. Then since e is neutral we have $e \top e^* = e^*$; and since e^* is neutral we have $e \top e^* = e$. Thus we have $e = e^*$ and so at most one such element exists in E.

> *Remark.* Because of the above result we can talk of *the* neutral element of a groupoid whenever such an element exists. In the additive notation the neutral element (when it exists) is denoted by 0 so that $(\forall x \in E)\ x+0=x=0+x$; and in the multiplicative notation we denote such an element (when it exists) by 1 so that $(\forall x \in E)\ x1=x=1x$.

We say that a law of composition $\top$ on a set E is **associative** if

$$(\forall x, y, z \in E) \quad x \top (y \top z) = (x \top y) \top z.$$

By a **semigroup** we shall mean a groupoid $(E, \top)$ in which $\top$ is associative.

The following are examples of semigroups:

$(\mathbf{N}, +)$; neutral element 0 [Theorems 8.16, 8.5];
$(\mathbf{N}, \cdot)$; neutral element 1 [Theorems 8.16, 8.4];
$(\mathbf{P}(E), \cap)$; neutral element E [property (c) of §1];
$(\mathbf{P}(E), \cup)$; neutral element $\varnothing$ [property (c) of §1];
(Map (E, E), $\circ$); neutral element id_E [Theorem 5.2];
$(\mathbf{N}\backslash\{0\}, +)$; no neutral element.

If x, y are elements of a semigroup $(E, \top)$ such that $x \top y = y \top x$ then we say that x and y **commute**. If every pair of elements of $(E, \top)$ commute then we say that $(E, \top)$ is **abelian** (or **commutative**). All of the semigroups listed above are abelian with the exception of the semigroup (Map (E, E), $\circ$); see §5.

Our aim now is to characterise, in the spirit of Theorem 9.3, the abelian semigroup $(\mathbf{N}, +)$. [Note that we do not bother with the abelian semigroup $(\mathbf{N}, \cdot)$ since multiplication can be defined recursively in terms of addition as we saw in Example 9.1.] For this purpose we require some additional notions concerning a particular type of mapping between groupoids.

Definition. Let $(E_1, \top_1)$ and $(E_2, \top_2)$ be groupoids. Then we shall say that a mapping $f: E_1 \to E_2$ is a **morphism** if

$$(\forall x, y \in E_1) \quad f(x \top_1 y) = f(x) \top_2 f(y).$$

Remark. It should be noted that the term **homomorphism** is widely used instead of simply **morphism**.

Note also that if we define the **cartesian product mapping** $f \times f : E_1 \times E_1 \to E_2 \times E_2$ by $(x, y) \mapsto (f(x), f(y))$ then to say that f is a morphism is equivalent to saying that the diagram

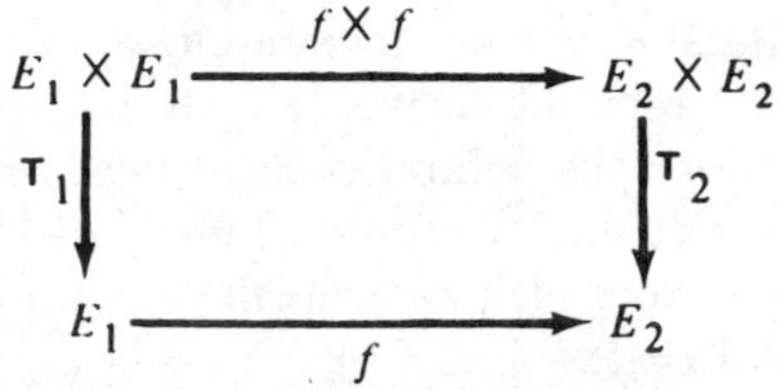

is **commutative** in that $f \circ \top_1 = \top_2 \circ (f \times f)$; for we have

$$(f \circ \top_1)(x, y) = f(x \top_1 y) \text{ and } [\top_2 \circ (f \times f)](x, y) = f(x) \top_2 f(y).$$

We mention this alternative definition to illustrate again the fact that all we are really interested in are sets and mappings; what the objects in the sets are is only of secondary importance to the ideas we are developing.

Example 11.4. For every $m \in \mathbf{N}$ the mapping $f_m : (\mathbf{N}, +) \to (\mathbf{N}, +)$ given by $f_m(n) = mn$ is a morphism; for, by Theorem 8.6, we have

$$(\forall n, p \in \mathbf{N}) \quad f_m(n+p) = m(n+p) = mn + mp = f_m(n) + f_m(p).$$

Example 11.5. The set **R** of real numbers forms a semigroup under addition and also a semigroup under multiplication. The mapping $f : (\mathbf{R}, +) \to (\mathbf{R}, \cdot)$ given by $f(x) = e^x$ (the **exponential map**) is a morphism since

$$(\forall x, y \in R) \quad f(x+y) = e^{x+y} = e^x e^y = f(x) f(y).$$

By a **monomorphism** we shall mean a morphism which is injective; by an **epimorphism** we shall mean a morphism which is surjective; and by an **isomorphism** we shall mean a morphism which is bijective.

Example 11.6. If $m \in \mathbf{N}$ with $m \neq 0$ then the morphism f_m of Example 11.4 is injective by Theorem 8.17 and so is a monomorphism.

We say that two semigroups are **isomorphic** if there is an isomorphism of one onto the other. We note that in Example 11.6 the set $\operatorname{Im} f_m$ is a semigroup under addition and this semigroup is isomorphic to the semigroup $(\mathbf{N}, +)$. This is a particular case of a general situation: *if E, F are semigroups and if $f : E \to F$ is a monomorphism then* $\operatorname{Im} f$ *is* *also a semigroup and the semigroups E and* Im f *are isomorphic.* The proof of this will be established later (see Theorem 12.9).

Let us now tie these notions in with that of an order. Suppose that $(E, \top)$ is a semigroup and that E is endowed with an order $\leqslant$. We shall say that $(E, \top, \leqslant)$ is an **ordered semigroup** if the law of composition and the order are related by the property: for all $x, y \in E$,

$$x \leqslant y \Rightarrow (\forall z \in E) \quad x \top z \leqslant y \top z \text{ and } z \top x \leqslant z \top y.$$

This condition may be expressed in the form: for every $z \in E$ the **left translation** $\lambda_z : E \to E$ given by $\lambda_z(x) = z \top x$ and the **right translation** $\rho_z : E \to E$ given by $\rho_z(x) = x \top z$ are each isotone.

Example 11.7. $(\mathbf{N}, +, \leqslant)$ is an abelian ordered semigroup (this follows from Theorems 8.4, 8.5, 8.7 and 8.16).

Combining the notions of order isomorphism and semigroup isomorphism, we say that an ordered semigroup $(E_1, \top_1, \leqslant_1)$ is **isomorphic** to an ordered semigroup $(E_2, \top_2, \leqslant_2)$ if there is a mapping $f: E_1 \to E_2$ which is both an order isomorphism and a semigroup isomorphism.

Our characterisation of the ordered abelian semigroup $(\mathbf{N}, +, \leqslant)$ is in terms of the following:

Definition. By a **naturally ordered semigroup** we shall mean a totally ordered abelian semigroup $(E, \top, \leqslant)$ such that

(1) $(E, \leqslant)$ is a naturally ordered set;
(2) the smallest element α of E is a neutral element for $\top$;
(3) for every $x \in E$ the successor of x is $x^+ = x \top \alpha^+$.

It is clear from this definition that $(\mathbf{N}, +, \leqslant)$ is a naturally ordered semigroup. That it is essentially the only one follows from the following result.

Theorem 11.2 *There is, to within ordered semigroup isomorphism, a unique naturally ordered semigroup.*

Proof. If $(E, \top, \leqslant)$ is a naturally ordered semigroup then by Theorem 9.3 there is an order isomorphism $g: \mathbf{N} \to E$. We show that this mapping is a semigroup isomorphism, from which the result will follow immediately. For this purpose, let m be an arbitrary element of $\mathbf{N}$ and let

$$H_m = \{n \in \mathbf{N};\ g(m+n) = g(m) \top g(n)\}.$$

Since α is the neutral element of E we have

$$g(m+0) = g(m) = g(m) \top \alpha = g(m) \top g(0)$$

and so $0 \in H_m$. Now since g is an order isomorphism it preserves successors in that $(\forall p \in \mathbf{N})\ g(p+1)=[g(p)]^+$; thus, for every $n \in H_m$,

$$\begin{aligned} g[m+(n+1)] = g[(m+n)+1] &= [g(m+n)]^+ \\ &= g(m+n) \top \alpha^+ \\ &= [g(m) \top g(n)] \top \alpha^+ \\ &= g(m) \top [g(n)]^+ = g(m) \top g(n+1) \end{aligned}$$

and so $n+1 \in H_m$. We deduce from the induction principle that $H_m = \mathbf{N}$. Since this holds for every $m \in \mathbf{N}$ we see that g is indeed a semigroup morphism; and being bijective it is a semigroup isomorphism.

Remark. When $\top$ is associative we may write $(x \top y) \top z$ and $x \top (y \top z)$ unambiguously as the multiple composite $x \top y \top z$. For a proper definition of multiple composites involving more than three elements and a proof that composites may be written unambiguously by omitting all brackets, we refer the reader to the details of Exercise 11.8.

Exercises for §11

1. Determine whether $(\mathbf{R}, \top)$ is a semigroup when $x \top y$ is given by

(a) $\sup\{x, y\}$; (b) $\inf\{x, 3\}$; (c) $x+y-xy$;
(d) $\sup\{n \in \mathbf{N};\ n \leqslant x+y\}$; (e) x^y; (f) $xy+1$.

2. Let $(E, \top)$ be a semigroup with neutral element e. An element $x^* \in E$ is said to be a **left inverse** of $x \in E$ if $x^* \top x = e$; and $x^{**} \in E$ is said to be a **right inverse** of $x \in E$ if $x \top x^{**} = e$. Show that if $x \in E$ has a left inverse x^* and a right inverse x^{**} then $x^* = x^{**}$.

3. Let $(E, \top)$ be an abelian semigroup in which every element is **idempotent** in that $x \top x = x$. Prove that the relation $\leqslant$ defined on E by

$$x \leqslant y \Leftrightarrow xy = x$$

is an order. Given any $a, b \in E$ show that $\inf\{a, b\}$ exists with respect to this order and is ab.

4. Let $(S, \top)$ be an abelian semigroup in which every element is idempotent (Exercise 3). For every $a \in S$ let $S_a = \{b \in S;\ (\exists x \in S)\ a \top x = b\}$. Prove that the mapping $f:(S, \top) \to (\mathbf{P}(S), \cap)$ given by $f(a) = S_a$ is a semigroup monomorphism.

[*Hint*. To show that f is injective, observe that if $S_a = S_b$ then $(\exists x, y \in S)\ a \top x = b,\ b \top y = a$; now simplify $[b \top (y \top x)] \top y$ in two ways.]

5. Let E be a non-empty set and define a law of composition $\circ$ on

$\mathbf{P}(E\times E)$ by $(G, H)\mapsto G\circ H$ where

$$(x, y)\in G\circ H \Leftrightarrow (\exists z\in E)\quad (x, z)\in G \text{ and } (z, y)\in H.$$

Show that $(\mathbf{P}(E\times E), \circ)$ is an ordered semigroup. Let $\Delta=\{(x, x);\ x\in E\}$ and, for every $G\in\mathbf{P}(E\times E)$, let $G^t=\{(y, x);\ (x, y)\in G\}$ and let R_G denote the relation on E given by $xR_Gy \Leftrightarrow (x, y)\in G$. Show that R_G is

(a) reflexive if and only if $\Delta\subseteq G$;
(b) symmetric if and only if $G=G^t$;
(c) transitive if and only if $G\circ G\subseteq G$.

Deduce that if R_G is an equivalence relation on E then G is idempotent in that $G\circ G=G$.

If $(G_\alpha)_{\alpha\in A}$ is a family of subsets of $E\times E$ prove that, for every $G\in\mathbf{P}(E\times E)$,

$$G\circ\bigcup_{\alpha\in A} G_\alpha=\bigcup_{\alpha\in A} G\circ G_\alpha \text{ and } G\circ\bigcap_{\alpha\in A} G_\alpha\subseteq\bigcap_{\alpha\in A} G\circ G_\alpha.$$

Show that equality need not hold in the second of these by considering $G=E\times E$, $A=\{1, 2\}$, $G_1=\Delta$ and $G_2=\complement(\Delta)$. Prove, in fact, that the following statements are equivalent:

(1) $(\forall H, K\in\mathbf{P}(E\times E))\quad G\circ(H\cap K)=(G\circ H)\cap(G\circ K)$;
(2) R_G is functional.

[*Hint.* (2) $\Rightarrow$ (1): Observe that it suffices to show that $G\circ(H\cap K)\supseteq(G\circ H)\cap(G\circ K)$.

(1) $\Rightarrow$ (2): Show that R_G is not functional if and only if $(G\circ\complement(\Delta))\cap(G\circ\Delta)\neq\varnothing$.]

Show also that, for every $G\in\mathbf{P}(E\times E)$, $(G^t)^t=G$ and that $G\mapsto G^t$ is an order isomorphism. Hence deduce that $\left(\bigcup_{\alpha\in A} G_\alpha\right)^t=\bigcup_{\alpha\in A} G^t_\alpha$ and $\left(\bigcap_{\alpha\in A} G_\alpha\right)^t=\bigcap_{\alpha\in A} G^t_\alpha$.

6. Let $(E, \top)$ be a semigroup. For any $x\in E$ consider the right translation $\rho_x: E\to E$ given by $\rho_x(y)=y\top x$. By Theorem 9.2 there is a unique mapping $g_x:\mathbf{N}\backslash\{0\}\to E$ such that

$$g_x(1)=x \text{ and } (\forall r\geqslant 1)\quad g_x(r+1)=g_x(r)\top x.$$

We define, for $n\geqslant 1$, $\overset{n}{\top}x=g_x(n)$. Thus, in a recursive manner, $\overset{1}{\top}x=x$, $\overset{2}{\top}x=x\top x, \ldots, \overset{n}{\top}x=\left(\overset{n-1}{\top}x\right)\top x$. In the case where E has a neutral element e we can extend each g_x to a unique mapping from $\mathbf{N}$ to E by setting $g_x(0)=e$ and $\overset{0}{\top}x=g_x(0)=e$. Prove that, for all $m, n\in\mathbf{N}$,

(1) $\overset{m+n}{\top}x=\left(\overset{m}{\top}x\right)\top\left(\overset{n}{\top}x\right)$.

[*Hint*. Given $m \in \mathbf{N}$ let S_m be the set of all $n \in \mathbf{N}$ for which (1) holds and use induction from the anchor point 1 or, in the case of a neutral element, 0.]

If $x, y \in E$ commute prove that, for all $m, n \in \mathbf{N}$,

(2) $\left(\overset{m}{\top} x\right) \top \left(\overset{n}{\top} y\right) = \left(\overset{n}{\top} y\right) \top \left(\overset{m}{\top} x\right)$;

(3) $\overset{n}{\top}(x \top y) = \left(\overset{n}{\top} x\right) \top \left(\overset{n}{\top} y\right)$;

(4) $\overset{n}{\top} e = e$.

[*Hint*. (2) Let $S = \{n \in \mathbf{N};\ y \top \left(\overset{n}{\top} x\right) = \left(\overset{n}{\top} x\right) \top y\}$ and prove by induction that $S = \mathbf{N}$. Now write y for x and $\overset{m}{\top} x$ for y.

(3) Use induction and (2).

(4) Use induction again.]

Note that when $\top$ is written as addition we have $\overset{0}{+} x = 0$ (when it exists), $\overset{1}{+} x = x$, $\overset{2}{+} x = x + x$, ..., $\overset{n}{+} x = \left(\overset{n-1}{+} x\right) + x$. In this case we write $\overset{n}{+} x$ as nx so that (1), (2), (3), (4) become

(1+) $(m+n)x = mx + nx$;

and, if $x + y = y + x$,

(2+) $mx + ny = ny + mx$;

(3+) $n(x+y) = nx + ny$;

(4+) $0x = 0$.

In the case wnere $\top$ is written multiplicatively we write $\overset{n}{\top} x$ as x^n so that the above become

(1·) $x^{m+n} = x^m x^n$;

and, if $xy = yx$,

(2·) $x^m y^n = y^n x^m$;

(3·) $(xy)^n = x^n y^n$;

(4·) $x^0 = 1$.

7. With the notation of Exercise 6, prove that if $(E, \top)$ is a semigroup (resp. with a neutral element) and if $x \in E$ then for all $m, n \in \mathbf{N}\backslash\{0\}$ (resp. $\mathbf{N}$)

$$\overset{nm}{\top} x = \overset{n}{\top}\left(\overset{m}{\top} x\right) = \overset{m}{\top}\left(\overset{n}{\top} x\right).$$

Note that in additive notation this reads $(nm)x = n(mx) = m(nx)$; and in multiplicative notation it reads $x^{nm} = (x^m)^n = (x^n)^m$.

[*Hint*. Let $y = \overset{m}{\mathsf{T}} x$ and let h be given by $h(n) = \overset{nm}{\mathsf{T}} x$. Show that $h(1) = y$ and, using Exercise 6, that $(\forall n \geqslant 1)\ h(n+1) = h(n) \mathsf{T} y$. Deduce that $h = g_y$ (where g_y is as in Exercise 6) and hence that $\overset{nm}{\mathsf{T}} x = \overset{n}{\mathsf{T}} y$.]

8. By an **n-ary operation** on a set E we mean a mapping $f: E^n \to E$. [The case $n=2$ gives a **binary operation** which is just a law of composition; the case $n=1$ gives a **unary operation** which is just a mapping from E to itself.]

Let T be a law of composition on E. Prove that there is one and only one sequence $(\mathsf{T}_r)_{r \geqslant 1}$ where each T_n is an n-ary operation on E and $\mathsf{T}_2 = \mathsf{T}$, such that

(1) $\mathsf{T}_1(x) = x$ for every $x \in E$;

(2) $\mathsf{T}_{n+1}(x_1, \ldots, x_n, x_{n+1}) = \mathsf{T}_n(x_1, \ldots, x_n) \mathsf{T} x_{n+1}$ for every $(n+1)$-tuple $(x_1, \ldots, x_{n+1}) \in E^{n+1}$.

[*Hint*. Define $G = \{g; (\exists n \geqslant 1)\ g: E^n \to E\}$ and let $f: G \to G$ be given by $g \mapsto f(g)$ where $f(g)(x_1, \ldots, x_n, x_{n+1}) = g(x_1, \ldots, x_n) \mathsf{T} x_{n+1}$. Use the principle of recursive definition (apply Theorem 9.3 to G and take $a = id_E$) to show that there is a unique sequence $(\mathsf{T}_r)_{r \geqslant 1}$ of elements of G such that $\mathsf{T}_1 = id_E$ and $\mathsf{T}_{n+1} = f(\mathsf{T}_n)$ for $n \geqslant 1$.]

For every n we write $\mathsf{T}_n(x_1, \ldots, x_n)$ as $\overset{n}{\underset{i=1}{\mathsf{T}}} x_i$ or $x_1 \mathsf{T} \ldots \mathsf{T} x_n$. In the additive case this becomes $\sum_{i=1}^{n} x_i$ or $x_1 + \ldots + x_n$; and in the multiplicative case it is written $\prod_{i=1}^{n} x_i$ or $x_1 x_2 \ldots x_n$.

Show by induction that if $x_i = x$ for $i = 1, \ldots, n$ then $\overset{n}{\underset{i=1}{\mathsf{T}}} x_i = \overset{n}{\mathsf{T}} x$ (so that we have a generalisation of the situation described in Exercise 11.6).

If T is associative prove that, for all $m, n \in \mathbf{N} \setminus \{0\}$,

$$\left(\overset{n}{\underset{i=1}{\mathsf{T}}} x_i \right) \mathsf{T} \left(\overset{m}{\underset{j=1}{\mathsf{T}}} x_{n+j} \right) = \overset{n+m}{\underset{i=1}{\mathsf{T}}} x_i.$$

[*Hint*. Fix n and use induction.]

Deduce, using the second principle of induction, that when T is associative composites may be written unambiguously by omitting all brackets.

If T is both associative and commutative prove that, for every bijection f on $\{1, 2, \ldots, n\}$,

$$\overset{n}{\underset{i=1}{\mathsf{T}}} x_i = \overset{n}{\underset{i=1}{\mathsf{T}}} x_{f(i)}.$$

[*Hint.* Observe by induction that $\left(\mathop{\top}\limits_{i=1}^{n} x_i\right) \top y = y \top \left(\mathop{\top}\limits_{i=1}^{n} x_i\right)$. For each bijection g on $\{1, 2, \ldots, n+1\}$ consider the element $\mathop{\top}\limits_{i=1}^{n+1} x_{g(i)} = x_{g(1)} \top \ldots \top x_{g(n+1)}$. Let t be such that $x_{n+1} = x_{g(t)}$ and note that, by the above observation,

$$\mathop{\top}\limits_{i=1}^{n+1} x_{g(i)} = \mathop{\top}\limits_{\substack{i=1\\ i\neq t}}^{n+1} x_{g(i)} \top x_{g(t)}.$$

Now use induction.]

§12. Groups; subgroups; group morphisms

Our immediate aim now is to set up the machinery which will provide a proper definition of the set **Z** of integers. Intuitively, we think of the set of integers as an infinite chain

$$\ldots < -3 < -2 < -1 < 0 < 1 < 2 < 3 < \ldots$$

which is an "extension" of **N** in that **N** is a proper subset of **Z** and that **Z** has two laws of composition (addition and multiplication) which, on the subset **N**, coincide with the corresponding laws of **N**. The essential difference between the algebraic properties of **Z** and those of **N**, in the intuitive sense, is that in **Z** any linear equation $a+x=b$ has a (unique) solution, namely $x=b-a$, whereas in **N** this is not the case. In this section our task will be to make these notations precise.

Suppose that $(E, \top)$ is a groupoid with neutral element e. Given $x \in E$ we say that $y \in E$ is a **left inverse** of x if $y \top x = e$; a **right inverse** of x if $x \top y = e$; and a **two-sided inverse** (or simply **inverse**) of x if $x \top y = e = y \top x$.

It should be noted that, in the case of a groupoid, inverses which may exist need not be unique. For example, let $E = \{e_1\ e_2, e_3\}$ and let a law of composition $\top$ be defined on E by means of the following **Cayley table**, the interpretation of which is that the element $e_i \top e_j$ is given by the entry appearing at the intersection of the ith row and the jth column:

$\top$	e_1	e_2	e_3
e_1	e_1	e_2	e_3
e_2	e_2	e_1	e_1
e_3	e_3	e_1	e_3

It is readily seen that e_1 is the neutral element of $(E, \top)$. Since $e_2 \top e_2 = e_1$ and $e_2 \top e_3 = e_1 = e_3 \top e_2$ we see that e_2 has two inverses, namely e_2 itself and e_3. Now we note that in this particular example the law $\top$ is not associative; for example,

$$(e_2 \top e_3) \top e_3 = e_1 \top e_3 = e_3;$$

$$e_2 \top (e_3 \top e_3) = e_2 \top e_3 = e_1.$$

In the presence of associativity, however, the possibility of multiple inverses cannot arise as the following result shows.

Theorem 12.1 *Let $(E, \top)$ be a semigroup with a neutral element. Then if $x \in E$ has both a left inverse x^* and a right inverse x^{**} we have $x^* = x^{**}$.*

Proof. Let e be the neutral element. Since $x^* \top x = e = x \top x^{**}$ we have, using the associativity of $\top$,

$$x^{**} = e \top x^{**} = (x^* \top x) \top x^{**} = x^* \top (x \top x^{**}) = x^* \top e = x^*.$$

Corollary. *In any semigroup with a neutral element, every element has at most one inverse (i.e., exactly one or none at all).*

> *Remark.* When dealing with an associative law $\top$ we shall denote the unique inverse of an element x by x^{-1} (when it exists) except when $\top$ is written additively, in which case we shall write $-x$.

An element which has an inverse will be called **invertible**. In any semigroup having a neutral element, the neutral element is clearly invertible and is its own inverse.

Example 12.1. In the semigroup $(\mathbf{N}, +)$ the only invertible element is 0; in the semigroup $(\mathbf{N}, \cdot)$ the only invertible element is 1. In the semigroup $(\mathrm{Map}(E, E), \circ)$ the elements which have left inverses are the injective mappings (Theorem 5.4); the elements having right inverses are the surjective mappings (Theorem 5.6); and the elements which have (two-sided) inverses are the bijective mappings (Theorem 5.9).

> *Remark.* Note that although two-sided inverses are necessarily unique, one-sided inverses need not be unique. By way of providing an illustrative example, we refer the reader to Exercise 5.10. In appreciating this example, the reader should bear in mind the result of Theorem 12.1.

Theorem 12.2 *Let* $(E, \top)$ *be a semigroup. If* $x_1, \ldots, x_n$ *are invertible elements of E then so also is* $x_1 \top \ldots \top x_n$*; moreover,*

$$(x_1 \top \ldots \top x_n)^{-1} = x_n^{-1} \top \ldots \top x_1^{-1}.$$

Proof. [The reader is referred to Exercise 11.8 for a proper definition of the composite element $x_1 \top \ldots \top x_n$.] We prove the result by induction. Clearly, the result holds for $n=1$. Since we shall require the result for $n=2$ we shall establish this as well. Suppose then that x_1, x_2 are invertible. Then

$$x_1 \top x_2 \top x_2^{-1} \top x_1^{-1} = x_1 \top e \top x_1^{-1} = x_1 \top x_1^{-1} = e;$$

$$x_2^{-1} \top x_1^{-1} \top x_1 \top x_2 = x_2^{-1} \top e \top x_2 = x_2^{-1} \top x_2 = e,$$

and so we see that $(x_1 \top x_2)^{-1}$ exists and is none other than $x_2^{-1} \top x_1^{-1}$. Suppose now that the result holds for n-fold composites and consider $x_1 \top \ldots \top x_n \top x_{n+1}$. Using the result for $n=2$ we have

$$\begin{aligned}(x_1 \top \ldots \top x_n \top x_{n+1})^{-1} &= [(x_1 \top \ldots \top x_n) \top x_{n+1}]^{-1} \\ &= x_{n+1}^{-1} \top (x_1 \top \ldots \top x_n)^{-1} \\ &= x_{n+1}^{-1} \top x_n^{-1} \top \ldots \top x_1^{-1}.\end{aligned}$$

Hence the result holds for $n+1$-fold composites and so is true for all $n \geqslant 1$ by the induction principle.

Definition. By a **group** we mean a semigroup with a neutral element in which every element is invertible. We say that a group is **abelian** if its law of composition is commutative (in other words, if every pair of elements commute).

Example 12.2. $(\mathbf{Z}, +)$ is an abelian group; the inverse of n is $-n$. $(\mathbf{Z}, \cdot)$ is not a group since the only elements with multiplicative inverses are 1 and -1. $(\mathbf{Q}, +)$ and $(\mathbf{Q}\backslash\{0\}, \cdot)$ are groups as are $(\mathbf{R}, +)$, $(\mathbf{R}\backslash\{0\}, \cdot)$, $(\mathbf{C}, +)$ and $(\mathbf{C}\backslash\{0\}, \cdot)$.

Example 12.3. By Theorem 5.10 the set Bij (E, E) of all bijections from a set E to itself forms a group with respect to composition of mappings. In the case where Card $E \geqslant 2$, this group is not abelian.

Example 12.4. The set $G=\{1, i, -1, -i\}$ where $i^2=-1$ forms an abelian group under multiplication. This is readily seen from the Cayley table

	1	i	-1	$-i$
1	1	i	-1	$-i$
i	i	-1	$-i$	1
-1	-1	$-i$	1	i
$-i$	$-i$	1	i	-1

The multiplication in G is associative since multiplication of complex numbers is associative. Clearly the neutral element is 1. The elements 1 and -1 are self-inverse and the elements $i, -i$ are inverses of each other.

Example 12.5. The set Map (**R**, **R**) forms an abelian group under the addition $+$ defined by $(f, g) \mapsto f+g$ where $(\forall x \in \mathbf{R})\,(f+g)(x)=f(x)+g(x)$. Associativity follows from the associativity of addition in **R**. The neutral element is the zero mapping (i.e., that which sends every element to 0); and the inverse of f is the mapping $-f$ given by $(\forall x \in \mathbf{R})$ $(-f)(x)=-f(x)$.

Example 12.6. Let $a, b \in \mathbf{R}$ with $a \neq 0$. We denote by $f_{a,b}$ the mapping from **R** to **R** given by $f_{a,b}(x)=ax+b$. Such a mapping is called an **affine transformation**. If Aff (**R**) denotes the set of all affine transformations and if $f_{a,b}$ and $f_{c,d}$ belong to this set then for every $x \in \mathbf{R}$ we have

$$(f_{a,b} \circ f_{c,d})(x)=f_{a,b}(cx+d)=acx+ad+b=f_{ac,\,ad+b}(x),$$

so that $f_{a,b} \circ f_{c,d}=f_{ac,\,ad+b}$. It follows that composition of mappings is a law of composition on Aff (**R**), for since $a \neq 0$ and $c \neq 0$ we have $ac \neq 0$. It follows that (Aff (**R**), $\circ$) is a semigroup and it is readily seen that $f_{1,0}=id_{\mathbf{R}}$ is a neutral element. Now for every $f_{a,b}$ we see that

$$f_{a,b} \circ f_{a^{-1},-ba^{-1}}=f_{1,0}=f_{a^{-1},-ba^{-1}} \circ f_{a,b}$$

and so $f_{a,b}$ is invertible with inverse $f_{a^{-1},-ba^{-1}}$. Thus (Aff (**R**), $\circ$) is a group.

The following result reflects the essential difference between **N** and **Z** as mentioned in an intuitive way above.

Theorem 12.3 *Let* $(E, \top)$ *be a semigroup. Then the following conditions are equivalent*:

(1) $(E, \top)$ *is a group*;

(2) $(\forall a, b \in E)(\exists x, y \in E)\quad x \top a=b, \quad a \top y=b.$

Proof. (1) ⇒ (2): If $(E, \top)$ is a group and $a, b \in E$ consider the elements $x=b \top a^{-1}$, $y=a^{-1} \top b$. These elements clearly satisfy (2).

(2) ⇒ (1): Let a be any fixed element of E. By (2) there exist $e_1, e_2 \in E$ such that $e_1 \top a=a$ and $a \top e_2=a$. Now let x be any element of E. Again by (2) there exist $y, z \in E$ such that $y \top a=x$ and $a \top z=x$. Now since

$$e_1 \top x=e_1 \top (a \top z)=(e_1 \top a) \top z=a \top z=x;$$
$$x \top e_2=(y \top a) \top e_2=y \top (a \top e_2)=y \top a=x,$$

it follows by choosing $x=e_2$ that $e_1 \top e_2=e_2$; and by choosing $x=e_1$ that $e_1 \top e_2=e_1$. Hence we have $e_1=e_2=e$, say; and from the above equalities e is a (hence the) neutral element of E. Now apply (2) to the elements x, e: there exist $y^*, z^* \in E$ such that $x \top y^*=e$ and $z^* \top x=e$. Applying Theorem 12.1 we deduce that $y^*=z^*=x^{-1}$. Consequently $(E, \top)$ is a group.

A fundamental property of every group is the following.

Theorem 12.4 *If $(E, \top)$ is a group then for every $a \in E$ the translations $\lambda_a: x \mapsto a \top x$ and $\rho_a: x \mapsto x \top a$ are bijections on E.*

Proof. Given any $b \in E$ there exist, by Theorem 12.3, elements $x, y \in E$ such that $\lambda_a(y)=b$ and $\rho_a(x)=b$; thus λ_a, ρ_a are surjective. Now since

$$a \top x=a \top y \Rightarrow a^{-1} \top a \top x=a^{-1} \top a \top y \Rightarrow e \top x=e \top y \Rightarrow x=y$$

we see that λ_a is also injective. Similarly ρ_a is injective.

> *Remark.* The injective properties of the translations λ_a, ρ_a may be expressed in the form $a \top x=a \top y \Rightarrow x=y$ and $x \top a=y \top a \Rightarrow x=y$. These are often referred to as the **left** and **right cancellation laws**.

If now $(E, \top)$ is a groupoid then we know that $\top$ is described by a mapping $f: E \times E \to E$. If F is a subset of E then in general all we can say about $f^{\to}(F \times F)$ is that $f^{\to}(F \times F) \subseteq E$. If, however, it should happen that $f^{\to}(F \times F) \subseteq F$ then we can clearly define a mapping $g: F \times F \to F$ (i.e., a law of composition on F) by writing $g(x, y)=f(x, y)$ for all $x, y \in F$. In this case we say that F is a **stable subset** of E with respect to $\top$ (or that it is **closed** under $\top$) and we say that the law of composition g on F is **induced by** (or **inherited from**) that on E. Thus, in terms of $\top$ we see that F is a stable subset of E if and only if it satisfies the property

$$(\forall x, y \in F) \quad x \top y \in F.$$

Example 12.7. Consider the law of composition $f: \mathbf{R} \times \mathbf{R} \to \mathbf{R}$ given by $f(x, y) = x - y$. Clearly $f^{\to}(\mathbf{Z} \times \mathbf{Z}) \subseteq \mathbf{Z}$ and so f induces a law of composition on $\mathbf{Z}$; put another way, $\mathbf{Z}$ is stable under subtraction. In contrast, we note that $f^{\to}(\mathbf{N} \times \mathbf{N}) \nsubseteq \mathbf{N}$ so $\mathbf{N}$ is not stable under subtraction.

If S is a semigroup then by a **subsemigroup** of S we shall mean a stable subset of S. If G is a group then by a **subgroup** of G we shall mean a stable subset H of G which is a group with respect to the law of composition induced on H by that of G.

Example 12.8. The set $\mathrm{Inj}(E, E)$ of all injections from a set E to itself forms a subsemigroup of the semigroup $\mathrm{Map}(E, E)$. $\mathbf{N}$ is a subsemigroup, but not a subgroup, of $\mathbf{Z}$.

Theorem 12.5 *All subgroups of a group G have the same neutral element as G.*

Proof. Let H be any subgroup of G and let e_H be the neutral element of H. If e_G denotes the neutral element of G then since $e_H \in G$ we have $e_H \top e_G = e_H$. But since e_H is the neutral element of H we have $e_H \top e_H = e_H$. Thus $e_H \top e_G = e_H \top e_H$ and so, by the left cancellation law, $e_G = e_H$.

The following extremely useful result gives a simple criterion for a subset H of a group G to be a subgroup of G.

Theorem 12.6 *A non-empty subset H of a group* $(G, \top)$ *is a subgroup if and only if*

$$(\forall x, y \in H) \quad x \top y^{-1} \in H.$$

Proof. Suppose first that H is a subgroup of G. Given $y \in H$ we see that the inverse y^{-1} of y in G belongs to H; for if y_H^{-1} denotes the inverse of y in the subgroup H then, by Theorem 12.5, we have $y_H^{-1} \top y = e_H = e_G = y^{-1} \top y$ whence $y_H^{-1} = y^{-1}$ by the right cancellation law. Since H is stable under $\top$ we therefore have $x \top y^{-1} \in H$ for all $x, y \in H$.

Conversely, suppose that H satisfies the given property. Choosing $y = x$ in the property gives $(\forall x \in H)\ x \top x^{-1} \in H$ and so we have $e \in H$. Now take $x = e$ in the property; we obtain $(\forall y \in H)\ y^{-1} \in H$, whence inverses of elements of H are also elements of H. Thus, given any $x, y \in H$ we see that $x, y^{-1} \in H$ whence, by the property, $x \top y = x \top (y^{-1})^{-1} \in H$. Consequently H is stable under $\top$. Since $\top$ is associative in G it is clearly associative in any stable subset. Combining these observations we see that H is indeed a subgroup of G.

Remarks. In additive notation we generally write $x+(-y)$ as $x-y$ so that the condition of the above result becomes $(\forall x, y \in H)$ $x - y \in H$. In multiplicative notation it becomes $(\forall x, y \in H)$ $xy^{-1} \in H$. Note that Theorem 12.6 is very useful in obtaining further examples of groups since, instead of verifying the group axioms each time, it is often possible to show that a given structure is a group by showing that it is a subgroup of some much larger group. The following examples illustrate this.

Example 12.9. For every $n \in \mathbf{Z}$ let $n\mathbf{Z}=\{np; p \in \mathbf{Z}\}$ be the set of all integer multiples of n. Alternatively, $n\mathbf{Z}$ is the class of 0 modulo the equivalence relation "mod n" (Example 6.5). That $(n\mathbf{Z}, +)$ forms a group follows from the fact that it is a subgroup of the group $(\mathbf{Z}, +)$; for if $x, y \in n\mathbf{Z}$ then $x=na$ and $y=nb$ for some $a, b \in \mathbf{Z}$ and so $x-y=n(a-b) \in n\mathbf{Z}$.

Example 12.10. Let α be a fixed real number and let $G=\{f \in \text{Map}(\mathbf{R}, \mathbf{R}); f(\alpha)=0\}$. Then G forms a group under addition of mappings. This follows from the fact that it is a subgroup of the group $(\text{Map}(\mathbf{R}, \mathbf{R}), +)$ of Example 12.5; for if $f, g \in G$ then $(f-g)(\alpha)=f(\alpha)-g(\alpha)=0-0=0$ and so $f-g \in G$.

Example 12.11. Let $D(\mathbf{R})$ denote the set of all mappings $f:\mathbf{R}\to\mathbf{R}$ which are differentiable. It is shown in analysis that if $f, g \in D(\mathbf{R})$ then $f-g \in D(\mathbf{R})$. Hence $(D(\mathbf{R}), +)$ is a subgroup of the group $(\text{Map}(\mathbf{R}, \mathbf{R}), +)$.

As an application of Theorem 12.6 we now prove:

Theorem 12.7 *The intersection of any family of subgroups of a group G is also a subgroup of G.*

Proof. Let $(A_i)_{i \in I}$ be a family of subgroups of the group $(G, \top)$. Then for all $x, y \in G$ we have

$$\begin{aligned} x, y \in \bigcap_{i \in I} A_i &\Rightarrow (\forall i \in I) \quad x, y \in A_i \\ &\Rightarrow (\forall i \in I) \quad x \top y^{-1} \in A_i \\ &\Rightarrow x \top y^{-1} \in \bigcap_{i \in I} A_i. \end{aligned}$$

The result is now immediate from Theorems 12.6 and 12.5.

Suppose now that X is any subset of a group G. Then the set of all subgroups of G which contain X is not empty (since G itself is in this set). By Theorem 12.7, the intersection of all the subgroups of G which contain X is also a subgroup of G and this subgroup clearly also contains X. It is then the smallest subgroup of G to contain X. It is called the **subgroup generated by** X and will be denoted by Gp X. In the case where $X=\emptyset$ it is clear that Gp $X=\{e\}$ which is the smallest subgroup of G. When $X\neq\emptyset$ we can describe Gp X in the following way.

Theorem 12.8 *If X is a non-empty subset of a group $(G, \top)$ let X^* denote the set of all elements of G of the form $x_1 \top x_2 \top \dots \top x_n$ where $n\geqslant 1$ and, for every i, either $x_i \in X$ or $x_i^{-1} \in X$. Then*

$$\text{Gp } X = X^*.$$

Proof. Since the inverse of $x_1 \top \dots \top x_n$ is $x_n^{-1} \top \dots \top x_1^{-1}$ it is immediate from Theorem 12.6 that X^* is a subgroup of G containing X. Since Gp X is the smallest subgroup with this property we then have Gp $X \subseteq X^*$. Suppose now that K is any subgroup of G which contains X. Then K contains the inverses of all the elements of X and consequently we see that $X^* \subseteq K$. Taking in particular $K=$ Gp X we obtain $X^* \subseteq$ Gp X and the result follows.

We recall now that if $(G_1, \top_1)$ and $(G_2, \top_2)$ are groupoids then by a **morphism** from G_1 to G_2 we mean a mapping $f: G_1 \to G_2$ such that $(\forall x, y \in G_1)\ f(x \top_1 y) = f(x) \top_2 f(y)$. We shall now consider morphisms between groups. We establish first the following elementary result.

Theorem 12.9 *If $(E_1, \top_1)$ and $(E_2, \top_2)$ are groupoids and if $f: E_1 \to E_2$ is a morphism then*

(1) Im f *is a stable subset of* E_2;

(2) *if E_1 is abelian then so also is* Im f;

(3) *if E_1 is a semigroup then so also is* Im f;

(4) *if e is neutral in E_1 then $f(e)$ is neutral in* Im f;

(5) *if y is an inverse of x in E_1 then $f(y)$ is an inverse of $f(x)$ in* Im f.

Proof. (1) If $a, b \in$ Im f then $a=f(x)$ and $b=f(y)$ for some $x, y \in E_1$. Since f is a morphism we have $a \top_2 b = f(x) \top_2 f(y) = f(x \top_1 y) \in$ Im f. Thus Im f is stable under $\top_2$.

(2) follows from the fact that if $x \top_1 y = y \top_1 x$ then we have $f(x) \top_2 f(y) = f(x \top_1 y) = f(y \top_1 x) = f(y) \top_2 f(x)$.

(3) follows from the fact that if $x \top_1 (y \top_1 z) = (x \top_1 y) \top_1 z$ then $f(x) \top_2 [f(y) \top_2 f(z)] = f(x) \top_2 f(y \top_1 z) = f[x \top_1 (y \top_1 z)] = f[(x \top_1 y) \top_1 z] = f(x \top_1 y) \top_2 f(z) = [f(x) \top_2 f(y)] \top_2 f(z)$.

(4) If $e \top_1 x = x = x \top_1 e$ for all $x \in E_1$ then for all $f(x) \in \operatorname{Im} f$ we have $f(e) \top_2 f(x) = f(e \top_1 x) = f(x) = f(x \top_1 e) = f(x) \top_2 f(e)$.

(5) follows from the fact that if $x \top_1 y = e = y \top_1 x$ then we have $f(x) \top_2 f(y) = f(x \top_1 y) = f(e) = f(y \top_1 x) = f(y) \top_2 f(x)$.

Corollary. *If G, H are groups and $f: G \to H$ is a morphism then* $\operatorname{Im} f$ *is a subgroup of H.*

If G, H are groups and $f: G \to H$ is a morphism then a particularly important subset of G is the set $\{x \in G; f(x) = e_H\}$. This set is called the **kernel** of f and is denoted by $\operatorname{Ker} f$. That $\operatorname{Ker} f \neq \emptyset$ follows by Theorems 12.9(4) and 12.5. It is clear that if $x, y \in \operatorname{Ker} f$ then, denoting the laws of G, H as $\top$ without confusion,

$$f(x \top y^{-1}) = f(x) \top f(y^{-1}) = f(x) \top [f(y)]^{-1} = e_H \top e_H = e_H$$

and so $x \top y^{-1} \in \operatorname{Ker} f$. Thus we see that $\operatorname{Ker} f$ *is a subgroup of G*. We have the following important criterion involving $\operatorname{Ker} f$ for a group morphism f to be a monomorphism (i.e., injective).

Theorem 12.10 *If G, H are groups and $f: G \to H$ is a morphism then the following conditions are equivalent*:
(1) f is *a monomorphism*;
(2) $\operatorname{Ker} f = \{e_G\}$.

Proof. (1) $\Rightarrow$ (2): Suppose that f is a monomorphism and let $x \in \operatorname{Ker} f$. Then we have $f(x) = e_H$. But by Theorem 12.9(4) we have $f(e_G) = e_H$. As f is injective we deduce that $x = e_G$ and so $\operatorname{Ker} f = \{e_G\}$.

(2) $\Rightarrow$ (1): If $\operatorname{Ker} f = \{e_G\}$ then, writing the laws of G and H both as $\top$ without confusion, it follows from Theorem 12.9(5) that

$$\begin{aligned} f(x) = f(y) &\Rightarrow f(x \top y^{-1}) = f(x) \top f(y^{-1}) = f(x) \top [f(y)]^{-1} = e_H \\ &\Rightarrow x \top y^{-1} = e_G \\ &\Rightarrow x = x \top y^{-1} \top y = e_G \top y = y. \end{aligned}$$

Thus f is injective and so is a monomorphism.

Example 12.12. Let $(G, \top)$ be a group. For every $a \in G$ let $\zeta_a: G \to G$ be given by $\zeta_a(x) = a \top x \top a^{-1}$. That ζ_a is a morphism follows from the equalities

$$\begin{aligned} \zeta_a(x \top y) = a \top (x \top y) \top a^{-1} &= (a \top x \top a^{-1}) \top (a \top y \top a^{-1}) \\ &= \zeta_a(x) \top \zeta_a(y). \end{aligned}$$

Now ζ_a is surjective (and hence an **epimorphism**) since for every $x \in G$ we have $\zeta_a(a^{-1} \top x \top a) = a \top (a^{-1} \top x \top a) \top a^{-1} = x$. Finally, if $x \in \operatorname{Ker} \zeta_a$

then $a \top x \top a^{-1} = e$ and so $x = a^{-1} \top (a \top x \top a^{-1}) \top a = a^{-1} \top e \top a = e$, showing that ζ_a is also a monomorphism. Hence ζ_a is an **isomorphism** (referred to as an **inner automorphism** on G).

We now give consideration to a type of group known as a **permutation group** and, as an illustration of the importance of the notion of isomorphism, we prove the celebrated Cayley theorem that every group is isomorphic to some permutation group.

If G, H are groups then by the notation $G \simeq H$ we shall mean that G and H are isomorphic.

By a **permutation** on a set E we mean simply a bijection $f: E \to E$. A group G is called a **permutation group** if, for some set E, G is a subgroup of the group (Bij (E, E), $\circ$).

Theorem 12.11 [Cayley] *Every group is isomorphic to some permutation group.*

Proof. Let $(G, \top)$ be a group and for every $a \in G$ let $\lambda_a: G \to G$ be the associated left translation. By Theorem 12.4 each λ_a is a bijection on G. Since, for all $a, b, x \in G$,

$$(\lambda_a \circ \lambda_b)(x) = \lambda_a(b \top x) = a \top b \top x = \lambda_{a \top b}(x)$$

we see that $\lambda_a \circ \lambda_b = \lambda_{a \top b}$ and so $F = \{\lambda_a;\, a \in G\}$ is a stable subset, and hence a subsemigroup, of the group (Bij (G, G), $\circ$). Now clearly $id_G = \lambda_e$ is the neutral element of this semigroup; and every element has an inverse, the inverse of λ_a being $\lambda_{a^{-1}}$. Thus F is a subgroup of Bij (G, G) and so is a permutation group. Consider now the mapping $f: (G, \top) \to (F, \circ)$ given by $f(a) = \lambda_a$. Since

$$(\forall a, b \in G) \quad f(a \top b) = \lambda_{a \top b} = \lambda_a \circ \lambda_b = f(a) \circ f(b)$$

we see that f is a morphism. It is also clear that f is surjective. Finally

$$a \in \operatorname{Ker} f \Leftrightarrow \lambda_a = id_G \Leftrightarrow (\forall x \in G) \quad a \top x = x \Leftrightarrow a = e_G$$

so that $\operatorname{Ker} f = \{e_G\}$ and, by Theorem 12.10, f is a monomorphism. Thus we see that $G \simeq F$.

For the remainder of this section we shall be concerned with properties of subgroups relative to morphisms.

Theorem 12.12 *Let G, H be groups and $f: G \to H$ a morphism. Then, for every subgroup J of G, $f^{\to}(J)$ is a subgroup of* $\operatorname{Im} f$; *and, for every subgroup K of H, $f^{\leftarrow}(K)$ is a subgroup of G containing* $\operatorname{Ker} f$.

Proof. If $x, y \in f^{\rightarrow}(J)$ then $x=f(a)$ and $y=f(b)$ for some $a, b \in J$ and so $x \top y^{-1}=f(a) \top [f(b)]^{-1}=f(a) \top f(b^{-1})=f(a \top b^{-1})$ whence $x \top y^{-1} \in f^{\rightarrow}(J)$ since J is a subgroup of G. Hence $f^{\rightarrow}(J)$ is a subgroup of $\operatorname{Im} f$.

Similarly, if $x, y \in f^{\leftarrow}(K)$ then $f(x), f(y) \in K$ and so, since K is a subgroup, $f(x \top y^{-1})=f(x) \top f(y^{-1})=f(x) \top [f(y)]^{-1} \in K$ and so $x \top y^{-1} \in f^{\leftarrow}(K)$. This shows that $f^{\leftarrow}(K)$ is a subgroup of G; moreover, since $\{e_H\} \subseteq K$ we have $\operatorname{Ker} f=f^{\leftarrow}\{e_H\} \subseteq f^{\leftarrow}(K)$.

For our next result we ask the reader to recall the definition of a *lattice* (see §7).

Theorem 12.13 *If $(G, \top)$ is a group then the set of subgroups of G, ordered by set inclusion, forms a lattice in which intersection is set-theoretic and union is given by*

$$A \vee B=\{x \in G;\, x=a_1 \top \ldots \top a_n,\, a_i \in A \cup B\}.$$

Proof. If A, B are subgroups of G then so also is $A \cap B$ by Theorem 12.7; moreover $A \cap B$ is clearly the greatest subgroup of G contained in both A and B, so we have $A \cap B=\inf\{A, B\}$ relative to the ordering by set inclusion.

Now $\sup\{A, B\}$ also exists, for it is simply the subgroup generated by $A \cup B$. Since A and B are subgroups we see from Theorem 12.8 that $\sup\{A, B\}$ coincides with $A \vee B$ defined above. Thus we see that the set of subgroups of G forms a lattice.

Theorem 12.14 *Let G, H be groups and $f: G \rightarrow H$ a morphism. Then, if J is a subgroup of G, we have*

$$f^{\leftarrow}[f^{\rightarrow}(J)]=J \vee \operatorname{Ker} f;$$

and, if K is a subgroup of H, we have

$$f^{\rightarrow}[f^{\leftarrow}(K)]=K \cap \operatorname{Im} f.$$

Proof. For the first part it suffices, by Theorems 5.8 and 12.12, to show that $f^{\leftarrow}[f^{\rightarrow}(J)] \subseteq J \vee \operatorname{Ker} f$. For this, let $t \in f^{\leftarrow}[f^{\rightarrow}(J)]$. Then we have $f(t)=f(j)$ for some $j \in J$ whence $t \top j^{-1} \in \operatorname{Ker} f$ and so $t=k \top j$ where $k \in \operatorname{Ker} f$ and $j \in J$. It follows that $t \in J \vee \operatorname{Ker} f$.

For the second part it suffices similarly to show that $f^{\rightarrow}[f^{\leftarrow}(K)] \supseteq K \cap \operatorname{Im} f$. Suppose then that $k \in K \cap \operatorname{Im} f$. Then $k=f(x)$ for some $x \in G$. Indeed since $k \in K$ we have $x \in f^{\leftarrow}(K)$ and so $k=f(x) \in f^{\rightarrow}[f^{\leftarrow}(K)]$.

Exercises for §12

1. Let $(E, \top)$ be a groupoid. Prove that $\top$ is associative if and only if, in the semigroup $(\mathrm{Map}(E, E), \circ)$, every left translation λ_a commutes with every right translation ρ_b. If $\top$ is associative prove that

(1) if $a \in E$ has an inverse with respect to $\top$ then λ_a and ρ_a are bijections;

(2) if λ_a and ρ_a are both surjective then E has a neutral element and a is invertible.

[*Hint* (2): If λ_a is surjective then $\lambda_a(e)=a$ for some $e \in E$. Show that if $x \in E$ then $x \top e = x$ (consider x' such that $\rho_a(x')=x$) so that e is a right neutral element for E. Show similarly the existence of a left neutral element.]

Suppose now that $\top$ is associative and that E has a neutral element e. Let $a_1, \ldots, a_n \in E$ be such that each λ_{a_i} is injective. Prove that if $a_1 \top \ldots \top a_n = e$ then, for every i, $a_{i+1} \top \ldots \top a_n \top a_1 \top \ldots \top a_i = e$. Deduce that if $a_1 \top \ldots \top a_n$ has an inverse then so also does every a_i.

2. Define a law of composition $\otimes$ on $\mathbf{R} \times \mathbf{R} \times \mathbf{R}$ by

$$(x, y, z) \otimes (x^*, y^*, z^*) = (xx^*, \alpha y^* + yx^*, \alpha z^* + zx^*)$$

where $\alpha = x + y + z$. Show that $\otimes$ is associative and that a neutral element exists. Show also that (x, y, z) has an inverse with respect to $\otimes$ if and only if $\alpha \neq 0$ and $x \neq 0$; and that the inverse is then $(1/x, -y/\alpha x, -z/\alpha x)$.

3. Let P be the set of primes $\{2, 3, 5, 7, 11, 13\}$. Define $f: P \times P \to \mathbf{Z}$ by

$$f(p, q) = \text{ the greatest prime factor of } p+q-2.$$

Show that f induces a law of composition on P. Writing this law as $\top$, construct the Cayley table for $\top$ and show that P has a neutral element and that every element of P has an inverse. Show, however, that $\top$ is not associative (so that $(P, \top)$ is not a group).

4. Let $(G, \top)$ be a finite groupoid. Prove that if G is a group then every element of G appears precisely once in each row and column of the Cayley table for $\top$. Show that the converse is not true in general. Prove that a finite semigroup is a group if and only if it is cancellative (in that every element is both left and right cancellative). The following is part of the Cayley table of a finite multiplicative group; fill in the missing entries:

	e	a	b	x	y	z
e	e	a	b	x	y	z
a	a		e	y		
b	b					
x	x	z				a
y	y					
z	z					

[*Hint.* Use the first part of the question to complete immediately three-quarters of the table. Then observe that either $x^2=e$ or $x^2=b$ and show that the latter is impossible since it yields $x \cdot x^2 \neq x^2 \cdot x$; similarly, either $z^2=e$ or $z^2=b$ and the latter is impossible.]

5. For every $z=x+iy \in \mathbf{C}$ define $|z|=\sqrt{(x^2+y^2)}$. Show that $G=\{z \in \mathbf{C};\ |z|=1\}$ is a group under multiplication (called the **circle group**).

6. If E is a non-empty set and $X, Y \in \mathbf{P}(E)$ define the **symmetric difference set** $X \Delta Y$ by $X \backslash Y \cup Y \backslash X$. Show that $(\mathbf{P}(E), \Delta)$ is a group.

7. Let $(G, \top)$ be a group and suppose that $a, b \in G$ are such that $a \neq b$, $a \neq e_G$, $b \neq e_G$ and $a \top b \neq b \top a$. Prove that the elements $e_G, a, b, a \top b, b \top a$ are distinct. Show that either $a \top a = e_G$ or $a \top a \notin \{e_G, a, b, a \top b, b \top a\}$. If $a \top a = e_G$ show that $a \top b \top a \notin \{e_G, a, b, a \top b, b \top a\}$. Hence deduce that every non-abelian group contains at least six elements.

If $f_1, \ldots, f_6$ are the mappings from $\mathbf{R} \backslash \{0, 1\}$ to $\mathbf{R}$ given by $f_1(x)=x$, $f_2(x)=\dfrac{1}{x}$, $f_3(x)=1-x$, $f_4(x)=\dfrac{x-1}{x}$, $f_5(x)=\dfrac{x}{x-1}$, $f_6(x)=\dfrac{1}{1-x}$ show that $f_1, \ldots, f_6$ forms, under composition of mappings, a non-abelian group with precisely six elements.

8. Let $G=\{(a, b) \in \mathbf{R} \times \mathbf{R};\ b \neq 0\}$. Show that G is a non-abelian group under the law of composition $\top$ given by $(a, b) \top (c, d) = (a+bc, bd)$. Show also that the subsets $H=\{(0, b); b \neq 0\}$, $K=\{(a, b); b \geqslant 0\}$ and

$L=\{(a, 1); a \in \mathbf{R}\}$ are subgroups of G. Are any of these subgroups abelian?

9. Show that $\left\{\dfrac{1+2m}{1+2n}; m, n \in \mathbf{Z}\right\}$ is a subgroup of $(\mathbf{Q}\backslash\{0\}, \cdot)$.

10. If p is a prime show that $\{m/p^r; m, r \in \mathbf{Z}, r \geqslant 0\}$ is a subgroup of $(\mathbf{Q}, +)$.

11. Let $S=\{(x, y) \in \mathbf{R}\times\mathbf{R}; x\neq 0, x+y\neq 0\}$. Prove that

$$(x_1, y_1) \top (x_2, y_2)=(x_1x_2, (x_1+y_1)(x_2+y_2)-x_1x_2)$$

defines an internal law of composition on S and that $(S, \top)$ is a group. Show also that the subset $P=\{(1, y) \in S; y>-1\}$ is a subgroup.

12. (a) Let G be a multiplicative group with neutral element e and let $(g, h) \mapsto g \top h = gh^{-1}$ be a second law of composition on G. Prove that, for all $g, h \in G$, (1) $g \top g = e$; (2) $g \top e = g$; (3) $e \top (g \top h) = h \top g$; (4) $gh = g \top (e \top h)$.

(b) Let a be a fixed element of a set S. Let $(x, y) \mapsto x \top y$ and $(x, y) \mapsto xy$ be two laws of composition on S such that, for all $x, y \in S$, (1) $x \top x = a$; (2) $x \top a = x$; (3) $a \top (x \top y) = y \top x$; (4) $xy = x \top (a \top y)$. Assuming that $(x, y) \mapsto xy$ is associative prove that S forms a group under this law. Prove also that every subset of S which is stable under $\top$ is a subgroup.

13. Let $G=\{f\in \text{Bij}\,(\mathbf{C}, \mathbf{C}); (\forall z, w \in \mathbf{C})\ |f(z)-f(w)|=|z-w|\}$. Show that $(G, \circ)$ is a subgroup of $(\text{Bij}\,(\mathbf{C}, \mathbf{C}), \circ)$.

14. For every group $(G, \top)$ define the **centre** of G to be the subset $Z(G)=\{a \in G; (\forall x \in G)\ a \top x = x \top a\}$. Prove that $Z(G)$ is an abelian subgroup of G.

15. If H_1, H_2 are subgroups of a group G prove that $H_1 \cup H_2$ is a subgroup of G if and only if either $H_1 \subseteq H_2$ or $H_2 \subseteq H_1$.

[*Hint.* $\Rightarrow$: If $H_1 \nsubseteq H_2$ and $H_2 \nsubseteq H_1$ choose elements $a \in H_2 \cap \complement_G(H_1)$ and $b \in H_1 \cap \complement_G(H_2)$ and show that $a \top b \notin H_1 \cup H_2$.]

16. Let $(G, \top)$ be a group and let $f: G \to H$ be a bijection from G to a set H. Show that there is one and only one law of composition ∇ on H such that (H, ∇) is a group with f an isomorphism.

[*Hint.* Consider $x \nabla y = f[f^{-1}(x) \top f^{-1}(y)]$.]

17. Define a law of composition $\top$ on $\mathbf{R}\backslash\{0\}\times\mathbf{R}$ by

$$(a, b) \top (c, d)=(ac, ad+b).$$

Show that under this law $\mathbf{R}\backslash\{0\}\times\mathbf{R}$ forms a group isomorphic to the group $(\text{Aff}\,(\mathbf{R}), \circ)$ of Example 12.6.

18. For every non-zero natural number p let $\mathbf{N}_p$ be the set $\{0, \ldots, p-1\}$. Define a law of composition $+_p$ on $\mathbf{N}_p$ (called **addition modulo** p) by setting $m +_p n = k$ where k is the unique element of $\mathbf{N}_p$ such that $m+n \equiv k \pmod{p}$. In other words, $m +_p n$ is the remainder term in the euclidean division of $m+n$ by p. Show that $(\mathbf{N}_p, +_p)$ is an abelian group.

Suppose now that $p \geqslant 3$ and let the vertices of a regular polygon Δ_p with p sides be labelled $0, 1, \ldots, p-1$ in an anti-clockwise manner. By a **symmetry** of Δ_p we mean a displacement of Δ_p in such a way that adjacent vertices are carried onto adjacent vertices. Observe that we can regard a symmetry of Δ_p as a bijection $f: \mathbf{N}_p \to \mathbf{N}_p$ such that, for every $t \in \mathbf{N}_p$, either $f(t +_p 1) = f(t) +_p 1$ or $f(t +_p 1) = f(t) -_p 1$.

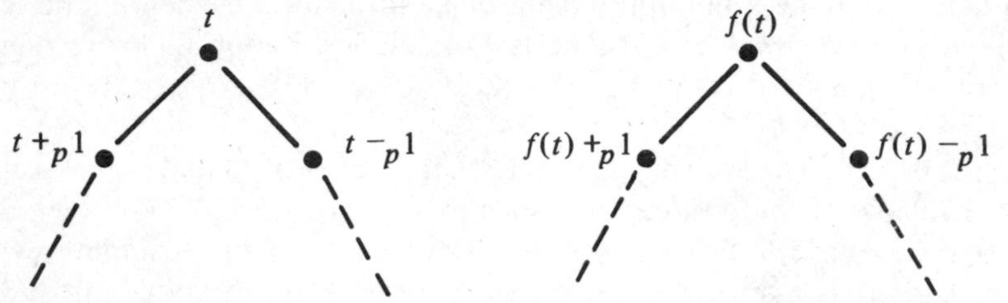

Show that if f is a symmetry on Δ_p and if $f(1) = f(0) +_p 1$ then $f(t +_p 1) = f(t) +_p 1$ for every $t \in \mathbf{N}_p$ whereas if $f(1) = f(0) -_p 1$ then $f(t +_p 1) = f(t) -_p 1$ for every $t \in \mathbf{N}_p$.

For every $t \in \mathbf{N}_p$ define the **rotation** r_t by $r_t(k) = t +_p k$ and the **reflection** s_t by $s_t(k) = t -_p k$. Prove that r_t and s_t are symmetries on Δ_p. Show further that if f is a symmetry on Δ_p then f is either a rotation or a reflection. Prove that, writing $f \circ f \circ \ldots \circ f$ (n terms) as f^n,

(1) $(\forall t \in \mathbf{N}_p)\quad r_t^p = id = s_t^2$;

(2) $(\forall t, n \in \mathbf{N}_p)\quad r_t \circ s_n = s_n \circ r_t^{-1}$.

Deduce that if D_p denotes the set of all symmetries on Δ_p then $(D_p, \circ)$ is a subgroup of the group $(\mathrm{Bij}\,(\mathbf{N}_p, \mathbf{N}_p), \circ)$. Show also that, for every $t \in \mathbf{N}_p$, $r_t = r_1^t$ and $s_t = r_t \circ s_0$. Hence deduce that every element of D_p can be expressed uniquely in the form $r_1^i \circ s_0^j$ where $i \in \mathbf{N}_p$ and $j \in \{0, 1\}$.

Suppose now that n is an integer $\geqslant 3$ and that G is a multiplicative group with neutral element e such that

$$(*)\quad \begin{cases} G \text{ contains elements } r, s \text{ with} \\ \quad (\alpha)\ s \neq e \text{ and } r^k \neq e \text{ if } 0 < k < n; \\ \quad (\beta)\ r^n = e = s^2; \\ \quad (\gamma)\ rs = sr^{-1}. \end{cases}$$

Show that $H=\{r^i s^j;\, i\in \mathbf{N}_n,\, j\in\{0,1\}\}$ is a subsemigroup of G containing e and that multiplication in H is given by

$$(r^i s^j)(r^a s^b)=\begin{cases} r^{i+a}s^b & \text{if } j=0;\\ r^{i-a}s^{j+b} & \text{if } j=1.\end{cases}$$

Show also that Card $H=2n$.

Deduce from the above that, for each integer $p\geqslant 3$, the group D_p is, to within isomorphism, the only group which consists of $2p$ elements and which satisfies (*). [The group D_p is called the **dihedral group of order 2p.**]

19. For every $n\in\mathbf{N}\backslash\{0\}$ let $\mathbf{E}_n=\{1,\ldots,n\}$. The permutation group (Bij $(\mathbf{E}_n,\mathbf{E}_n),\circ$) is called the **symmetric group of degree n** and is denoted by S_n. By Exercise 9.9, S_n has $n!$ elements. For each $\rho\in S_n$ define a binary relation R_ρ on S_n by

$$x\equiv y(R_\rho)\Leftrightarrow(\exists t\in\mathbf{Z})\quad x=\rho^t(y).$$

Show that R_ρ is an equivalence relation. The ρ-classes are called the **orbits** of ρ. If X is an orbit of ρ prove that there exists $k\in\mathbf{N}$ such that, for every $x\in X$,

$$X=\{x,\rho(x),\ldots,\rho^{k-1}(x)\}.$$

[*Hint.* Since S_n is finite, given $x\in X$ there is a smallest integer k such that $\rho^k(x)=x$, so that $\{x,\rho(x),\ldots,\rho^{k-1}(x)\}\subseteq X$. If $y\in X$ then $y=\rho^t(x)$ for some t; in the case where $t>k$ use euclidean division to show that $y=\rho^r(x)$ for some r with $0\leqslant r<k$.]

Write down the orbits of the permutations $f, g\in S_6$ described by

i	1	2	3	4	5	6
↓	↓	↓	↓	↓	↓	↓
$f(i)$	3	1	2	4	6	5

i	1	2	3	4	5	6
↓	↓	↓	↓	↓	↓	↓
$g(i)$	4	5	3	1	6	2

Determine also the orbits of $f\circ g$ and $g\circ f$.

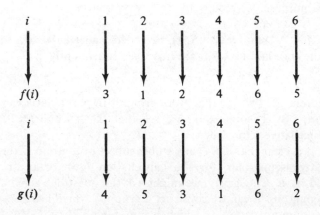

An orbit X is called **non-trivial** if Card $X \geqslant 2$. A permutation $\rho \in S_n$ is called a **cycle** if it has precisely one non-trivial orbit. Two cycles are said to be **disjoint** if their non-trivial orbits are disjoint. Prove that disjoint cycles commute.

Let $\rho \in S_n$ be such that $\rho \neq id$. Let $X_1, \ldots, X_t$ be the non-trivial orbits of ρ and for $i=1, \ldots, t$ let $\rho_i \in S_n$ be given by

$$\rho_i(x) = \begin{cases} \rho(x) & \text{if } x \in X_i; \\ x & \text{if } x \in \complement_{E_n}(X_i). \end{cases}$$

Prove that ρ is the product of the disjoint cycles $\rho_1, \ldots, \rho_n$.

Express the permutations f, g above as products of disjoint cycles.

A cycle whose non-trivial orbit is of cardinality 2 is called a **transposition**. Prove that every cycle is a product of transpositions.

20. Let p be a fixed integer greater than 1. For every $n \in \mathbf{N}$ let n^* be the greatest multiple of p which is less than or equal to n; i.e., n^* is given by $n^* = kp \leqslant n < (k+1)p$. Endow $\mathbf{N}$ with the law of composition $(m, n) \mapsto m \oplus n = m + n^*$. Prove that $(\mathbf{N}, \oplus)$ is a semigroup and that $n \in \mathbf{N}$ is idempotent (i.e., $n \oplus n = n$) if and only if $0 \leqslant n \leqslant p-1$. Deduce that the semigroup $(\mathbf{N}, \oplus)$ is the union of p disjoint isomorphic groups.

§13. Embedding a cancellative abelian semigroup in a group; Z

Our aim now is to give a formal construction of the set $\mathbf{Z}$ of integers. For this purpose, we require the following notion.

Definition. A semigroup $(S, \top)$ is called **cancellative** if, for every $x \in S$, the translations λ_x, ρ_x are injective; equivalently, if

$$(\forall x, y, z \in S) \quad x \top y = x \top z \Rightarrow y = z \text{ and } y \top x = z \top x \Rightarrow y = z.$$

By Theorem 12.4 we see that every group is in particular a cancellative semigroup. By Theorem 8.17 we see that $(\mathbf{N}, +)$ and $(\mathbf{N}\backslash\{0\}, \cdot)$ are each cancellative semigroups.

Now it is clear that if E is any stable subset of a group G then E is a cancellative semigroup, for the cancellation laws of G are clearly inherited by E. Conversely, we might ask the question: given a cancellative semigroup E can we construct a group G which has E as a stable subset (or, more naturally, which has a stable subset F which is isomorphic to E)? Clearly there may be many possible solutions to this problem; for example, in an intuitive way (since they have not been formally constructed as yet) the additive groups **Z**, **Q**, **R**, and **C** each possess subsets isomorphic to the additive semigroup **N**. However, if we impose the condition that G be generated by F, so that $G = \text{Gp}\, F$, then we clearly obtain the most economical solution in that G is then the "smallest" group with the required property.

In this section we shall solve this problem for the case of a given abelian cancellative semigroup $(S, \top)$ with neutral element e. By taking $(S, \top)$ to be $(\mathbf{N}, +)$ we can then give a proper definition of **Z** and derive some important properties. In tackling this problem, we have to be able to construct a group from the material at hand; this we shall do using the concept of an equivalence relation and it is precisely now that we begin to see the power of this important notion.

We begin by solving the following problem: given a set E on which there is defined an equivalence relation R and a law of composition $\top$, can we define a law of composition on the quotient set E/R in such a way that the canonical surjection $\natural_R : E \to E/R$ becomes a morphism?

Theorem 13.1 *If E is a set endowed with a law of composition* $\top$ *and an equivalence relation R then the following are equivalent:*

(1) *there is a unique law of composition* $\top_R$ *on* E/R *such that* $\natural_R$ *is a morphism* [i.e., $(\forall x, y \in E)\ x/R \top_R y/R = (x \top y)/R$];

(2) *for all* $x, x^*, y, y^* \in E$,

$$(x \equiv x^*(R) \text{ and } y \equiv y^*(R)) \Rightarrow x \top y \equiv x^* \top y^*(R).$$

Proof. Consider the diagram

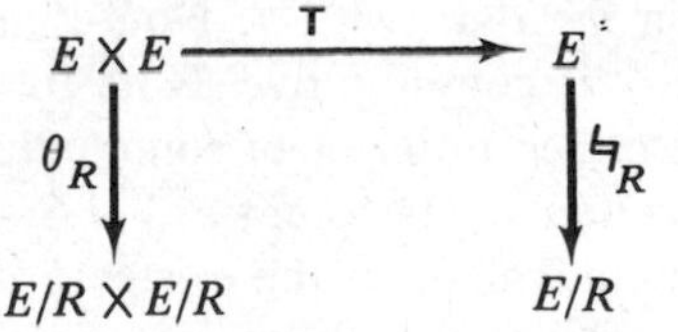

in which θ_R is the mapping given by $\theta_R(x, y) = (x/R, y/R)$ for all $(x, y) \in E \times E$. It is clear that (1) is equivalent to saying that there exists a unique mapping $\top_R : E/R \times E/R \to E/R$ which makes the diagram commutative, in that $\top_R \circ \theta_R = \natural_R \circ \top$. Applying Theorem 5.3 we deduce immediately that (1) and (2) are equivalent.

Remark. Note that condition (2) of the above result expresses the fact that, given any $x^* \in x/R$ and any $y^* \in y/R$, the elements

$x^* \top y^*$ and $x \top y$ belong to the same class modulo R. If this property holds, we say that R is **compatible with** $\top$. The unique law $\top_R$ so defined on the quotient set E/R is called the **law induced by** $\top$ relative to the compatible equivalence relation R.

In practice it is often convenient to use the following result when testing for the compatibility of an equivalence relation with respect to a given law of composition.

Theorem 13.2 *Let $(E, \top)$ be a groupoid and let R be an equivalence relation on E. Then R is compatible with $\top$ if and only if, for all $a, a^* \in E$,*

(1) $a \equiv a^* \ (R) \Rightarrow (\forall x \in E) \quad x \top a \equiv x \top a^* \ (R)$;

(2) $a \equiv a^* \ (R) \Rightarrow (\forall x \in E) \quad a \top x \equiv a^* \top x \ (R)$.

Proof. Suppose that (1) and (2) hold and let $a \equiv a^* \ (R)$, $b \equiv b^* \ (R)$. Then by (2) with $x = b$ we have $a \top b \equiv a^* \top b \ (R)$ and by (1) with $x = a^*$ we deduce from $b \equiv b^* \ (R)$ that $a^* \top b \equiv a^* \top b^* \ (R)$. The transitivity of R now shows that R is compatible with $\top$.

Conversely, suppose that R is compatible with $\top$. Then we have

$$(a \equiv a^* \ (R) \text{ and } b \equiv b^* \ (R)) \Rightarrow a \top b \equiv a^* \top b^* \ (R).$$

Taking $b = b^* = x$, this reduces to (2); and taking $a = a^* = x$ it reduces to (1) with b instead of a.

Remark. When condition (1) above is satisfied we say that R is **compatible on the left** with $\top$; and when (2) holds we say that R is **compatible on the right** with $\top$. Note that (1) and (2) are the same whenever $\top$ is commutative. Note further that (1) and (2) may also be described in terms of commutative diagrams; in fact a simple application of Theorem 5.3 shows that condition (1), for example, is equivalent to the existence of a unique mapping $h: E/R \to E/R$, making the following diagrams (one for each $x \in E$) commutative:

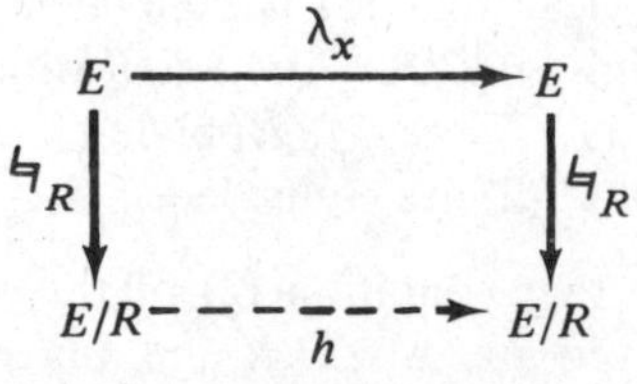

Replacing λ_x by ρ_x gives condition (2). We remark that each of these is a particular case of Exercise 6.14.

Example 13.1. Consider the relation "mod n" defined on **Z** as in Example 6.5. We have

$$\begin{aligned} x \equiv y \pmod{n} &\Rightarrow (\exists k \in \mathbf{Z})\ x-y=kn \\ &\Rightarrow (\forall z \in \mathbf{Z})(\exists k \in \mathbf{Z})\ zx-zy=z(x-y)=zkn \\ &\Rightarrow (\forall z \in \mathbf{Z})\ zx \equiv zy \pmod{n} \end{aligned}$$

and so the relation "mod n" is compatible with multiplication. We also have

$$\begin{aligned} x \equiv y \pmod{n} &\Rightarrow (\forall z \in \mathbf{Z})(\exists k \in \mathbf{Z})\quad (x+z)-(y+z)=x-y=kn \\ &\Rightarrow (\forall z \in \mathbf{Z})\quad x+z \equiv y+z \pmod{n} \end{aligned}$$

so that "mod n" is also compatible with addition. Thus we can define a multiplication $\cdot_n$ and an addition $+_n$ (the induced laws) on the quotient set. In the case where $n=3$, for example, the following are the Cayley tables for $\cdot_3$ and $+_3$ (in which, for convenience, we write $x/$mod 3 as $[x]$):

$+_3$	[0]	[1]	[2]
[0]	[0]	[1]	[2]
[1]	[1]	[2]	[0]
[2]	[2]	[0]	[1]

$\cdot_3$	[0]	[1]	[2]
[0]	[0]	[0]	[0]
[1]	[0]	[1]	[2]
[2]	[0]	[2]	[1]

Remark. A compatible equivalence relation is often called a **congruence relation.**

Theorem 13.3 *Let R be an equivalence relation on a groupoid* $(G, \top)$ *such that R is compatible with* $\top$. *If* $\top$ *is associative* (*resp. commutative*) *then so also is the law* $\top_R$ *induced on* G/R.

Proof. The proof is immediate from Theorem 12.9(3), (2) on taking $E_1=G$, $E_2=G/R$ and $f=\natural_R$.

We shall now return to our general problem and show how these results can be used to establish the existence of and, up to isomorphism, the uniqueness of a smallest abelian group G which contains an isomorphic copy of a given cancellative abelian semigroup S with neutral element e. Observing that G contains an isomorphic copy of S if and only if there is a semigroup monomorphism $f: S \to G$, we introduce the following concept.

Definition. Let S be a cancellative abelian semigroup. By a **group of quotients** of S we shall mean an abelian group G together with a (semigroup) monomorphism $f: S \to G$ such that, for any group H and any (semigroup) monomorphism $g: S \to H$, there is a unique (group) monomorphism $h: G \to H$ such that the diagram

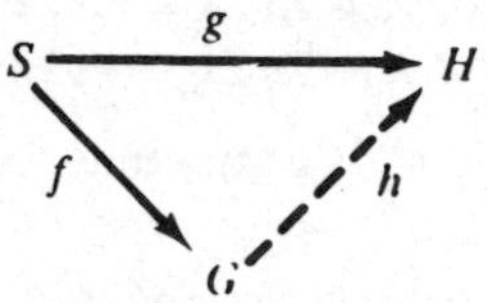

is commutative.

Theorem 13.4 *If S is a cancellative abelian semigroup with a neutral element then there exists, to within isomorphism, a unique group of quotients (G, f) of S. Moreover, G is generated by* $\operatorname{Im} f$.

Proof. Throughout the proof we shall denote almost all laws of composition by the same symbol $\top$. This will cause no confusion since the context will make it clear which law is involved.

It is readily seen that the prescription $(x, y) \top (x', y') = (x \top x', y \top y')$ defines a law of composition on $S \times S$ with respect to which $S \times S$ is also a cancellative abelian semigroup. We shall define an equivalence relation R on $S \times S$ such that R is compatible with $\top$ and $(S \times S)/R$ is a group under the induced law; this will turn out to be the group we are seeking. Consider the relation R defined on $S \times S$ by

$$(x, y) \equiv (x', y')\,(R) \Leftrightarrow x \top y' = x' \top y.$$

It is clear that R is reflexive and symmetric. It is also transitive since

$$\left.\begin{aligned}(x, y) &\equiv (x', y')(R)\\ (x', y') &\equiv (x'', y'')(R)\end{aligned}\right\} \Rightarrow x \top y' = x' \top y \text{ and } x' \top y'' = x'' \top y'$$

$$\begin{aligned}
&\Rightarrow x \top y' \top y'' = x' \top y \top y'' \text{ and } x' \top y'' \top y = x'' \top y' \top y\\
&\Rightarrow x \top y' \top y'' = x'' \top y' \top y \qquad (S \text{ abelian})\\
&\Rightarrow x \top y'' \top y' = x'' \top y \top y' \qquad (S \text{ abelian})\\
&\Rightarrow x \top y'' = x'' \top y \qquad (S \text{ cancellative})\\
&\Rightarrow (x, y) \equiv (x'', y'')\,(R).
\end{aligned}$$

Thus R is an equivalence relation on $S \times S$. That R is compatible with $\top$

follows from the observation

$$\left.\begin{array}{l}(x, y) \equiv (x', y')\,(R)\\ (x_1, y_1) \equiv (x_1', y_1')\,(R)\end{array}\right\} \Rightarrow x \top y' = x' \top y \text{ and } x_1 \top y_1' = x_1' \top y_1$$

$$\Rightarrow x \top x_1 \top y' \top y_1' = x' \top x_1 \top y \top y_1 \qquad (S \text{ abelian})$$

$$\Rightarrow (x \top x_1, y \top y_1) \equiv (x' \top x_1', y' \top y_1')\,(R).$$

We can therefore define a law of composition $\top_R$ on $G = (S \times S)/R$ by

$$(x, y)/R \top_R (x', y')/R = (x \top x', y \top y')/R.$$

Since $\top$ is commutative and associative, the same is true of $\top_R$ by Theorem 13.3 and so $(G, \top_R)$ is an abelian semigroup. Now for every $a \in S$ we have, from the definition of R,

$$(x, y)/R \top_R (a, a)/R = (x \top a, y \top a)/R = (x, y)/R.$$

Thus, for every $a \in S$, the class $(a, a)/R$ is a (hence the) neutral element e_G of G. Since now

$$(x, y)/R \top_R (y, x)/R = (x \top y, y \top x)/R = (x \top y, x \top y)/R = e_G$$

we see that $(x, y)/R$ has an inverse in G, namely $(y, x)/R$. This then shows that $(G, \top_R)$ is an abelian group.

We now define $f: S \to G$ by the prescription $f(x) = (x, e)/R$ where e is the neutral element of S. It is easily verified that f is injective. Now since

$$f(x) \top_R f(y) = (x, e)/R \top_R (y, e)/R = (x \top y, e)/R = f(x \top y)$$

we see that f is also a morphism whence it is a monomorphism.

We shall now show that (G, f) is a group of quotients of S. For this purpose, let H be any group and let $g: S \to H$ be a monomorphism. Then we have to show the existence of a unique monomorphism $h: G \to H$ such that $h \circ f = g$. Now since

$$(x_1, y_1)/R = (x_2, y_2)/R \Rightarrow y_2 \top x_1 = y_1 \top x_2$$

$$\Rightarrow g(y_2) \top g(x_1) = g(y_1) \top g(x_2)$$

$$\Rightarrow g(x_1) \top [g(x_2)]^{-1} = g(y_1) \top [g(y_2)]^{-1},$$

the last implication following from the fact that Im g is abelian, we can define a mapping $h: G \to H$ by the prescription

$$h[(x, y)/R] = g(x) \top [g(y)]^{-1}.$$

It is readily verified that h is a morphism. That it is injective is a consequence of the fact that g is injective; for

$$e_H = h[(x, y)/R] = g(x) \top [g(y)]^{-1} \Rightarrow g(x) = g(y) \Rightarrow x = y$$

and so Ker $h=\{e_H)$ whence h is a monomorphism by Theorem 12.10. For every $x \in S$ we have

$$(h \circ f)(x) = h[(x, e)/R] = g(x) \top [g(e)]^{-1} = g(x),$$

since $g(e)$ is the neutral element of H (by Theorems 12.9 and 11.1), so $h \circ f = g$. To establish the uniqueness of h, suppose that $k: G \to H$ is also a monomorphism satisfying $k \circ f = g$. Then for all $x, y \in S$ we have

$$\begin{aligned} k[(x, y)/R] &= k[(x, e)/R \top_R (e, y)/R] \\ &= k[f(x) \top_R [f(y)]^{-1}] \\ &= (k \circ f)(x) \top [(k \circ f)(y)]^{-1} \qquad \text{by Theorem 12.9(5)} \\ &= g(x) \top [g(y)]^{-1} \\ &= h[(x, y)/R] \end{aligned}$$

from which it follows that $k = h$.

We have thus established the existence of a group of quotients of S. To establish the uniqueness, to within isomorphism, of (G, f), let (G^*, f^*) also be a group of quotients of S. Then, by definition, there are unique monomorphisms j, k such that the following diagrams are commutative:

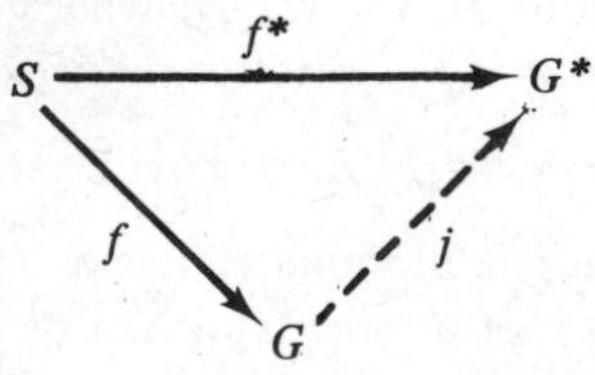

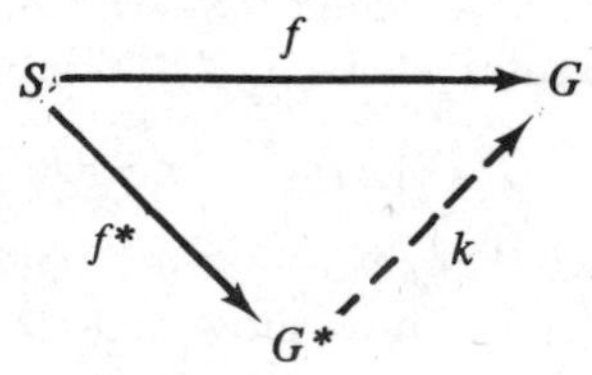

It follows that $k \circ j$ is a monomorphism such that the diagram

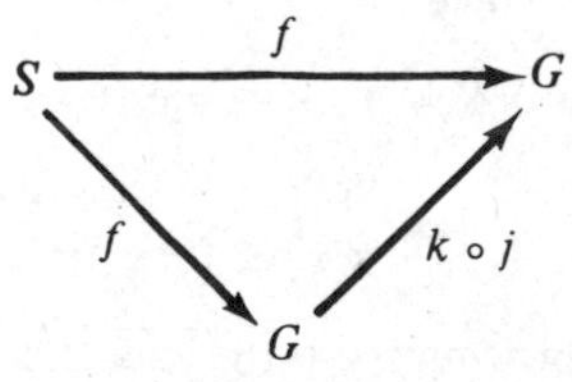

is commutative. Now, on the one hand, since (G, f) is a group of quotients of S there is precisely one monomorphism which makes this last diagram commutative; and, on the other hand, it is clear that id_G is such a morphism. Consequently we must have $k \circ j = id_G$. Likewise, we can show that $j \circ k = id_{G^*}$ whence it follows that j is an isomorphism whose inverse is k.

Finally, we have to show that G is generated by $\operatorname{Im} f$. For this purpose, consider the diagram

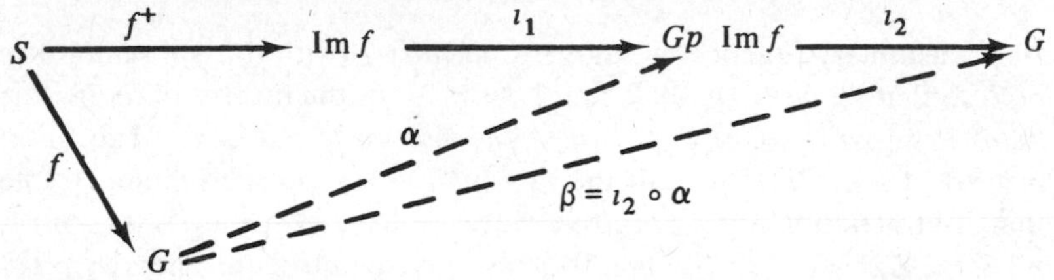

where $f^+: S \to \operatorname{Im} f$ is the monomorphism given by $x \mapsto f(x)$ and ι_1, ι_2 are the canonical injections. Since (G, f) is a group of quotients of S there exists, on the one hand, a unique monomorphism $\alpha: G \to \operatorname{Gp} \operatorname{Im} f$ such that $\alpha \circ f = \iota_1 \circ f^+$ and so the mapping $\beta = \iota_2 \circ \alpha: G \to G$ is a monomorphism such that $\beta \circ f = \iota_2 \circ \iota_1 \circ f^+$; on the other hand, only one such monomorphism β can exist and clearly id_G is such a morphism. Thus we have $\iota_2 \circ \alpha = \beta = id_G$ and consequently the canonical injection ι_2 is also surjective. We therefore see that $\operatorname{Gp} \operatorname{Im} f$ coincides with G.

As we have seen earlier, $(\mathbf{N}, +)$ is a cancellative abelian semigroup with a neutral element. By the above result there is therefore an abelian group, unique to within isomorphism, which is generated by a subsemigroup isomorphic to $\mathbf{N}$. As a model of this group we shall choose that which is obtained as in the above proof with $(S, \top)$ replaced by $(\mathbf{N}, +)$. We call this model the **group of integers** and denote it henceforth by $(\mathbf{Z}, +)$.

It is instructive to recall from the above proof that $\mathbf{Z}$ is therefore obtained from $\mathbf{N} \times \mathbf{N}$ by considering the equivalence relation R given by

$$(m, n) \equiv (m', n')\ (R) \Leftrightarrow m + n' = m' + n.$$

The equivalence classes are depicted in the following diagram:

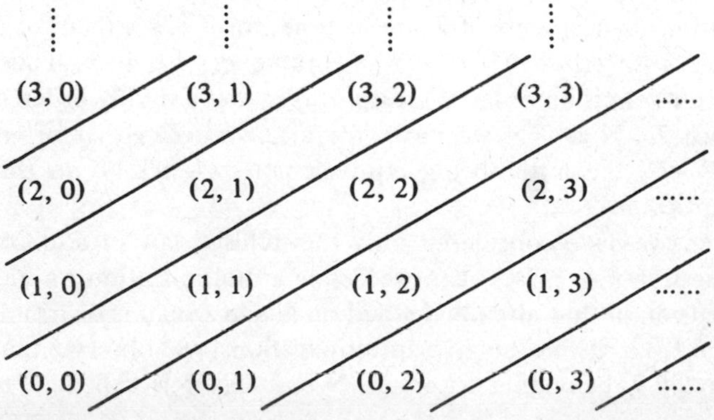

The isomorphism of **N** with a subset of **Z** is given by

$$f(n)=(n, 0)/R.$$

As is customary practice, we agree to *identify* **N** with this subsemigroup of **Z** so that we write $(n, 0)/R$ simply as n. Since the inverse of $(n, 0)/R$ is $(0, n)/R$ we write $(0, n)/R$ as $-n$. As is also customary, we write $n-m$ instead of $n+(-m)$ and call the law $(n, m) \mapsto n-m$ **subtraction**. [Note that subtraction is not associative since we have $n-(m-p)=(n-m)+p$.] Since **Z** is generated by **N** it follows immediately from Theorem 12.8 that every element of **Z** can be written in the form $n-m$ where $m, n \in \mathbf{N}$. If now we write $-\mathbf{N}=\{-n;\, n \in \mathbf{N}\}$ then since $n-m= -(m-n)$ we see that every element of **Z** is of the form p or $-p$ where $p \in \mathbf{N}$, so that $\mathbf{Z}=\mathbf{N} \cup -\mathbf{N}$. Moreover, $\mathbf{N} \cap -\mathbf{N}=\{0\}$; for if $x \in \mathbf{N} \cap -\mathbf{N}$ then $x=p= -q$ where $p, q \in \mathbf{N}$ and since $p+q=0$ we have $p=0=q$ whence $x=0$.

Definition. By an **ordered group** we shall mean an ordered semigroup which is a group.

Since **N** is an ordered semigroup it is natural to examine whether we can order **Z**. To show how **Z** can be made into an ordered group, consider the relation $\leqslant$ defined on **Z** by

$$m \leqslant n \Leftrightarrow n-m \in \mathbf{N}.$$

It is reflexive since for every $m \in \mathbf{Z}$ we have $m-m=0 \in \mathbf{N}$; it is anti-symmetric since if $m \leqslant n$ and $n \leqslant m$ then $n-m \in \mathbf{N}$ and $m-n \in \mathbf{N}$, the latter giving $n-m= -(m-n) \in -\mathbf{N}$, whence $n-m \in \mathbf{N} \cap -\mathbf{N}=\{0\}$ and so $n=m$; finally, it is transitive since if $m \leqslant n$ and $n \leqslant p$ then $n-m \in \mathbf{N}$ and $p-n \in \mathbf{N}$ whence $p-m=(p-n)+(n-m) \in \mathbf{N}$ and so $m \leqslant p$. Thus $\leqslant$ is an order on **Z** and it is clear from its definition that this order is an extension of the natural order on **N**. With respect to this order $\leqslant$ on **Z**, translations by elements of **Z** are isotone; for if $m \leqslant n$ then for all $p \in \mathbf{Z}$ we have $n+p-(m+p)=n-m \in \mathbf{N}$ and so $m+p \leqslant n+p$. Thus we see that $(\mathbf{Z}, +, \leqslant)$ is an ordered group. It is in fact a totally ordered group; for since $\mathbf{Z}=\mathbf{N} \cup -\mathbf{N}$ we have, for all $m, n \in \mathbf{Z}$, either $m-n \in \mathbf{N}$ or $m-n \in -\mathbf{N}$, the latter being equivalent to $n-m \in \mathbf{N}$, so that either $m \leqslant n$ or $n \leqslant m$.

So far, we have considered only the induced law of addition on **Z**. The question now arises: can we define a multiplication on **Z** which is an extension of that already defined on **N** and which is distributive over addition? To define such a multiplication, we observe that, every element of **Z** belonging to either **N** or $-\mathbf{N}$, it is sufficient to define

$(-a)b$, $a(-b)$ and $(-a)(-b)$ when $a, b \in \mathbf{N}$ since ab is already defined in **N**. Now for every $a \in \mathbf{N}$ we have $a+(-a)=0$ so the distributive law of multiplication over addition, if it held in **Z**, would give

$$(\forall a, b \in \mathbf{N}) \quad \begin{cases} 0=0b=[a+(-a)]b=ab+(-a)b; \\ 0=b0=b[a+(-a)]=ba+b(-a), \end{cases}$$

whence we would have $(-a)b=-(ab)=b(-a)$. It would then follow that

$$(\forall a, b \in \mathbf{N}) \quad 0=-(0b)=0(-b)=[a+(-a)](-b)=a(-b)+(-a)(-b)$$

whence we would have $(-a)(-b)=-[a(-b)]=-[-(ab)]=ab$. We therefore **define** the products $(-a)b$, $a(-b)$ and $(-a)(-b)$ by

$$(\forall a, b \in \mathbf{N}) \quad (-a)b=a(-b)=-(ab);\ (-a)(-b)=ab.$$

It is then an easy matter to show that the multiplication thus defined on **Z** is commutative, associative and distributive over addition; we simply consider all possible cases arising: e.g., to check that $xy=yx$ for all $x, y \in \mathbf{Z}$ consider the cases (i) $x \in \mathbf{N}, y \in -\mathbf{N}$, (ii) $x \in -\mathbf{N}, y \in -\mathbf{N}$ noting that it is not necessary to consider the cases (iii) $x \in -\mathbf{N}, y \in \mathbf{N}$ [which follows from (ii) by symmetry] or (iv) $x \in \mathbf{N}, y \in \mathbf{N}$ [since the result already holds in **N**].

It is readily seen that 1 is the neutral element for multiplication on **Z** and that if we write $\mathbf{Z}^*$ for $\mathbf{Z}\setminus\{0\}$ then every $n \in \mathbf{Z}^*$ is multiplicatively cancellative. Finally we note that if $p \in \mathbf{Z}$ then the multiplicative translation associated with p is isotone if $p \in \mathbf{N}$ and is antitone if $p \in -\mathbf{N}$; in other words, given any $m, n \in \mathbf{Z}$ we have

$$m \leqslant n \Rightarrow \begin{cases} (\forall p \geqslant 0) & mp \leqslant np; \\ (\forall p \leqslant 0) & mp \geqslant np. \end{cases}$$

Concerning the order of **Z** we also have the following extensions of properties of the order of **N**. First we note that since the order of **N** is archimedean (Theorem 8.18) the same is true of that of **Z**. We also have the following analogue of Theorem 8.19.

Theorem 13.5 **[Euclidean division]** *Given any $a, b \in \mathbf{Z}$ with $b>0$ there exist $q, r \in \mathbf{Z}$ with $0 \leqslant r < b$ such that $a=bq+r$.*

Proof. If $a \geqslant 0$ then the result holds by Theorem 8.19. Suppose therefore that $a<0$. Then $-a>0$ and so, again by Theorem 8.19, there exists $q^*>0$ such that $q^*b \leqslant -a < (q^*+1)b$. Now if $q^*b=-a$ define $q=-q^*$ so that $qb=a$; and if $q^*b<-a$ define $q=-(q^*+1)$ so that

$-(q+1)b < -a < -qb$ whence $qb < a < (q+1)b$. Thus, in either eventuality, $qb \leqslant a < (q+1)b$; and defining $r = a - qb$ it is clear that $0 \leqslant r < b$.

Remark. It is clear from the above proof that, just as in Theorem 8.19, q and r are *unique*.

By way of application of Theorem 13.5, we shall now determine the form of every subgroup of the group $(\mathbf{Z}, +)$.

Theorem 13.6 *Every subgroup of the additive abelian group* $\mathbf{Z}$ *is of the form* $n\mathbf{Z} = \{np;\ p \in \mathbf{Z}\}$ *for some* $n \in \mathbf{Z}$.

Proof. We recall that every $n\mathbf{Z}$ is a subgroup of $\mathbf{Z}$ (Example 12.9). Suppose now that H is any subgroup of $\mathbf{Z}$. If $H = \{0\}$ then clearly $H = 0\mathbf{Z}$. Suppose then that $H \neq \{0\}$ and let $n \neq 0$ be an element of H. Since H is a subgroup we have $-n \in H$ and so, whether the given integer n is positive or not, H contains positive integers. Let $H_+ = \{x \in H;\ x > 0\}$. Then by the well-ordering of $\mathbf{N}$ we see that H_+ has a smallest element, a say. Now by Theorem 13.5, for any $x \in H$ there exist $q, r \in \mathbf{Z}$ such that $x = aq + r$ with $0 \leqslant r < a$. Since $a \in H$ we clearly have $aq = a + \ldots + a \in H$ and since also $x \in H$ we deduce that $r = x - aq \in H$. If $r \neq 0$, this contradicts the minimality of a. Hence we must have $r = 0$ in which case $x = aq \in a\mathbf{Z}$. Since x was chosen arbitrarily in H we then have $H \subseteq a\mathbf{Z}$. But for every $n \in \mathbf{Z}$ we clearly have $an = a + \ldots + a \in H$ from which there results the reverse inclusion $a\mathbf{Z} \subseteq H$.

Definition. Let G be an ordered group. Then by the **positive cone** of G we shall mean the subsemigroup $G_+ = \{x \in G;\ x \geqslant e_G\}$.

We now have the following characterisation of the totally ordered abelian group $(\mathbf{Z}, +, \leqslant)$.

Theorem 13.7 *The group* $\mathbf{Z}$ *is, to within ordered group isomorphism, the only totally ordered abelian group whose positive cone is a naturally ordered semigroup.*

Proof. It is clear from what has gone before that $\mathbf{Z}$ fits the description, its positive cone being $\mathbf{N}$. Suppose now that $(H, \top)$ is a totally ordered abelian group with H_+ a naturally ordered semigroup. By Theorems 9.3 and 11.2 there is an order isomorphism $g: \mathbf{N} \to H_+$, namely that given by $g(n) = e_H \top \ldots \top e_H$ (n terms). For convenience we shall write $\top$ as addition, so that we have $g(n) = ne_H$. We can extend g to a morphism from $(\mathbf{Z}, +)$ to $(H, +)$ by defining $g(-n) = (-n)e_H$ for every $-n \in -\mathbf{N}$. To show that $g: \mathbf{Z} \to H$ thus defined is surjective, it suffices to observe that if $x < e_H$ then $-x > e_H$ and so $-x = g(n)$ for some $n \in \mathbf{N}$ whence $x = -g(-n) = g(n)$. To show that g is an isomorphism of ordered groups we observe that if $n, m \in \mathbf{Z}$ are such that $n < m$ then $m - n \in \mathbf{N}\backslash\{0\}$ and so $g(m) - g(n) = g(m-n) \in H_+\backslash\{e_H\}$ whence we have $g(n) < g(m)$; we now apply Theorem 7.2. Thus we see that $(H, +, \leqslant)$ is isomorphic to $(\mathbf{Z}, +, \leqslant)$.

We end this section by considering the cardinality of **Z**. We recall that in Theorem 10.3 we obtained the surprising result that $\mathbf{N} \times \mathbf{N}$ is denumerable. We shall now show that the same is true of **Z**. Roughly speaking, therefore, **Z** has "the same number of elements" as **N**, another result which, phrased in this imprecise manner, contradicts intuition.

Theorem 13.8 *$\mathbf{Z}$ is denumerable.*

Proof. We have seen that $\mathbf{Z} = \mathbf{N} \cup -\mathbf{N}$. Thus **Z** is the union of two countable sets and so, by Theorem 10.4, is also countable. Being infinite, it is therefore denumerable.

Exercises for §13

1. Consider the equivalence relation R defined on **Z** by

$$x \equiv y\,(R) \Leftrightarrow x^2 \equiv y^2 \pmod 6.$$

Prove that R is compatible with multiplication but is not compatible with addition. Construct the Cayley table for $(\mathbf{Z}/R, \cdot_R)$ and determine whether or not $(\mathbf{Z}/R, \cdot_R)$ is a group.

2. Show that the equivalence relation M defined on **Z** by

$$m \equiv n\,(M) \Leftrightarrow |\,m\,| = |\,n\,|$$

is compatible with multiplication but is not compatible with addition. Show also that the semigroups $(\mathbf{Z}/M, \cdot_M)$ and $(\mathbf{N}, \cdot)$ are isomorphic.

3. Observing that $10 \equiv 1 \pmod 9$, use the fact that the relation "mod 9" is compatible with both addition and multiplication on **Z** (Example 13.1) to show that, if $x = b_r \ldots b_1 b_0$ is any positive integer expressed in the usual decimal notation (Exercise 9.4), then x is divisible by 9 if and only if $b_0 + \ldots + b_r$ is divisible by 9. Deduce in a similar way that

(a) x is divisible by 3 if and only if $b_0 + \ldots + b_r$ is divisible by 3;

(b) x is divisible by 11 if and only if $b_0 - b_1 + \ldots + (-1)^r b_r$ is divisible by 11;

(c) x is divisible by 12 if and only if $b_0 - 2b_1 + 4(b_2 + \ldots + b_r)$ is divisible by 12.

[*Hint*. (b) $10 \equiv -1 \pmod{11}$; (c) $10 \equiv -2 \pmod{12}$ and $10^n \equiv 4 \pmod{12}$ for $n \geqslant 2$.]

4. Let E be a set with Card $E \geqslant 3$. If $\top$ is a law of composition on E with respect to which *every* equivalence relation on E is compatible, prove that either (a) $(\forall x, y \in E)\ x \top y = x$; or (b) $(\forall x, y \in E)\ x \top y = y$; or (c) $(\exists a \in E)(\forall x, y \in E)\ x \top y = a$.

[*Hint*. First show that each of (a), (b), (c) yields a law of composition on E with respect to which every equivalence relation on E is compatible. Now suppose that $(\forall z \in E)\ z \top z = z$. Let $x, y \in E$ be such that $x \neq y$ and let R be the equivalence relation associated with the partition $\{\{x, y\}, \complement_E\{x, y\}\}$. Use the compatibility of R to show that either $x \top y = x$ or $x \top y = y$. If $x \top y = x$ let S be the equivalence relation associated with the partition $\{\{y\}, \complement_E\{y\}\}$; use the compatibility of S to show that $(\forall z \neq y)\ z \top y \neq y$ and hence that $z \top y = z$ for all $z \in S$, so that (a) holds. Establish (b) similarly by supposing that $x \top y = y$. Suppose now that $(\exists z \in E)\ z \top z \neq z$. In this case consider the equivalence relation P associated with the partition $\{\{z \top z\}, \complement_E\{z \top z\}\}$; use the compatibility of P to show that $x \top y = z \top z$ for all $x, y \neq z \top z$. Now consider the equivalence relation Q associated with the partition $\{\{z, z \top z\}, \complement_E\{z, z \top z\}\}$ and show that, for every $x \in E$, $(z \top z) \top x$ is either z or $z \top z$; then argue that the possibility z is excluded.]

5. Define the relation S on $\mathbf{R}$ by

$$x \equiv y\ (S) \Leftrightarrow x - y \in \mathbf{Z}.$$

Show that S is an equivalence relation on $\mathbf{R}$ which is compatible with addition. Show also that $\mathbf{R}/S$ is a group under the induced law. Now let $a \in \mathbf{R} \backslash \mathbf{Q}$ and define $f_a : \mathbf{Z} \to]0, 1[$ by $f_a(m) = am - [am]$ where $[am]$ is the greatest integer less than or equal to am (Example 4.8). Show that f_a is injective. If $\natural_0$ denotes the restriction to $]0, 1[$ of the canonical epimorphism $\natural : \mathbf{R} \to \mathbf{R}/S$ show that $\natural_0 \circ f_a$ is a monomorphism. Deduce that $\mathbf{Z}$ is isomorphic to a subgroup of $\mathbf{R}/S$.

6. Show that Theorem 13.4 holds without the assumption that S has a neutral element.

[*Hint*. Show that, for all $x, z, z^* \in S$, $(x \top z, z) \equiv (x \top z^*, z^*)\ (R)$ so that $f : S \to G$ may be defined by $f(x) = (x \top z, z)/R$.]

7. What results from applying Theorem 13.4 to the semigroup $(\mathbf{N} \backslash \{0\}, \cdot)$?

8. Let m, n be non-zero integers and let d be their greatest common

divisor. [This notion will be considered formally later.] Prove that there exist $x, y \in \mathbf{Z}$ such that $d = mx + ny$.

[*Hint.* Let $S_{m,n} = \{mx + ny;\ x, y \in \mathbf{Z}\}$. Observe that $S_{m,n} \cap \mathbf{N}\backslash\{0\} \neq \emptyset$ since, for example, it contains $|m|$. Since $\mathbf{N}$ is well-ordered, $S_{m,n} \cap \mathbf{N}\backslash\{0\}$ has a least element, say $d = mx_1 + ny_1$. Let $m = qd + r$ $(0 \leqslant r < d)$ be the euclidean division of m by d and observe that $r = m - qd = m(1 - qx_1) + n(-qy_1)$ so that $r > 0$ implies $r \in S_{m,n} \cap \mathbf{N}\backslash\{0\}$ thus contradicting the minimality of d. Conclude that d is a common divisor of m, n and show that every common divisor of m, n divides d.]

9. Let m, n be natural numbers with $n \neq 0$. Define the binary relation $R_{m,n}$ on $\mathbf{N}$ by

$$x \equiv y(R_{m,n}) \Leftrightarrow \begin{cases} \textit{either } x = y \\ \textit{or } m \leqslant x < y \textit{ and } n \textit{ divides } y - x \\ \textit{or } m \leqslant y < x \textit{ and } n \textit{ divides } x - y. \end{cases}$$

Show that $R_{m,n}$ is an equivalence relation on $\mathbf{N}$ which is compatible with addition. Exhibit, for example, the $R_{0,4}$-classes and the $R_{2,3}$-classes. Conversely, let R be an equivalence relation on $\mathbf{N}$ which is compatible with addition. Prove that if R is non-trivial (in that the R-classes are not all singletons) then there exist $m, n \in \mathbf{N}$ with $n \neq 0$ such that $x \equiv y(R) \Leftrightarrow x \equiv y(R_{m,n})$.

[*Hint.* Let m be the smallest integer for which m/R is not a singleton and let $m+n$ be the "successor" of m in m/R.]

§14. Compatible equivalence relations on groups; quotient groups; isomorphism theorems; cyclic groups

In the previous section we introduced the notion of an equivalence relation on a groupoid $(G, \top)$ which was compatible with $\top$, together with the notions of left and right compatibility (which coincide whenever $\top$ is commutative). If we now strengthen the structure of G so that it becomes a group, we can say a lot more.

Theorem 14.1 *Let $(G, \top)$ be a group and let R be a binary relation on G. Then the following are equivalent:*

(1) *R is an equivalence relation which is compatible on the left [resp. right] with $\top$;*

(2) *there is a subgroup H of G such that, for all $x, y \in G$,*

$$x \equiv y(R) \Leftrightarrow x^{-1} \top y \in H \qquad [resp.\ x \equiv y(R) \Leftrightarrow y \top x^{-1} \in H].$$

Proof. (1) $\Rightarrow$ (2): Suppose that R is an equivalence relation which is compatible on the left with $\top$. Then, e being the neutral element of G, we have

$$x \equiv y(R) \Leftrightarrow x^{-1} \top y \in e/R.$$

[In fact, if $x \equiv y(R)$ then by the left compatibility of R we have $e = x^{-1} \top x \equiv x^{-1} \top y(R)$; and conversely $x^{-1} \top y \equiv e(R)$ gives $x = x \top e \equiv x \top x^{-1} \top y = y(R)$.] It therefore suffices to show that e/R is a subgroup of G; and this follows from Theorem 12.6 and the implications

$$\begin{cases} y \equiv e(R) \Rightarrow y^{-1} = y^{-1} \top e \equiv y^{-1} \top y = e(R); \\ x, y \equiv e(R) \Rightarrow x, y^{-1} \equiv e(R) \Rightarrow x \top y^{-1} \equiv x \top e = x \equiv e(R). \end{cases}$$

(2) $\Rightarrow$ (1): Suppose now that H is a subgroup of G such that $x \equiv y(R) \Leftrightarrow x^{-1} \top y \in H$. That R is an equivalence relation on G follows from

(a) $(\forall x \in G)\quad x^{-1} \top x = e \in H$, so $x \equiv x(R)$;

(b) if $x \equiv y(R)$ then $x^{-1} \top y \in H$ whence, H being a subgroup, we have $y^{-1} \top x = (x^{-1} \top y)^{-1} \in H$ and hence $y \equiv x(R)$;

(c) if $x \equiv y(R)$ and $y \equiv z(R)$ then $x^{-1} \top y \in H$ and $y^{-1} \top z \in H$ whence $x^{-1} \top z = (x^{-1} \top y) \top (y^{-1} \top z) \in H$ and so $x \equiv z(R)$.

That R is compatible on the left with $\top$ follows from the fact that if $x \equiv y(R)$ then $x^{-1} \top y \in H$ and so $(\forall z \in G)\, (z \top x)^{-1} \top (z \top y) = x^{-1} \top y \in H$ whence $(\forall z \in G)\; z \top x \equiv z \top y(R)$.

A similar proof holds for right compatible equivalences.

Corollary 1. *If $(G, \top)$ is a group then the following sets are equipotent:*

A: the set of partitionings of G associated with equivalence relations which are compatible on the left with $\top$;

B: the set of partitionings of G associated with equivalence relations which are compatible on the right with $\top$;

C: the set of subgroups of G.

Proof. Define $\theta: A \to C$ by the prescription $\theta(f_R) = e/R$ where f_R is the partitioning associated with the left compatible equivalence relation R. Then θ is injective; for if R, S are such that $e/R = e/S$ then we have

$$x \equiv y(R) \Leftrightarrow x^{-1} \top y \in e/R = e/S \Leftrightarrow x \equiv y(S)$$

whence R is equivalent to S and so $f_R = f_S$. To show that θ is also surjective, let H be any subgroup of G and note that if R_H is defined by $x \equiv y(R_H) \Leftrightarrow x^{-1} \top y \in H$ then R_H is a left-compatible equivalence relation on G. Taking $x = e$, we obtain $e/R_H = H$. This then shows that θ is a bijection whence A and C are equipotent; in a similar way, so also are B, C.

Corollary 2. *Let $(G, \top)$ be a group and let R be an equivalence relation on G which is compatible on the left [resp. right] with $\top$. Then for every $x \in G$ we have*

$$x/R = \{x \top h;\ h \in e/R\} \qquad [\textit{resp. } x/R = \{h \top x;\ h \in e/R\}].$$

Moreover, every equivalence class modulo R is equipotent to the subgroup e/R.

Proof. Writing $H = e/R$ we have, in the case where R is compatible on the left,

$$\begin{aligned} y \in x/R \Leftrightarrow x^{-1} \top y \in H &\Leftrightarrow (\exists h \in H)\ x^{-1} \top y = h \\ &\Leftrightarrow (\exists h \in H)\ y = x \top h \end{aligned}$$

and so, for every $x \in G$, we have $x/R = \{x \top h;\ h \in e/R\}$. Given $x \in G$, consider now the mapping $f_x : e/R \to x/R$ given by $f_x(h) = x \top h$. It is clear from the above that f_x is surjective. Since the left translation λ_x is injective we see that f_x is also injective. Thus f_x is a bijection and so e/R and x/R are equipotent.

Definition. If G is a finite group then by the **order** of G we shall mean simply the cardinality of G. We shall denote this as usual by Card G; other common notation is $| G |$ or $\#(G)$.

An important consequence of the above results is the following.

Theorem 14.2 [**Lagrange**] *In a finite group G the order of every subgroup is a divisor of the order of G.*

Proof. Let H be a subgroup of G and, for every $x \in G$, denote (by abuse of notation) the set $\{x \top h;\ h \in H\}$ by $x \top H$. If we define the relation R_H on G by $x \equiv y(R_H) \Leftrightarrow x^{-1} \top y \in H$ then we know that $H = e/R_H$ and, by Corollary 2 of Theorem 14.1, $x/R_H = x \top H$. Thus we see that $\{x \top H;\ x \in G\}$ is a partition of G. Moreover, again by Corollary 2 of Theorem 14.1, each of the sets $x \top H$ is equipotent to H. Consequently, if $p = \text{Card}\ \{x \top H;\ x \in G\}$, we have Card $G = p$ Card H and so Card H divides Card G.

Remark. If H is a subgroup of a group G then the sets $x \top H$ are often called the **left cosets of G modulo H**.

Suppose now that H is a subgroup of the group $(G, \top)$. Let R_H be the left compatible equivalence relation given by

$$x \equiv y(R_H) \Leftrightarrow x^{-1} \top y \in H$$

and let S_H be the right compatible equivalence relation given by

$$x \equiv y\,(S_H) \Leftrightarrow y \top x^{-1} \in H.$$

As observed above, we have $x^{-1} \top y \in H \Leftrightarrow y \in x \top H$; and similarly $y \top x^{-1} \in H \Leftrightarrow y \in H \top x$. In the case where G is abelian it is clear that R_H and S_H are equivalent. In general, however, this is not the case. For example, the group of Exercise 12.7 has the Cayley table

$\circ$	f_1	f_2	f_3	f_4	f_5	f_6
f_1	f_1	f_2	f_3	f_4	f_5	f_6
f_2	f_2	f_1	f_6	f_5	f_4	f_3
f_3	f_3	f_4	f_1	f_2	f_6	f_5
f_4	f_4	f_3	f_5	f_6	f_2	f_1
f_5	f_5	f_6	f_4	f_3	f_1	f_2
f_6	f_6	f_5	f_2	f_1	f_3	f_4

and in this group we see that $H = \{f_1, f_2\}$ is a subgroup; moreover,

$$H \circ f_4 = \{f_1 \circ f_4, f_2 \circ f_4\} = \{f_4, f_5\} \text{ and } f_4 \circ H = \{f_4 \circ f_1, f_4 \circ f_2\} = \{f_4, f_3\}.$$

Now we note that the left compatible equivalence R_H is not in general right compatible; for example, in the above group we have $f_4^{-1} \circ f_3 = f_6 \circ f_3 = f_2 \in H$ so that $f_4 \equiv f_3(R_H)$, but $(f_4 \circ f_3)^{-1} \circ (f_3 \circ f_3) = f_5 \circ f_1 = f_5 \notin H$ and so $f_4 \circ f_3 \not\equiv f_3 \circ f_3(R_H)$. Thus in general we cannot define a law of composition on the quotient set G/R_H which is related to $\top$ in the desirable way of Theorem 13.1 (namely in such a way that the canonical surjection $\natural$ is a morphism). However, if the subgroup H has the property that R_H and S_H are *equivalent* then it *is* possible to do so, for each will then be compatible on both sides with respect to $\top$. We are thus led to make the following definition.

Definition. A subgroup H of a group $(G, \top)$ is said to be **normal** (or **invariant**) if and only if R_H and S_H are equivalent; in other words, if and only if, $(\forall x \in G)\ x \top H = H \top x$.

Remark. If we write $x \top H \top x^{-1}$ for $\{x \top h \top x^{-1};\ h \in H\}$ then H is normal if and only if $(\forall x \in G)\ H = x \top H \top x^{-1}$. In fact, suppose that H is normal. Then for every $x \in G$ we have $x \top H \top x^{-1} = H \top x \top x^{-1} = H \top e = H$. Conversely, if $H = x \top H \top x^{-1}$ for every $x \in G$ than clearly $H \top x = x \top H \top x^{-1} \top x = x \top H \top e = x \top H$ and so H is normal. In fact, to show that H is a normal subgroup we need only show that $x \top H \top x^{-1}$ *is included in* H for every $x \in G$. For suppose this condition holds. Then if x is any element of G we have $x \top H \top x^{-1} \subseteq H$ and likewise $x^{-1} \top H \top (x^{-1})^{-1} \subseteq H$ and from the second of these it follows at once that $H \subseteq x \top H \top x^{-1}$.

Example 14.1. *If G, H are groups and $f: G \to H$ is a morphism then* $\operatorname{Ker} f$ *is a normal subgroup of G.* In fact, let K denote the subgroup $\operatorname{Ker} f$. To show that it is normal, let $x \in G$ and note that, for any $k \in K$, if $y = x \top k \top x^{-1}$ then $f(y) = f(x) \top f(k) \top f(x^{-1}) = f(x) \top e_H \top [f(x)]^{-1} = e_H$ so that we have $x \top K \top x^{-1} \subseteq K$.

Theorem 14.3 *Let $(G, \top)$ be a group and let R be a binary relation on G. Then the following are equivalent*:

(1) *R is an equivalence relation which is compatible with $\top$;*

(2) *there is a normal subgroup H of G such that, for all $x, y \in G$,*

$$x \equiv y(R) \Leftrightarrow x^{-1} \top y \in H \Leftrightarrow y \top x^{-1} \in H.$$

Proof. (1) $\Rightarrow$ (2): By virtue of Theorem 14.1 it suffices to show that the subgroup $H = e/R$ is normal. This is immediate from Corollary 2 of Theorem 14.1 since $(\forall x \in G)\ x \top H = x/R = H \top x$.

(2) $\Rightarrow$ (1): This is immediate from Theorem 14.1.

Corollary. *The set of normal subgroups of a group G is equipotent to the set of partitionings of G associated with compatible equivalence relations on G.*

Proof. This is immediate from Corollary 1 of Theorem 14.1.

Remark. If R is a compatible equivalence relation on the group $(G, \top)$ then we know that G/R is a group under the induced law $\top_R$. If H is the normal subgroup of G which corresponds (under the bijection of the previous corollary) to the partitioning associated with R then we shall henceforth commit the standard abuse of notation by writing G/H instead of G/R. We call G/H the **quotient group** (or **factor group**) **of** G **by the normal subgroup** H.

Example 14.2. As we have seen in Theorem 13.6, every subgroup of the abelian group $(\mathbf{Z}, +)$ is of the form $n\mathbf{Z}$ for some $n \in \mathbf{Z}$ and is necessarily normal. The relation "mod n" is described by $x \equiv y \pmod{n}$ $\Leftrightarrow x - y \in n\mathbf{Z}$ and so yields the quotient group $\mathbf{Z}/n\mathbf{Z}$. The elements of this group are the equivalence classes $x/n\mathbf{Z}$ and, by Corollary 2 of Theorem 14.1, we have

$$x/n\mathbf{Z} = x + n\mathbf{Z} = \{\ldots, x-2n, x-n, x, x+n, x+2n, \ldots\}.$$

Clearly $\mathbf{Z}/n\mathbf{Z} = \{0/n\mathbf{Z}, 1/n\mathbf{Z}, \ldots, (n-1)/n\mathbf{Z}\}$ and so is an additive abelian group of order n.

We shall now establish an important analogue of Theorem 5.3 in the case of groups and group morphisms.

Theorem 14.4 *Consider the diagram of groups and group morphisms*

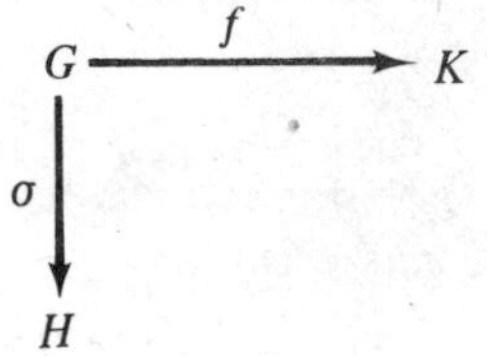

where σ is surjective. The following conditions are then equivalent:
(1) *there is a unique morphism $f_* : H \to K$ such that $f = f_* \circ \sigma$;*
(2) $\mathrm{Ker}\, \sigma \subseteq \mathrm{Ker}\, f$.
Moreover, if these conditions are satisfied then f_* is
(a) *a monomorphism if and only if* $\mathrm{Ker}\, \sigma = \mathrm{Ker}\, f$;
(b) *an epimorphism if and only if f is an epimorphism.*

Proof. We write the laws of G, H, K as $\top$ without confusion. By Theorem 5.3 the existence of a mapping $f_* : H \to K$ such that $f = f_* \circ \sigma$ is equivalent to the property

$$(*) \qquad (\forall x, y \in G) \qquad \sigma(x) = \sigma(y) \Rightarrow f(x) = f(y).$$

Now since σ is a morphism we have

$$\begin{aligned}\sigma(x) = \sigma(y) &\Leftrightarrow e_H = \sigma(x) \top [\sigma(y)]^{-1} = \sigma(x) \top \sigma(y^{-1}) = \sigma(x \top y^{-1}) \\ &\Leftrightarrow x \top y^{-1} \in \mathrm{Ker}\, \sigma\end{aligned}$$

and likewise, since f is a morphism, $f(x) = f(y) \Leftrightarrow x \top y^{-1} \in \mathrm{Ker}\, f$. Thus, if (*) above holds then on taking $y = e_G$ we obtain $x \in \mathrm{Ker}\, \sigma \Rightarrow x \in \mathrm{Ker}\, f$. It follows therefore that (1) $\Rightarrow$ (2).

Conversely, if (2) holds then clearly

$$(\forall x, y \in G) \qquad \sigma(x)=\sigma(y) \Rightarrow f(x)=f(y)$$

and so, by Theorem 5.3, there exists a mapping $f_*: H \to K$ such that $f=f_* \circ \sigma$. That f_* is unique follows from the fact that σ is surjective and hence right cancellative. To show that f_* is a morphism we again use the fact that σ is surjective. Given $x, y \in H$ let $x^*, y^* \in G$ be such that $\sigma(x^*)=x$ and $\sigma(y^*)=y$; then since σ, f are morphisms we have

$$\begin{aligned} f_*(x \top y) = f_*[\sigma(x^*) \top \sigma(y^*)] &= f_*[\sigma(x^* \top y^*)] \\ &= f(x^* \top y^*) \\ &= f(x^*) \top f(y^*) \\ &= f_*[\sigma(x^*)] \top f_*[\sigma(y^*)] \\ &= f_*(x) \top f_*(y). \end{aligned}$$

This shows that f_* is also a morphism and hence that (2) $\Rightarrow$ (1).

Suppose now that (1) and (2) hold. By Theorem 12.10, f_* is a monomorphism if and only if $\operatorname{Ker} f_* = \{e_H\}$. Now $\operatorname{Ker} f_* = \sigma^{\rightarrow}[\operatorname{Ker} f]$; for, on the one hand, if $y \in \sigma^{\rightarrow}[\operatorname{Ker} f]$ then $y=\sigma(x)$ where $f(x)=e_K$ so that $f_*(y)=f_*[\sigma(x)]=f(x)=e_K$ and so $y \in \operatorname{Ker} f_*$ and, on the other hand, if $y \in \operatorname{Ker} f_*$ then choosing $y^* \in G$ such that $\sigma(y^*)=y$ we have $f(y^*)=f_*[\sigma(y^*)]=f_*(y)=e_K$ so that $y=\sigma(y^*) \in \sigma^{\rightarrow}[\operatorname{Ker} f]$. It follows, therefore, that f_* is a monomorphism if and only if $\sigma^{\rightarrow}[\operatorname{Ker} f]=\{e_H\}$, i.e., if and only if $\operatorname{Ker} f \subseteq \operatorname{Ker} \sigma$, i.e., if and only if $\operatorname{Ker} f = \operatorname{Ker} \sigma$ (since we are supposing that (1) holds). This then establishes (a). As for (b), we observe that, since σ is surjective,

$$\operatorname{Im} f_* = f_*^{\rightarrow}(H) = f_*^{\rightarrow}[\operatorname{Im} \sigma] = \operatorname{Im} f_* \circ \sigma = \operatorname{Im} f$$

whence it is clear that f_* is surjective if and only if f is surjective.

Corollary 1. *If G, G^* are groups and $f: G \to G^*$ is a morphism then in the diagram*

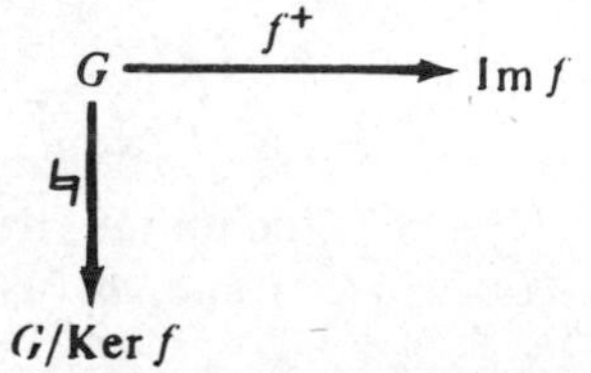

where $f^+: G \to \operatorname{Im} f$ is given by $x \mapsto f(x)$, there is a unique isomorphism $f_*: G/\operatorname{Ker} f \to \operatorname{Im} f$ such that $f_* \circ \natural = f^+$.

Proof. Recall by Example 14.1 that $\operatorname{Ker} f$ is a normal subgroup of G so we can form the quotient group $G/\operatorname{Ker} f$. Let us find the kernel of the canonical epimorphism $\natural : G \to G/\mathrm{Ker}\, f$. Since

$$x \in \mathrm{Ker}\, \natural \Leftrightarrow x/\mathrm{Ker}\, f = e_G/\mathrm{Ker}\, f \Leftrightarrow x = x \top e_G^{-1} \in \mathrm{Ker}\, f$$

we see immediately that $\mathrm{Ker}\, \natural = \mathrm{Ker}\, f$. We can therefore apply Theorem 14.4 to the given diagram to deduce the existence of a unique monomorphism $f_* : G/\mathrm{Ker}\, f \to \mathrm{Im}\, f$ such that $f_* \circ \natural = f^+$; and since f^+ is surjective we see from property (b) of Theorem 14.4 that so also is f_*. Thus f_* is an isomorphism.

Corollary 2. *Every group morphism can be written as the composite of a monomorphism, an isomorphism and an epimorphism.*

Proof. This is immediate from Corollary 1 and Theorem 6.7. [Note that, by the standard abuse of notation, $G/\mathrm{Ker}\, f$ is written for G/R where $x \equiv y(R) \Leftrightarrow x \top y^{-1} \in \mathrm{Ker}\, f \Leftrightarrow f(x) = f(y) \Leftrightarrow f^+(x) = f^+(y) \Leftrightarrow x \equiv y\ (R_{f^+})$.]

Corollary 3. *Let G, G^* be groups and let H, H^* be normal subgroups of G, G^* respectively. If $g : G \to G^*$ is a morphism then the following conditions are equivalent:*

(1) $g^{\rightarrow}(H) \subseteq H^*$;

(2) *there is a unique morphism $g_* : G/H \to G^*/H^*$ making the following diagram commutative:*

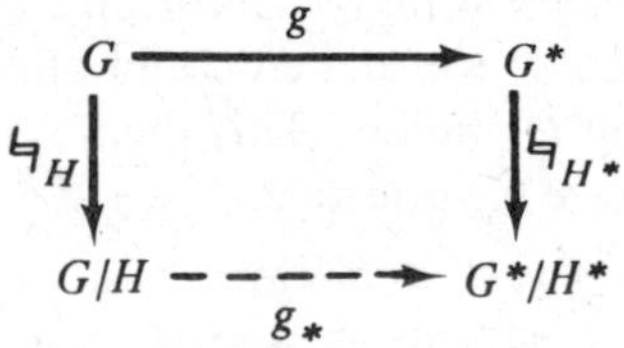

Proof. Applying Theorem 14.4 with $f = \natural_{H^*} \circ g$ and $\sigma = \natural_H$ we see that (2) holds if and only if $\mathrm{Ker}\, \natural_H \subseteq \mathrm{Ker}\, \natural_{H^*} \circ g$. Now

$$x \in \mathrm{Ker}\, \natural_H \Leftrightarrow x/H = e_G/H \Leftrightarrow x = x \top e_G^{-1} \in H$$

and so $\mathrm{Ker}\, \natural_H = H$. Likewise, we have

$$x \in \mathrm{Ker}\, \natural_{H^*} \circ g \Leftrightarrow \natural_{H^*}[g(x)] = e_{g^*}/H^* \Leftrightarrow g(x) \in \mathrm{Ker}\, \natural_{H^*} = H^*$$

and so $\mathrm{Ker}\, \natural_{H^*} \circ g = g^{\leftarrow}(H^*)$. The condition $\mathrm{Ker}\, \natural_H \subseteq \mathrm{Ker}\, \natural_{H^*} \circ g$ is therefore equivalent to $H \subseteq g^{\leftarrow}(H^*)$ and, by Theorem 5.1(v), this is equivalent to (1).

Corollary 4. *Let G be a group and let H be a normal subgroup of G. If $g : G \to G^*$ is a morphism from G to a group G^* then $g^{\rightarrow}(H)$ is a normal subgroup of* Im g *and the following conditions are equivalent:*

(1) $\mathrm{Ker}\, g \subseteq H$;

(2) *there is a unique isomorphism* $g_*: G/H \to \operatorname{Im} g/g^{\rightarrow}(H)$ *making the following diagram commutative*:

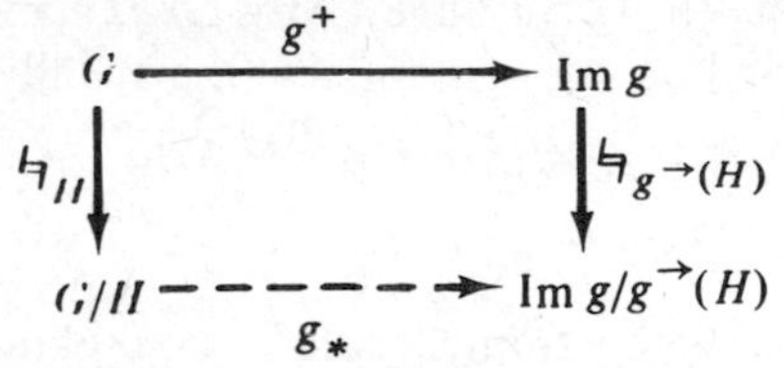

Proof. By Theorem 12.12, $g^{\rightarrow}(H)$ is a subgroup of $\operatorname{Im} g$. Since H is normal in G we have $(\forall x \in G)\ H = x \top H \top x^{-1}$. It follows readily that

$$(\forall x \in G) \qquad g^{\rightarrow}(H) = g(x) \top g^{\rightarrow}(H) \top g(x^{-1}) = g(x) \top g^{\rightarrow}(H) \top [g(x)]^{-1}$$

and so $g^{\rightarrow}(H)$ is normal in $\operatorname{Im} g$. Applying Corollary 3 to the above diagram (taking $G^* = \operatorname{Im} g$ and $H^* = g^{\rightarrow}(H)$), we see that there exists a unique morphism g_* which makes the rectangle commutative. Now since $\natural_{g^{\rightarrow}(H)} \circ g^+$ is surjective in this diagram it follows by Theorem 14.4(b) that g_* is surjective. It follows from these observations that (2) holds if and only if g_* is injective; and by Theorem 14.4(a) this is the case if and only if $\operatorname{Ker} \natural_H = \operatorname{Ker} \natural_{g^{\rightarrow}(H)} \circ g^+$. Now as was shown in the proof of Corollary 3 we have $\operatorname{Ker} \natural_H = H$ and $\operatorname{Ker} \natural_{g^{\rightarrow}(H)} \circ g^+ = (g^+)^{\leftarrow}[g^{\rightarrow}(H)] = g^{\leftarrow}[g^{\rightarrow}(H)]$. Thus we can see that (2) holds if and only if $H = g^{\leftarrow}[g^{\rightarrow}(H)]$. By Theorem 12.14 this is equivalent to $H = H \vee \operatorname{Ker} g$ which is the case if and only if $\operatorname{Ker} g \subseteq H$.

Corollary 5. *Let G be a group and let H, K be normal subgroups of G with $K \subseteq H$. Then*

(1) *K is normal in H;*

(2) *H/K is normal in G/K;*

(3) *there is a unique isomorphism* $\theta: G/H \to (G/K)/(H/K)$ *such that the following diagram is commutative*:

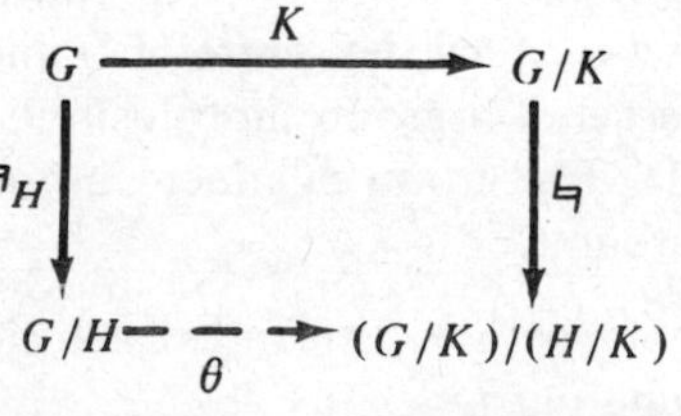

Proof. (1) Since K is normal in G we have $(\forall x \in G)\ K = x \top K \top x^{-1}$. It follows that $(\forall x \in H)\ K = x \top K \top x^{-1}$ and so K is normal in H.

(2) Since H is normal in G we have $(\forall x \in G)\ H = x \top H \top x^{-1}$. By (1) we can form H/K; and on taking quotients it is readily seen that

$$(\forall x \in G) \qquad H/K = x/K \top H/K \top (x/K)^{-1}$$

which shows that H/K is normal in G/K.

(3) This follows immediately from Corollary 4 on taking $g = \natural_K : G \to G/K$ and noting that $\natural_K^{\to}(H) = H/K$ and $\operatorname{Ker} \natural_K = K \subseteq H$.

Example 14.3. For each positive integer p let $\mathbf{N}_p = \{0, \ldots, p-1\}$. Define addition modulo p on $\mathbf{N}_p$ by setting $m +_p n = k$ where k is the unique element of $\mathbf{N}_p$ such that $m + n \equiv k \pmod{p}$. It is readily seen that $(\mathbf{N}_p, +_p)$ is an abelian group (see Exercise 12.18). Define $f : \mathbf{Z} \to \mathbf{N}_p$ by $f(m) = m^*$ where $m^* \in \mathbf{N}_p$ is such that $m \equiv m^* \pmod{p}$. Since $(m+n)^* \equiv m+n \equiv m^* + n^* \pmod{p}$ we see that $(m+n)^* = m^* +_p n^*$ and so f is a morphism. It is clear that f is surjective. Moreover, we have $\operatorname{Ker} f = \{m \in Z;\ m^* = 0\} = p\mathbf{Z}$. Thus, by Corollary 1 to Theorem 14.4, the group $\mathbf{N}_p$ is isomorphic to the group $\mathbf{Z}/p\mathbf{Z}$.

Example 14.4. Let G be a group and let $I(G)$ be the set of all inner automorphisms on G (see Example 12.12). Recall that the inner automorphism ζ_a associated with $a \in G$ is given by $\zeta_a(x) = a \top x \top a^{-1}$. It is readily seen that $\zeta_a \circ \zeta_b = \zeta_{a \top b}$ for all $a, b \in G$ and so $I(G)$ is a semigroup under composition of mappings in which the neutral element is $\zeta_e = id_G$ and the inverse of ζ_a is $\zeta_{a^{-1}}$. Thus $(I(G), \circ)$ is a group. Consider now the mapping $f : G \to I(G)$ given by $f(a) = \zeta_a$. That f is a morphism follows from $f(a \top b) = \zeta_{a \top b} = \zeta_a \circ \zeta_b$. Clearly, f is surjective. By the fundamental isomorphism of Corollary 1 to Theorem 14.4 we then have $G/\operatorname{Ker} f \simeq I(G)$. Let us now describe $\operatorname{Ker} f$. We have

$$\begin{aligned} a \in \operatorname{Ker} f \Leftrightarrow \zeta_a = \zeta_e &\Leftrightarrow (\forall x \in G) \qquad a \top x \top a^{-1} = x \\ &\Leftrightarrow (\forall x \in G) \qquad a \top x = x \top a \end{aligned}$$

and so $\operatorname{Ker} f$ is simply the set of all $a \in G$ which commute with every element of G. This set is called the **centre** of G and is denoted by $Z(G)$. Since $Z(G)$ is the kernel of a group morphism it is a normal subgroup of G (Example 14.1). The group of inner automorphisms $I(G)$ is thus isomorphic to the group $G/Z(G)$.

From what has gone before, we can extract the following characterisation of normal subgroups.

Theorem 14.5 *A subset H of a group G is a normal subgroup of G if and only if H is the kernel of a morphism f from G to a group G^*.*

Proof. If H is a normal subgroup of G then, as was seen above, $H = \text{Ker}\, \natural_H$. Conversely, if H is the kernel of a morphism $f: G \to G^*$ from G to a group G^* then H is a normal subgroup of G by Example 14.1.

By way of another application of Theorem 14.4 we shall now give a complete description of a particular type of group. We say that a group $(G, \top)$ is **cyclic** if there exists $a \in G$ such that $G = \text{Gp}\{a\}$, such an element being called a **generator** of G. By Theorem 12.8 we see that G is cyclic if and only if there exists $a \in G$ such that every $x \in G$ can be written $x = \overset{n}{\top} a$ or $x = \overset{n}{\top} a^{-1}$. [For a proper definition of $\overset{n}{\top} a$ the reader is referred to Exercise 11.6.]

Example 14.5. The additive group $\mathbf{Z}$ is cyclic; 1 is a generator. Note that -1 is also a generator, so that generators need not be unique. Likewise, the additive group $\mathbf{Z}/n\mathbf{Z}$ is cyclic; $1/n\mathbf{Z}$ is a generator.

We shall now show that the cyclic groups of Example 14.5 are, to within isomorphism, the only cyclic groups.

Theorem 14.6 *If $(G, \top)$ is a cyclic group then either $(G, \top) \simeq (\mathbf{Z}, +)$ or, for some non-zero integer p, $(G, \top) \simeq (\mathbf{Z}/p\mathbf{Z}, +)$.*

Proof. Let x be a generator of G. Then the morphism $f: \mathbf{Z} \to G$ given by $(\forall n \in \mathbf{Z})\ f(n) = \overset{n}{\top} x$ is surjective. Now $\text{Ker}\, f$ is a subgroup of $\mathbf{Z}$ and so, by Theorem 13.6, is of the form $p\mathbf{Z}$ for some $p \in \mathbf{Z}$; and we can choose $p \geqslant 0$ since $p\mathbf{Z} = (-p)\mathbf{Z}$. We consider two cases:

(1) $p = 0$: In this case $\text{Ker}\, f = 0\mathbf{Z} = \{0\}$ and so f is injective by Theorem 12.10. Thus f is an isomorphism and so $G \simeq \mathbf{Z}$.

(2) $p > 0$: In this case we apply Theorem 14.4 to the diagram

$$\begin{array}{ccc} \mathbf{Z} & \xrightarrow{\ f\ } & G \\ {\scriptstyle \natural}\big\downarrow & & \\ \mathbf{Z}/p\mathbf{Z} & & \end{array}$$

Since $\text{Ker}\, \natural = p\mathbf{Z} = \text{Ker}\, f$ we obtain the existence of an isomorphism $f_*: \mathbf{Z}/p\mathbf{Z} \to G$ such that $f_* \circ \natural = f$.

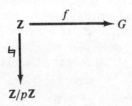

Corollary 1. *All cyclic groups are abelian.*

Corollary 2. *Any two cyclic groups of the same order are isomorphic.*

Corollary 3. *The additive group* **Z** *is, to within isomorphism, the only infinite cyclic group.*

If $(G, \top)$ is a group then by Theorem 12.8, for every $x \in G$, the subgroup $\mathrm{Gp}\{x\}$ generated by $\{x\}$ is given by $\mathrm{Gp}\{x\} = \{\overset{n}{\top} x;\ n \in \mathbf{Z}\}$ and is cyclic. Consider the morphism $f: \mathbf{Z} \to \mathrm{Gp}\{x\}$ given by $f(n) = \overset{n}{\top} x$. If $\mathrm{Gp}\{x\}$ is finite then by Theorem 14.6 we have $\mathrm{Gp}\{x\} \simeq \mathbf{Z}/n\mathbf{Z}$ where $n\mathbf{Z} = \mathrm{Ker}\, f = \{p \in \mathbf{Z};\ f(p) = \overset{p}{\top} x = e_G\}$, so that $\mathrm{Card}\ \mathrm{Gp}\{x\} = \mathrm{Card}\ \mathbf{Z}/n\mathbf{Z} = n$ where n, being the smallest positive integer in $n\mathbf{Z}$ is the smallest positive integer such that $\overset{n}{\top} x = e_G$.

Definition. For every $x \in G$ we define the **order** of x by $\mathrm{o}(x) = \mathrm{Card}\ \mathrm{Gp}\{x\}$. Thus, if $\mathrm{o}(x)$ is finite, it is the smallest positive integer p such that $\overset{p}{\top} x = e_G$.

Theorem 14.7 *Let G be a finite group of order n. Then for every $x \in G$ the order of x is a divisor of n; moreover $\overset{n}{\top} x = e_G$.*

Proof. Since G is finite so is $\mathrm{Gp}\{x\}$. The first part is an immediate consequence of Lagrange's Theorem (Theorem 14.2). As for the second part, if $\mathrm{o}(x) = m$ and $n = dm$ then

$$\overset{n}{\top} x = \overset{dm}{\top} x = \overset{d}{\top} e_G = e_G.$$

The problem of determining all the subgroups of a given group is one which, to date, is far from being solved. A notable exception occurs, however, in the case of cyclic groups, as we shall now show. For typographical reasons we shall write the law of composition as multiplication in the following result, in which we restrict our attention to finite cyclic groups since we know that all infinite cyclic groups are isomorphic to the additive group **Z** and its subgroup structure is known (Theorem 13.6).

Theorem 14.8 *Let $G = \mathrm{Gp}\{g\}$ be a finite cyclic group of order n. For every positive divisor m of n there is a unique subgroup of order m,* *namely the subgroup* $\mathrm{Gp}\{g^{n/m}\}$. *Moreover, there is a bijection from the set of positive divisors of n to the set of subgroups of G.*

Proof. If $1 \leqslant r < m$ then $(n/m)r < n$ and so, since the generator g of G is of order n, we have $e_G \neq g^{(n/m)r} = (g^{n/m})^r$. Now since $(g^{n/m})^m = g^n = e_G$ by Theorem 14.7, it follows that $o(g^{n/m}) = m$. Let H be a subgroup of G of order m. If $g^t \in H$ then we have $g^{tm} = (g^t)^m = e_G$ and so $tm \in n\mathbf{Z}$ (since $n\mathbf{Z} = \mathrm{Ker} f$ where $f: \mathbf{Z} \to \mathrm{Gp}\{g\}$ is the morphism described above). Consequently n divides tm and hence n/m divides t. It follows that $g^t \in \mathrm{Gp}\{g^{n/m}\}$. Since every element of G, and hence every element of H, is a power of g we deduce that $H \subseteq \mathrm{Gp}\{g^{n/m}\}$ and since each of these subgroups is of order m we conclude that $H = \mathrm{Gp}\{g^{n/m}\}$. Finally, it follows by Lagrange's Theorem that the mapping ζ from the set of positive divisors of n to the set of subgroups of G described by $\zeta(m) = \mathrm{Gp}\{g^{n/m}\}$ is a bijection.

Corollary. *Every subgroup of a cyclic group is cyclic.*

We end this section by considering subgroups of quotient groups. These are linked to subgroups of the original group in a way which is a special case of the following result, in which we denote the set of subgroups of a group G by $S(G)$.

Theorem 14.9 *Let* G, G^* *be groups and let* $f: G \to G^*$ *be an epimorphism. If* $A = \{H \in S(G);\ \mathrm{Ker} f \subseteq H\}$ *then the mapping* $\theta_f: A \to S(G^*)$ *given by the prescription* $\theta_f(H) = f^{\to}(H)$ *is a bijection.*

Proof. Let H^* be any subgroup of G^*. Then, by Theorem 12.12, $f^{\leftarrow}(H^*)$ is a subgroup of G containing $\mathrm{Ker} f$. Moreover, since f is an epimorphism we have, by Theorem 12.14, $f^{\to}[f^{\leftarrow}(H^*)] = H^*$. This then shows that θ_f is surjective. To show that it is also injective, suppose that $H, K \in A$ with $f^{\to}(H) = f^{\to}(K)$. Then by Theorem 12.14 we have $H = f^{\leftarrow}[f^{\to}(H)] = f^{\leftarrow}[f^{\to}(K)] = K$ whence θ_f is injective.

Corollary. *If H is a normal subgroup of a group G then the prescription* $K \mapsto \natural_H^{\to}(K)$ *yields a bijection from the set of subgroups K of G such that* $H \subseteq K$ *to the set of subgroups of the quotient group* G/H. *Moreover, this bijection preserves normal subgroups.*

Proof. It suffices to take $G^* = G/H$ and $f = \natural_H$ in Theorem 14.9. That normal subgroups are preserved follows from Corollary 4 of Theorem 14.4.

Remark. The above corollary is often referred to as the **correspondence theorem for groups.**

Exercises for §14

1. For every element x of a group G let ζ_x be the associated inner automorphism (Example 12.12). If H is a subgroup of G prove that H is normal if and only if H is invariant under every inner automorphism, in that $(\forall x \in G)\ \zeta_x^{\rightarrow}(H) \subseteq H$. [Hence the terminology *invariant* which is often used instead of *normal*.]

2. Given a group $(G, \top)$ define the **commutator** $[x, y]$ of $x, y \in G$ by

$$[x, y] = x \top y \top x^{-1} \top y^{-1}.$$

If H is a normal subgroup of G and either x or y belongs to H, show that $[x, y] \in H$.

[*Hint.* Use Exercise 1.]

Now define $\delta(G) = \text{Gp}\{[x, y];\ x, y \in G\}$. Show that $\delta(G)$ is a normal subgroup of G (called the **derived group** of G).

[*Hint.* Observe that $[x, y]^{-1} = [y, x]$ and that, for every inner automorphism ζ_a, $\zeta_a([x, y]) = [\zeta_a(x), \zeta_a(y)]$; now use Exercise 1.]

If H is a normal subgroup of G prove that G/H is abelian if and only if $\delta(G) \subseteq H$.

[*Hint.* Show that G/H is abelian if and only if $(\forall x, y \in G)\ x \top y \top H = y \top x \top H$ and that this is equivalent to $[x, y] \in H$.]

3. If H is a subgroup of a group G define the **normaliser** of H in G by

$$N(H) = \{a \in G;\ a \top H \top a^{-1} = H\}.$$

Prove that $N(H)$ is a subgroup of G and is the largest subgroup of G which contains H as a normal subgroup. Deduce that H is a normal subgroup of G if and only if $N(H) = G$.

4. Let H be a normal subgroup of a multiplicative group G and let $a \in G$ be of order n. If k is the least integer such that $a^k \in H$ prove that k divides n.

[*Hint.* Observe that k is the order of a/H in G/H.]

5. Let G be a multiplicative group with the property that $(\exists n \in \mathbf{N}\backslash\{0\})$ $(\forall x, y \in G)\ (xy)^n = x^n y^n$. Prove that the subsets $G^n = \{x^n;\ x \in G\}$ and $G_n = \{x \in G;\ x^n = e_G\}$ are normal subgroups of G and that $G^n \simeq G/G_n$.

[*Hint.* Consider the mapping $G \to G^n$ given by $x \mapsto x^n$.]

6. Let G be an infinite cyclic group with generator g. Show that g^{-1} is also a generator and that these are the only generators.

7. Let G be a group in which every element is of order 2. Prove that G is abelian.

8. Prove that all groups of prime order are cyclic.

9. Use Exercises 7 and 8 to show that all groups of order less than 6 are abelian (cf. Exercise 12.7).
[*Hint*. In the case where Card $G=4$ use Lagrange's Theorem.]

10. In the dihedral group D_p of order $2p$ as described in Exercise 12.18, show that $N=\{e, r, r^2, \ldots, r^{p-1}\}$ is a normal subgroup and that $D_p/N \simeq \mathbf{Z}/2\mathbf{Z}$.

11. If $(G, \top)$ is a group and H, K are non-empty subsets of G define

$$H \top K=\{h \top k;\ h \in H, k \in K\}.$$

Show by means of an example that if H and K are subgroups of G then $H \top K$ need not be a subgroup.

[*Hint*. Consider, for example, the group of Exercise 12.7, the Cayley table for which is given following Theorem 14.2. Let $H=\{f_1, f_2\}$, $K=\{f_1, f_3\}$. Observe that $H \top K$ is not a stable subset.]

If H, K are subgroups of G and at least one of them is normal (in particular, if G is abelian) prove that

$$H \top K=K \top H=H \vee K.$$

12. Let G be a group and let H, K be finite subgroups of G. If $f: H \times K \to H \top K$ is given by $f(h, k)=h \top k$ and if R_f is the equivalence relation associated with f, prove that $(H \times K)/R_f$ and $H \top K$ are equipotent. Show also that, for every $(h, k) \in H \times K$,

$$(h, k)/R_f=\{(h \top x, x^{-1} \top k);\ x \in H \cap K\}$$

and deduce that

$$\text{Card } H \top K=\frac{\text{Card } H \text{ Card } K}{\text{Card } H \cap K}.$$

13. A sequence of groups and morphisms

$$\ldots \to A_{i-1} \xrightarrow{f_{i-1}} A_i \xrightarrow{f_i} A_{i+1} \to \ldots$$

is said to be **exact** if $\operatorname{Im} f_i=\operatorname{Ker} f_{i+1}$ for every i. [In other words, at each group in the sequence the image of the input morphism is the kernel of the output morphism.] Denoting the trivial group by $\{0\}$, the trivial monomorphism by $\{0\} \to A$ and the trivial epimorphism by

$B \to \{0\}$, prove that if A, B are groups and $f: A \to B$ is a morphism then f is (1) a monomorphism if and only if $\{0\} \to A \xrightarrow{f} B$ is exact; (2) an epimorphism if and only if $A \xrightarrow{f} B \to \{0\}$ is exact; (3) an isomorphism if and only if $\{0\} \to A \xrightarrow{f} B \to \{0\}$ is exact. Deduce that for every group G and every normal subgroup H of G there is an exact sequence $\{0\} \to H \to G \to G/H \to \{0\}$.

14. If H, K are normal subgroups of the group $(G, \top)$ show that there is a diagram of the form

$$
\begin{array}{ccccccccc}
 & & \{0\} & & \{0\} & & \{0\} & & \\
 & & \downarrow & & \downarrow & & \downarrow & & \\
\{0\} & \longrightarrow & H \cap K & \longrightarrow & K & \longrightarrow & K/(H \cap K) & \longrightarrow & \{0\} \\
 & & \downarrow & & \downarrow & & \downarrow & & \\
\{0\} & \longrightarrow & H & \longrightarrow & G & \longrightarrow & G/H & \longrightarrow & \{0\} \\
 & & \downarrow & & \downarrow & & \downarrow & & \\
\{0\} & \longrightarrow & H/(H \cap K) & \longrightarrow & G/K & \longrightarrow & G/(H \top K) & \longrightarrow & \{0\} \\
 & & \downarrow & & \downarrow & & \downarrow & & \\
 & & \{0\} & & \{0\} & & \{0\} & &
\end{array}
$$

which is commutative (in that, following the arrows, any two composite mappings with the same departure set and the same arrival set are equal) and each row and column is exact. Deduce that

$$(H \top K)/H \simeq K/(H \cap K).$$

[*Hint.* Use Corollary 3 of Theorem 14.4 to establish the bottom left-hand and top right-hand rectangles, then use Exercise 13 to complete the top left-hand rectangle. To complete the diagram, show how to define a morphism from G/K to $G/(H \top K)$ which will make the bottom row exact. For the last part take $G = H \top K$ and use Exercise 13 (3).]

15. A group G is said to be **simple** if it has no normal subgroups other than $\{e_G\}$ and G. Prove that every finite cyclic group whose order is a prime is simple.

[*Hint.* Use Lagrange's Theorem.]

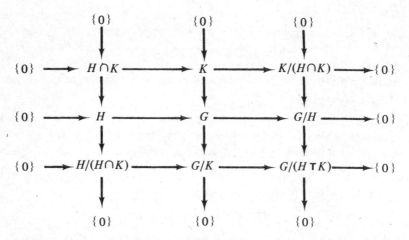

16. Let G, H be groups the laws of which are both written multiplicatively. Define a multiplication on $G \times H$ by

$$(g_1, h_1)(g_2, h_2) = (g_1h_1, g_2h_2).$$

Show that $G \times H$ is a group with respect to this law (called the **direct product group** of G, H). Prove that $G \times \{e_H\}$ and $\{e_G\} \times H$ are normal subgroups of $G \times H$ and establish the isomorphisms

$$(G \times H)/(G \times \{e_H\}) \simeq H; \quad (G \times H)/(\{e_G\} \times H) \simeq G.$$

[*Hint.* Consider the projections $(g, h) \mapsto h$ and $(g, h) \mapsto g$.]

Show that every element of $G \times \{e_H\}$ commutes with every element of $\{e_G\} \times H$ and that every element of $G \times H$ can be expressed uniquely as the product of an element of $G \times \{e_H\}$ by an element of $\{e_G\} \times H$.

Prove that if H, K are subgroups of a group G such that every element of H commutes with every element of K and every element of G can be expressed in a unique way as a product of an element of H by an element of K then the groups G and $H \times K$ are isomorphic.

17. If G, H are additive groups prove that the set $\mathrm{Mor}(G, H)$ of morphisms $f: G \to H$ is a group under the addition $(f, g) \mapsto f+g$ where $(\forall x \in G)\ (f+g)(x) = f(x) + g(x)$.

Show that for every positive integer n the set $H_n = \{h \in H;\ nh = 0\}$ is a subgroup of H and that the following are equivalent:

(1) $h \in H_n$;

(2) there is a unique $\theta_h \in \mathrm{Mor}(\mathbf{Z}/n\mathbf{Z}, H)$ such that $\theta_h(1/n\mathbf{Z}) = h$.

Deduce that $\mathrm{Mor}(\mathbf{Z}/n\mathbf{Z}, H) \simeq H_n$. Hence show that if m, n are positive integers then

$$\mathrm{Mor}(\mathbf{Z}/n\mathbf{Z}, \mathbf{Z}/m\mathbf{Z}) \simeq \mathbf{Z}/d\mathbf{Z}$$

where d is the greatest common divisor of m, n.

[*Hint.* For the last part observe that $(\mathbf{Z}/m\mathbf{Z})_n$ is cyclic and that the order of this group is the order of any generator. Now prove that $t/m\mathbf{Z} \in (\mathbf{Z}/m\mathbf{Z})_n \Leftrightarrow n(t/m\mathbf{Z}) = 0/m\mathbf{Z} \Leftrightarrow d(t/m\mathbf{Z}) = 0/m\mathbf{Z}$ (use Exercise 13.8).]

18. Let A, B be subgroups of a group $(G, \top)$. Then G is said to be the **semi-direct product** of A and B if and only if

(1) A is normal;

(2) $A \vee B = G$;

(3) $A \cap B = \{e_G\}$.

If G is the semi-direct product of A and B prove that there is a morphism, described by $b \mapsto \phi_b$, from B to the group of automorphisms of A such that

$$(\forall a \in A)(\forall b \in B) \qquad b \top a = \phi_b(a) \top b.$$

[*Hint.* Since A is normal it is invariant under every inner automorphism of G (Exercise 1). For every $b \in B$, let ζ_b denote the inner automorphism on G associated with b; then $\phi_b: A \to A$ given by $\phi_b(a) = \zeta_b(a)$ for every $a \in A$ is an automorphism on A.]

Show also that every $g \in G$ can be written uniquely in the form $g = a \top b$ where $a \in A$, $b \in B$.

[*Hint.* Use Exercise 11.]

Deduce that $G/A \simeq B$.

[*Hint.* Show that the mapping $\theta: G \to B$ given by $\theta(a \top b) = b$ is a morphism with kernel A.]

19. Let A, B be multiplicative groups and suppose that there is given a morphism, described by $b \mapsto \phi_b$, from B to the group of automorphisms of A. Show that $A \times B$, equipped with the law of composition

$$(a, b)(a', b') = (a\phi_b(a'), bb')$$

is a group. Show further that this group is the semi-direct product of subgroups A', B' isomorphic respectively to A, B.

[*Hint.* Take $A' = A \times \{e_B\}$, $B' = \{e_A\} \times B$.]

20. Let $\alpha, \beta \in \text{Bij}(\mathbf{N}, \mathbf{N})$ be given by $\alpha(n) = n+1$, $\beta(n) = -n$. Prove that $\beta \circ \beta = id_{\mathbf{N}}$ and $(\forall n \in \mathbf{N})\ \alpha^n \circ \beta = \beta \circ \alpha^{-n}$ where $\alpha^0 = id_{\mathbf{N}}$. Deduce that the subgroup of $\text{Bij}(\mathbf{N}, \mathbf{N})$ generated by $\{\alpha, \beta\}$ is $K = \{\alpha^n, \alpha^n \circ \beta;\ n \in \mathbf{N}\}$. Given any $p, k \in \mathbf{N}$ let $H_{p,k} = \{\alpha^k, \alpha^{p+\lambda k} \circ \beta;\ \lambda \in \mathbf{N}\}$ and $J_k = \{\alpha^{\lambda k};\ \lambda \in \mathbf{N}\}$. Show that $H_{p,k} \cup \{id_{\mathbf{N}}\}$ and J_k are subgroups of K. Prove also that every subgroup H of K is of the form $H_{p,k} \cup \{id_{\mathbf{N}}\}$ or J_k for some $p, k \in \mathbf{N}$.

[*Hint.* If H contains no multiple of β then we have $H \subseteq \text{Gp}\{\alpha\} \simeq \mathbf{Z}$. Use Theorem 13.6 and the fact that $J_n \simeq \mathbf{Z}/n\mathbf{Z}$ to deduce that $H = J_k$ for some $k \in \mathbf{N}$. Now suppose that H contains a multiple of β. If the only multiples of β in H are powers of β then clearly $H = \{id_{\mathbf{N}}, \beta\} = H_{0,0}$. Suppose then that H contains elements of the form $\alpha^t \circ \beta$ with $t \neq 0$. Let p be the least positive integer such that $\alpha^p \circ \beta \in H$. Show that if $\alpha^t \circ \beta \in H$ where $t \in \mathbf{N}$ then $\alpha^{t-p} \in H$ (consider $\alpha^t \circ \beta \circ \alpha^p \circ \beta$). Now let k be the least positive integer such that $\alpha^k \in H$. Show that if $\alpha^r \in H$ then r is a multiple of k (use euclidean division). Deduce that if $\alpha^t \circ \beta \in H$ then $t = p + \lambda k$ for some $\lambda \in \mathbf{N}$. Consequently, $H \subseteq H_{p,k} \cup \{id_{\mathbf{N}}\}$.]

Show that every subgroup J_k is normal in K and the quotient group K/J_k is a dihedral group of order $2k$.

[*Hint.* Show that $a=\alpha/J_k$, $b=\beta/J_k$ generate K/J_k and use Exercise 12.18.]

§15. Rings; subrings; compatible equivalences on rings; ideals; ring morphisms

We now turn our attention to sets which are endowed with two laws of composition. Guided by the properties already established for **Z**, we make the following definition.

Definition. By a **ring** we shall mean a set A on which there are defined two laws of composition (conventionally written as addition and multiplication) such that

(1) A is an abelian group under addition;

(2) A is a semigroup under multiplication;

(3) multiplication is distributive over addition, in that

$$(\forall x, y, z \in A) \qquad x(y+z)=xy+xz, \quad (y+z)x=yx+zx.$$

We shall often write a ring in the form $(A, +, \cdot)$.

The additive neutral element of a ring A is denoted by 0 and is called the **zero** of A. A ring need not possess a neutral element with respect to multiplication, however; but whenever it does, this element is necessarily unique, is denoted by 1 and is called the **identity element**. A ring which has an identity element is called a **unitary ring** (or a **ring with an identity**). Although the addition in a ring is, by definition, commutative the same need not be true of multiplication. However, if the multiplication is commutative then we say that the ring is a **commutative ring**.

Immediate consequences of the above definition are the following.

Theorem 15.1 *If A is a ring then, for all $x, y \in A$,*

(1) $x0=0=0x$;

(2) $x(-y)=(-x)y=-(xy)$;

(3) $(-x)(-y)=xy$.

Proof. (1) Since $x0+0=x0=x(0+0)=x0+x0$ it follows by the cancellation law for addition that $0=x0$; similarly $0x=0$.

(2) Since $x(-y)+xy=x[(-y)+y]=x0=0$ we see that $x(-y)$ is the additive inverse of xy, so $x(-y)=-(xy)$; similarly $(-x)y=-(xy)$.

(3) By (2) we have $(-x)(-y)=-[(-x)y]=-[-(xy)]=xy$.

Theorem 15.2 *Let A be a ring. Then*

(1) $(\forall m, n \in \mathbf{Z})(\forall x \in A) \quad (m+n)x=mx+nx$;

(2) $(\forall m, n \in \mathbf{Z})(\forall x \in A) \quad (mn)x=m(nx)=n(mx)$;

(3) $(\forall m \in \mathbf{Z})(\forall x, y \in A) \quad m(xy)=(mx)y=x(my)$.

If, moreover, A is unitary then

(4) $(\forall m \in \mathbf{Z})(\forall x \in A)(\exists y \in A) \quad mx=yx$

(*i.e., every integer multiple is a ring multiple*).

Proof. We recall that $mx=x+\ldots+x$ (m summands). It is clear that for all $m, n \in \mathbf{N}$ we have $(m+n)x=mx+nx$ and $(mn)x=m(nx)=n(mx)$. Since $(-m)x=-(mx)$ it is therefore clear that (1) and (2) hold. To establish (3), we observe first that if $m=0$ the result is trivial. Suppose by way of induction that $m(xy)=(mx)y=x(my)$ for some $m \in \mathbf{N}$; then we have

$$(m+1)xy=m(xy)+xy=\begin{cases} x(my)+xy=x(my+y)=x[(m+1)y]; \\ (mx)y+xy=(mx+x)y=[(m+1)x]y. \end{cases}$$

The result therefore holds for all $m \in \mathbf{N}$ by the induction principle. Now $\mathbf{Z}=\mathbf{N} \cup -\mathbf{N}$ and by taking $n=-m$ in (1) we obtain $(\forall m \in \mathbf{N})(\forall x \in A)$ $(-m)x=-(mx)$. Hence the equality $m(xy)=(mx)y$ holds for all $m \in \mathbf{Z}$; similarly so does $m(xy)=x(my)$. Finally, to establish (4) we observe that, by (3),

$$(\forall m \in \mathbf{Z})(\forall x \in A) \qquad mx=m(1x)=(m1)x$$

whence the result follows with $y=m1 \in A$.

Example 15.1. $(\mathbf{Z}, +, \cdot)$ is a commutative unitary ring.

Example 15.2. $(2\mathbf{Z}, +, \cdot)$ is a commutative ring without an identity.

Example 15.3. The group $(\mathrm{Map}(\mathbf{R}, \mathbf{R}), +)$ of Example 12.5 becomes a commutative unitary ring if we define multiplication by $(f, g) \mapsto fg$ where $(\forall x \in \mathbf{R})$ $(fg)(x)=f(x)g(x)$. [This multiplication is not to be confused with composition of mappings.]

Example 15.4. Let $(G, +)$ be an abelian group and let End (G) be the set of endomorphisms on G (i.e., morphisms from G to itself). Define addition and multiplication on End (G) by $(f, g) \mapsto f+g$ and $(f, g) \mapsto f \circ g$. Then it is readily seen that $f+g$ and $f \circ g$ are in End (G) and that (End (G), $+$, $\circ$) is a unitary ring which is not in general commutative.

Remark. Note that the **trivial ring** $\{0\}$ is unitary; in this case we have $0=1$. In a non-trivial unitary ring A we have $0 \neq 1$; for if we had $0=1$ then for every $a \in A$ we would have $a=a1=a0=0$.

By a **subring** of a ring A we shall mean a subset S of A which is stable with respect to both the laws of A and is a ring with respect to the inherited laws. Note that the distributivity of multiplication over addition is automatically inherited by any subset which is both additively and multiplicatively stable. By the additive version of Theorem 12.6 we therefore have the following characterisation of subrings.

Theorem 15.3 *A non-empty subset S of a ring A is a subring of A if and only if*

$$(\forall x, y \in S) \qquad x-y \in S \text{ and } xy \in S.$$

As an immediate application of this result we can determine all the subrings of the ring $\mathbf{Z}$.

Theorem 15.4 *Every subring of the ring $\mathbf{Z}$ is of the form $n\mathbf{Z}$ for some $n \in \mathbf{Z}$ and conversely.*

Proof. We know by Theorem 13.6 that every subgroup of the group $\mathbf{Z}$ is of the form $n\mathbf{Z}$ for some $n \in \mathbf{Z}$. Now if $p, q \in n\mathbf{Z}$ then $p=na$, $q=nb$ for some $a, b \in \mathbf{Z}$ and so $pq=n^2ab \in n\mathbf{Z}$. Thus we see, by Theorem 15.3, that $n\mathbf{Z}$ is a subring of the ring $\mathbf{Z}$. It follows that every subring of $\mathbf{Z}$ is of the form $n\mathbf{Z}$ (since every subgroup is of this form).

Given a subring S of a ring A the relation R_S defined on A by

$$x \equiv y \,(R_S) \Leftrightarrow x-y \in S$$

is, as we know, an equivalence relation which is compatible with addition in A. In general, however, it is not compatible with multiplication; for if $a \in A$ and $x \equiv y\,(R_S)$ then $ax-ay=a(x-y)$ and $xa-ya=(x-y)a$ need not in general be elements of S. To discuss the compatibility of an equivalence relation with respect to both the laws of composition of a ring A we therefore require some additional terminology.

Let S be a semigroup in which the law of composition is written as multiplication. Define a multiplication $(X, Y) \mapsto XY$ on $\mathbf{P}(S)$ by

$$XY=\begin{cases}\{xy;\, x \in X,\, y \in Y\} & \text{if } X, Y \neq \varnothing;\\ \varnothing & \text{otherwise.}\end{cases}$$

It is readily verified that this multiplication, called the **induced multiplication** on $\mathbf{P}(S)$, is associative. We shall often write xY instead of $\{x\}Y$ and Xy instead of $X\{y\}$.

Definition. Let S be a (multiplicative) semigroup. By a **left ideal** [resp. **right ideal**] of S we shall mean a non-empty subset I of S such that $SI \subseteq I$ [resp. $IS \subseteq I$]; in other words $sx \in I$ [resp. $xs \in I$] for all $s \in S$ and all $x \in I$. By a **left ideal** [resp. **right ideal**] of a ring $(A, +, \cdot)$ we shall mean a subgroup I of $(A, +)$ which is a left ideal [resp. right ideal] of the semigroup $(A, \cdot)$.

Theorem 15.5 *Let A be a ring and let R be a binary relation on A. Then the following are equivalent:*

(1) *R is an equivalence relation which is compatible with addition and compatible on the left* [*resp. right*] *with multiplication*;

(2) *there is a left ideal* [*resp. right ideal*] *I of A such that, for all $x, y \in A$, $x \equiv y\ (R) \Leftrightarrow x - y \in I$.*

Proof. (1) $\Rightarrow$ (2): If (1) holds then by Theorem 14.3 there is an additive subgroup $I = 0/R$ of A such that, for all $x, y \in A$, $x \equiv y\ (R) \Leftrightarrow x - y \in I$. Given any $x \in 0/R$ and any $a \in A$ we have, by the left compatibility of R with respect to multiplication, $ax \equiv a0 = 0\ (R)$ and so $ax \in 0/R$. It follows that $AI \subseteq I$ and so I is a left ideal of the ring A.

(2) $\Rightarrow$ (1): If (2) holds then since I is in particular an additive subgroup of A we see from Theorem 14.3 that R is compatible with addition. Since I is a left ideal we also have, for all $x, y \in A$,

$$\begin{aligned} x \equiv y\ (R) &\Rightarrow x - y \in I \Rightarrow (\forall a \in A)\ ax - ay = a(x-y) \in AI \subseteq I \\ &\Rightarrow (\forall a \in A)\ ax \equiv ay\ (R) \end{aligned}$$

and so R is compatible on the left with multiplication.

We define a **two-sided ideal** (or simply **ideal**) of a ring A to be a subset I of A which is both a left and right ideal of A. We then deduce the following results.

Corollary 1. *If A is a ring and R is a binary relation on A then the following are equivalent*:

(1) *R is an equivalence relation which is compatible with both addition and multiplication*;

(2) *there is a (two-sided) ideal I of A such that, for all $x, y \in A$, $x \equiv y\ (R) \Leftrightarrow x - y \in I$.*

Corollary 2. *The set of (two-sided) ideals of a ring A is equipotent to the set of partitionings of A associated with equivalence relations on A which are compatible with both addition and multiplication.*

In dealing with rings and ideals we shall commit the same abuse of notation as we did in dealing with groups and normal subgroups in that, if R is an equivalence relation on a ring A which is compatible with both addition and multiplication and if I is the ideal of A which corresponds (under the bijection of the previous corollary) to the partitioning associated with R, we shall write A/I instead of A/R and call A/I the **quotient ring** (or **residue class ring**) of A by the (*two-sided*) ideal I.

Example 15.5. For every $n \in \mathbf{Z}$ the set $n\mathbf{Z}$ is a subring of $\mathbf{Z}$ by Theorem 15.4. Now if $x \in n\mathbf{Z}$ then $x=np$ for some $p \in \mathbf{Z}$ and so $(\forall y \in \mathbf{Z})$ $xy=npy \in n\mathbf{Z}$. Thus we see that $n\mathbf{Z}$ is an ideal of $\mathbf{Z}$. It follows that $\mathbf{Z}/n\mathbf{Z}$ is a ring which is moreover unitary since $1/n\mathbf{Z}$ is clearly an identity element.

For typographical reasons we shall now make the convention that, whenever possible, in dealing with several rings at once we shall denote their laws of addition all by the same symbol $+$ and their laws of multiplication all by juxtaposition (or sometimes a dot for clarity). The context (and by now the reader's expertise in handling laws of composition) will make it clear to which laws reference is being made.

We now come to consider mappings from one ring to another and in particular the notion of a ring morphism.

Definition. If A, A^* are rings then a mapping $f: A \to A^*$ is called a **ring morphism** if $f:(A, +) \to (A^*, +)$ is a group morphism and $f:(A, \cdot) \to (A^*, \cdot)$ is a semigroup morphism; equivalently, if

$$(\forall x, y \in A) \qquad f(x+y)=f(x)+f(y) \text{ and } f(xy)=f(x)f(y).$$

Example 15.6. For every ring A, id_A is a ring morphism.

Example 15.7. If A is a ring and I is an ideal of A then A/I is a ring under the induced laws and the canonical surjection $\natural_I: A \to A/I$ is a ring morphism, called the **canonical epimorphism**.

Example 15.8. If A, B are rings and $f: A \to B$ is given by $(\forall x \in A)$ $f(x)=0$ then clearly f is a ring morphism, called the **zero morphism**.

The following result is an extension of Theorem 12.12 for group morphisms.

Theorem 15.6 *If A, A^* are rings and $f: A \to A^*$ is a ring morphism then* $\operatorname{Im} f$ *is a subring of A^* and* $\operatorname{Ker} f$ *is a two-sided ideal of A.*

Proof. It is clear that $\operatorname{Im} f$ is a subring of A^* and that $\operatorname{Ker} f$ is a subgroup of A. Now if $x \in A$ is such that $f(x)=0$ then for every $y \in A$ we have $f(yx)=f(y)f(x)=f(y)0=0$ whence we see that $\operatorname{Ker} f$ is a left ideal of A; similarly, it is a right ideal.

We shall leave to the reader the task of showing that Theorem 14.4, its corollaries and Theorem 14.5 have analogues in the case of rings and ring morphisms. [Essentially, it suffices to check that, when the groups are replaced by rings and the normal subgroups are replaced by two-sided ideals, all the mappings in question are also semigroup morphisms.] We mention in particular the *fundamental ring isomorphism* $A/\operatorname{Ker} f \simeq \operatorname{Im} f$ *associated with a ring morphism* $f: A \to A^*$. Also, the Corollary to Theorem 14.9 has a ring-theoretic analogue (called the **correspondence theorem for rings**), namely that *if I is an ideal of the ring A then there is a bijection from the set of ideals J such that $I \subseteq J$ to the set of ideals of the quotient ring A/I.*

Exercises for §15

1. Let A be a ring and let End (A) be the ring of endomorphisms of the abelian group $(A, +)$ (see Example 15.4). For every $a \in A$ define $\lambda_a: A \to A$ by $\lambda_a(x)=ax$. Prove that $\zeta: A \to \operatorname{End}(A)$ described by $\zeta(a)=\lambda_a$ is a ring morphism. Deduce the ring analogue of Cayley's Theorem, namely that every unitary ring is isomorphic to a ring of endomorphisms.

2. Show that $(\mathbf{P}(E), \Delta, \cap)$ is a ring where Δ is defined as in Exercise 12.6. If F is a subset of E show that $\mathbf{P}(F)$ is an ideal of the ring $\mathbf{P}(E)$ and establish the ring isomorphism $\mathbf{P}(E)/\mathbf{P}(F) \simeq \mathbf{P}(\complement_E(F))$.
[*Hint.* Consider the mapping given by $X \mapsto X \cap \complement_E(F)$.]

3. If A is a unitary ring and $x, y \in A$ commute (i.e., $xy=yx$) prove the **Binomial Theorem** that, for all $n \in \mathbf{N}$,

$$(x+y)^n = \sum_{r=0}^{n} \binom{n}{r} x^r y^{n-r}.$$

[*Hint.* Use induction and the properties of binomial coefficients given in Exercise 9.10.]

4. Let A be a commutative ring. For every $a \in A$ define the **annihilator** of a by $\operatorname{Ann}(a)=\{x \in A;\ xa=0\}$. Show that $\operatorname{Ann}(a)$ is an ideal of A. In the ring $\mathbf{Z}/12\mathbf{Z}$ determine Ann $(3/12\mathbf{Z})$, Ann $(5/12\mathbf{Z})$, Ann $(8/12\mathbf{Z})$.

5. An element a of a ring A is said to be **nilpotent** if there is a positive integer n such that $a^n=0$. Suppose that A is commutative and unitary. Prove that the set N of nilpotent elements of A is an ideal of A.

[*Hint.* If $a^n=0=b^m$ consider $(a-b)^{n+m}$ and use Exercise 3.]

Show also that the quotient A/N contains no non-zero nilpotent elements.

6. If I is an ideal of a unitary ring A define

$$\sqrt{I}=\{x \in A; (\exists n \in \mathbf{N})\ x^n \in I\}.$$

Prove that $\sqrt{I}$ is an ideal of A containing I. Show also that if I, J are ideals of A then

(a) $I \subseteq J \subseteq \sqrt{I} \Rightarrow \sqrt{J}=\sqrt{I}$;

(b) $\sqrt{(I \cap J)}=\sqrt{I} \cap \sqrt{J}$.

7. An ideal I of a ring A is said to be **regular** if the quotient ring A/I is unitary. Prove that

(1) every ideal of a unitary ring is regular;

(2) every ideal which contains a regular ideal is regular;

(3) the intersection of two regular ideals is regular.

[*Hint.* (2) Use the ring analogue of the Corollary to Theorem 14.9.
(3) If $e/I, f/J$ are the identity elements of A/I, A/J respectively, consider $(e+f-ef)/(I \cap J)$.]

8. Prove that the intersection of any family of ideals of a ring A is also an ideal of A. If I and J are ideals of A prove that the ideal generated by $I \cup J$ (i.e., the smallest ideal of A which contains I and J) is given by

$$I+J=\{x+y;\ x \in I,\ y \in J\}.$$

[*Hint.* Use Theorem 12.13.]

9. If I is an ideal of the ring A, prove that the quotient ring A/I is commutative if and only if I contains the ideal generated by the elements of the form $xy-yx$.

10. Show that for every subset X of a ring A the ideal generated by X is $\mathrm{Gp}\{X \cup AX \cup XA \cup AXA\}$. What does this become when A is unitary?

11. Let A, B be rings. Show that $A \times B$, equipped with the laws given by

$$(a, b)+(a', b')=(a+a', b+b'), \quad (a, b)(a', b')=(aa', bb')$$

is also a ring (called the **cartesian product ring**). Establish ring-theoretic analogues of Exercises 14.13, 14.14. Show that, associated with any two ideals I, J of a ring A there is an exact sequence of rings and ring morphisms

$$\{0\} \to I \cap J \to A \xrightarrow{f} A/I \times A/J \xrightarrow{g} A/(I+J) \to \{0\}.$$

[*Hint.* Define $f(x)=(x/I, -x/J)$ and $g(x/I, y/J)=(x+y)/(I+J)$.]

Hence show that

$$(I+J)/(I \cap J) \simeq (I+J)/I \times (I+J)/J \simeq I/(I \cap J) \times J/(I \cap J).$$

12. If A, B are unitary rings and $f: A \to B$ is a ring epimorphism prove that

(a) $f(1_A)=1_B$;

(b) if a is an invertible element of A then $f(a)$ is invertible in B and $[f(a)]^{-1}=f(a^{-1})$.

13. Prove that if $f: \mathbf{Z} \to \mathbf{Z}$ is a ring morphism which is not the zero morphism then $f=id_{\mathbf{Z}}$.

[*Hint.* Show that $(\forall n \in \mathbf{Z})\ f(n)=nf(1)$ and use Exercise 12.]

14. Let R be a ring and let $f, g: R \to \mathbf{Z}$ be ring epimorphisms. Prove that $f=g$ if and only if $\operatorname{Ker} f = \operatorname{Ker} g$.

[*Hint.* $\Leftarrow$: Consider the canonical decompositions of f, g (the ring analogue of Corollary 2 to Theorem 14.4) and use Exercise 13.]

§16. Integral domains; division rings; fields

If we compute the Cayley tables for the ring $\mathbf{Z}/4\mathbf{Z}$, we obtain the following (in which we write $[m]$ for $m/4\mathbf{Z}$):

$+_4$	[0]	[1]	[2]	[3]
[0]	[0]	[1]	[2]	[3]
[1]	[1]	[2]	[3]	[0]
[2]	[2]	[3]	[0]	[1]
[3]	[3]	[0]	[1]	[2]

4	[0]	[1]	[2]	[3]
[0]	[0]	[0]	[0]	[0]
[1]	[0]	[1]	[2]	[3]
[2]	[0]	[2]	[0]	[2]
[3]	[0]	[3]	[2]	[1]

We observe a curious fact in the multiplication table: namely that although $[2] \neq [0]$ we have $[2][2]=[0]$. We are thus led to consider the following notion.

Definition. If the product of two non-zero elements a, b or a non-trivial ring A is the zero element then a and b are called **divisors of zero**. (Note that we do not count 0 as a divisor of zero.)

Theorem 16.1 *If A is a ring then the following are equivalent*:

(1) *A has no zero divisors*;

(2) *$A \backslash \{0\}$ is a (multiplicative) subsemigroup of A*;

(3) *the (multiplicative) cancellation laws hold in $A \backslash \{0\}$*.

Proof. The equivalence of (1) and (2) is obvious. To show that (1) $\Rightarrow$ (3), suppose that $a, b, x \in A$ are such that $x \neq 0$ and $ax = bx$. Then we have $0 = ax - bx = (a-b)x$ whence, by (1), $a-b=0$ and $a=b$. As for (3) $\Rightarrow$ (1), let a, b be such that $a \neq 0$, $b \neq 0$. Suppose that $ab=0$; then $ab=0b$ and the cancellation law gives the contradiction $a=0$.

We thus see that, by imposing on A the property of not having any zero divisors, we strengthen the semigroup structure of A in such a way that $A \backslash \{0\}$ becomes a cancellative subsemigroup of A.

Definition. By an **integral domain** we shall mean a non-trivial unitary ring with no zero divisors.

It is natural to investigate integral domains A for which the cancellative subsemigroup $A \backslash \{0\}$ is a group. This gives rise to the following notion.

Definition. By a **division ring** we shall mean an integral domain in which every non-zero element has a multiplicative inverse. By a **field** we mean a commutative division ring.

As the following result shows, the ideal structure of a division ring (and hence of a field) is particularly simple.

Theorem 16.2 *A unitary ring A is a division ring if and only if it has no left or right ideals other than $\{0\}$ and A.*

Proof. Suppose that A is a division ring and let I be a non-zero left ideal of A. Given any non-zero $x \in I$ we note that x^{-1} exists. For every $y \in A$ we then have $y = y1 = yx^{-1}x \in yx^{-1}I \subseteq I$ whence we deduce that $I = A$. Similarly, if J is a non-zero right ideal of A then $J = A$.

Conversely, suppose that $\{0\}$ and A are the only left or right ideals of A. Given $a \in A \backslash \{0\}$, note that the set $Aa = \{xa;\ x \in A\}$ is a left ideal of A which is distinct from $\{0\}$ since $a = 1a \in Aa$. By the hypothesis we therefore have $Aa = A$ and so there exists $y \in A$ such that $ya = 1$. Hence a has a left inverse. Arguing in a similar way with the right ideal $aA = \{ax;\ x \in A\}$ we see that a also has a right inverse. It follows by Theorem 12.1 that a is invertible. Consequently $A \backslash \{0\}$ is a group and so A is a division ring.

This result yields the following fact concerning morphisms.

Theorem 16.3 *If $f: D \to A$ is a morphism from a division ring D to a ring A then either* $\text{Im} f = \{0\}$ *or else* $\text{Im} f$ *is a division ring and f is a monomorphism.*

Proof. Since $\text{Ker} f$ is an ideal of D it follows from Theorem 16.2 that either $\text{Ker} f = \{0\}$ or $\text{Ker} f = D$. In the latter case we have $f(x) = 0$ for every $x \in D$ and so $\text{Im} f = \{0\}$; in the former case f is clearly a monomorphism and $\text{Im} f \simeq D/\text{Ker} f = D/\{0\} = D$ is a division ring.

We shall now introduce particularly important types of ideal and show how they may be characterised by properties of the corresponding quotient ring.

Definition. By a **maximal ideal** of a ring A we shall mean an ideal $I \neq A$ such that if M is any ideal of A with $I \subseteq M \subset A$ then $I = M$. By a **prime ideal** of A we shall mean an ideal $I \neq A$ such that, for all $x, y \in A$, if $xy \in I$ then either $x \in I$ or $y \in I$.

Theorem 16.4 *Let A be a commutative unitary ring and let I be an ideal of A. Then*

(1) *I is maximal if and only if A/I is a field;*

(2) *I is prime if and only if A/I is an integral domain.*

Proof. (1) By the ring-theoretic analogue of the Corollary to Theorem 14.9, there is a bijection from the set of ideals which contain I to the set of ideals of the quotient ring A/I. Now it is clear that I is a maximal ideal if and only if the set of ideals which contain I is $\{I, A\}$. The above bijection shows that this is equivalent to A/I having only two ideals, namely A/I itself and $\{0\}/I$. It follows by Theorem 16.2 that this is equivalent to A being a field.

(2) That A/I is an integral domain is equivalent to saying that it has no zero divisors which in turn is equivalent to saying that for all $x, y \in A$ if $x/I . y/I = 0/I$ then $x/I = 0/I$ or $y/I = 0/I$. This is clearly equivalent to saying that I is a prime ideal.

Corollary. *In a commutative unitary ring every maximal ideal is prime.*

Proof. Every field is an integral domain.

Example 16.1. Let A be the set of all mappings $f: [0, 1] \to \mathbf{R}$ which are continuous. It is shown in analysis that $(A, +, \cdot)$ is a commutative unitary ring (being a subring of the ring described in Example 15.3). Consider the mapping $\zeta: A \to \mathbf{R}$ given by $\zeta(f) = f(\alpha)$ where α is a fixed element of $[0, 1]$. This mapping ζ is a ring epimorphism and so

$\mathbf{R} = \text{Im } \zeta \simeq A/\text{Ker } \zeta$. Since $\mathbf{R}$ is a field it follows from Theorem 16.4 that Ker $\zeta = \{f \in A; f(\alpha) = 0\}$ is a maximal ideal of A.

We end this section with another application of Zorn's axiom (see §10).

Theorem 16.5 **[Krull]** *If A is a unitary ring then every ideal I of A with $I \neq A$ is contained in a maximal ideal of A.*

Proof. Denote by $I(A)$ the set of all ideals of A. Given any ideal I of A define

$$F_I = \{J \in I(A);\ I \subseteq J \subset A\}.$$

Clearly $F_I \neq \emptyset$ since it contains I. Now let $\{J_\beta; \beta \in B\}$ be a totally ordered subset of F_I. We note that $\bigcup_{\beta \in B} J_\beta \neq A$; for if we had equality then the identity element 1 of A would belong to at least one J_β whence we would have $A = A1 \subseteq AJ_\beta \subseteq J_\beta \subset A$, a contradiction. Suppose now that $x, y \in \bigcup_{\beta \in B} J_\beta$. Then there exist $\alpha, \gamma \in B$ such that $x \in J_\alpha$ and $y \in J_\gamma$. Since $\{J_\beta; \beta \in B\}$ is totally ordered we have either $J_\alpha \subseteq J_\gamma$ or $J_\gamma \subseteq J_\alpha$. Suppose, without loss of generality, that $J_\alpha \subseteq J_\gamma$; then we have $x - y \in J_\gamma \subseteq \bigcup_{\beta \in B} J_\beta$. In a similar way we have $xa, ax \in J_\gamma$ for every $a \in A$. Hence we see that $\bigcup_{\beta \in B} J_\beta$ is an ideal of A, distinct from A. Thus F_I is inductively ordered and we can apply Zorn's axiom to obtain a maximal element M of F_I. M is then a maximal ideal of A containing I.

Exercises for §16

1. Prove that every finite integral domain is a division ring.

[*Hint.* Use Exercise 12.4.]

2. Show that $A = \{m + n\sqrt{3};\ m, n \in \mathbf{Z}\}$ is a subring of $\mathbf{R}$. If the binary relation T is defined on A by

$$m + n\sqrt{3} \equiv p + q\sqrt{3}(T) \Leftrightarrow m \equiv p \pmod 2,\ n \equiv q \pmod 2$$

show that T is an equivalence relation on A which is compatible with both addition and multiplication. What is Card A/T? By considering the Cayley tables for A/T show that A/T has only one divisor of zero and only one ideal different from $\{0\}/T$ and A/T.

3. Let A be a ring and define laws of addition and multiplication on $A \times \mathbf{Z}$ by

$$(x, m) + (y, n) = (x + y, m + n), \quad (x, m)(y, n) = (xy + my + nx, mn).$$

Show that in this way $A \times \mathbf{Z}$ becomes a unitary ring. Show also that the mapping $f: A \to A \times \mathbf{Z}$ given by $h(x) = (x, 0)$ is a ring monomorphism with $\operatorname{Im} f$ an ideal of $A \times \mathbf{Z}$. In the case where $A = 2\mathbf{Z}$ show that $A \times \mathbf{Z}$ has divisors of zero whereas A does not.

4. Let A, A^* be rings and let $f: A \to A^*$ be a ring morphism. If I is a prime ideal of A such that $\operatorname{Ker} f \subseteq I$ prove that $f^{\to}(I)$ is a prime ideal of A^*. Prove conversely that if J is a prime ideal of $\operatorname{Im} f$ then $f^{\leftarrow}(J)$ is a prime ideal of A containing $\operatorname{Ker} f$. If K is a maximal ideal of A prove that $f^{\to}(K)$ is a maximal ideal of $\operatorname{Im} f$ and, conversely, if L is a maximal ideal of $\operatorname{Im} f$ prove that $f^{\leftarrow}(L)$ is a maximal ideal of A.

5. By a **boolean ring** we mean a non-trivial ring A in which every element x is idempotent (in that $x^2 = x$). If A is a boolean ring prove that
(a) $(\forall x \in A)\ x = -x$;
(b) A is commutative.

[*Hint.* (a) $x + x$ is idempotent; (b) $x + y$ is idempotent.]

If I is an ideal of a boolean ring A prove that the following are equivalent:
(1) I is prime; (2) I is maximal; (3) $A/I \simeq \mathbf{Z}/2\mathbf{Z}$.

[*Hint.* (1) $\Rightarrow$ (3): Observe that in a boolean ring we have $xy(x+y) = 0$; apply this to the quotient ring A/I.]

6. For a commutative unitary ring A define $\operatorname{Rad}_J(A)$, the **Jacobson radical** of A, to be the intersection of all the maximal ideals of A. If I is an ideal of A prove that the following are equivalent:
(1) $I \subseteq \operatorname{Rad}_J(A)$;
(2) every element of $1/I$ is invertible in A.

[*Hint.* (1) $\Rightarrow$ (2): Suppose that $a \in I$ is such that $1 + a$ is not invertible. Use Theorem 16.5 to show that there is a maximal ideal M containing $1 + a$ and derive from (1) the contradiction that $1 \in M$.
(2) $\Rightarrow$ (1): If $I \nsubseteq \operatorname{Rad}_J(A)$ then there is a maximal ideal M of A with $I \nsubseteq M$. Observe that for some $a \in I$ we have $M + Aa = A$ so that $1 = m + xa$ for some $m \in M$, $x \in A$. Derive the contradiction that m is invertible.]

Hence show that the following are equivalent:
(3) $a \in \operatorname{Rad}_J(A)$;
(4) $(\forall x \in A)\quad 1 + xa$ is invertible in A.

[*Hint.* Take $I = Aa$ where $a \in \operatorname{Rad}_J(A)$.]

Deduce from the above that $\text{Rad}_J(A/\text{Rad}_J(A))=\{0\}$.

[*Hint.* Use the equivalence of (3), (4) twice to show that if $R=\text{Rad}_J(A)$ then $a/R \in \text{Rad}_J(A/R) \Rightarrow (\forall x \in A)(\exists y \in A)\ y+xay$ is invertible in A. Now use the equivalence of (3), (4) again to show that $a \in R$.]

7. A ring A is said to be a **Jacobson ring** if, for every $x \in A$, there is an integer $n \geqslant 2$ such that $x^n=x$. If A is a unitary Jacobson ring prove that

(1) no non-zero element is nilpotent (Exercise 15.5);
(2) if $e \in A$ is idempotent ($e^2=e$) then $(xe-exe)^2=0=(ex-exe)^2$;
(3) for every $x \in A$ some power of x is idempotent;
(4) every left (right) ideal of A is two-sided;
(5) $\text{Rad}_J(A)=\{0\}$.

[*Hint.* (3) If $x^p=x$ consider $(x^{p-1})^2$.

(4) If $x^p=x$ use (1), (2), (3) to show that, for every $x \in A$, $ax=xb$ where $b=x^{p-2}ax$.

(5) Suppose that $x \neq 0$ belongs to $\text{Rad}_J(A)$, the intersection of all the maximal ideals of A. If $x^p=x$ then the idempotent $e=x^{p-1}$ also belongs to $\text{Rad}_J(A)$. Show that $1-e$ does not belong to any proper ideal of A and deduce that $(1-e)A=A$. Hence derive the contradiction $e=0$.]

8. If A is a unitary ring and I is an ideal of A with $I \neq A$, show that no element of I is invertible. Use Theorem 16.5 to show that an element of A is invertible if and only if it does not belong to any maximal ideal.

9. Let A be a commutative unitary ring having precisely one maximal ideal M. (Such a ring is called a **local ring**.) If $a \in A$ is idempotent prove that either $a=0$ or $a=1$.

[*Hint.* Let $a^2=a$ with $a \notin \{0, 1\}$. Show that the elements a, $1-a$ are zero divisors and hence non-invertible. Now use Exercise 8.]

10. Let Q denote the set of all expressions $a_0+a_1i+a_2j+a_3k$ where every $a_i \in \mathbf{R}$ and define $a_0+a_1i+a_2j+a_3k$ to be equal to $b_0+b_1i+b_2j+b_3k$ if and only if $a_i=b_i$ for every i. Define addition and multiplication on Q by

$$(a_0+a_1i+a_2j+a_3k)+(b_0+b_1i+b_2j+b_3k) = (a_0+b_0)+(a_1+b_1)i+(a_2+b_2)j+(a_3+b_3)k$$

and

$$\begin{aligned}(a_0+a_1i+a_2j+a_3k)(b_0+b_1i+b_2j+b_3k) &= (a_0b_0-a_1b_1-a_2b_2-a_3b_3)+(a_0b_1+a_1b_0+a_2b_3-a_3b_2)i \\ &\quad +(a_0b_2+a_2b_0+a_3b_1-a_1b_3)j+(a_0b_3+a_3b_0+a_1b_2-a_2b_1)k.\end{aligned}$$

[Note that the formula for multiplication may be obtained by multiplying out the product as though they were real numbers and using the relations $i^2=j^2=k^2=-1$, $ij=-ji=k$, $jk=-kj=i$, $ki=-ik=j$.] Show that in this way Q becomes a non-commutative ring with zero element $0+0i+0j+0k$ and identity element $1+0i+0j+0k$. If the element $a_0+a_1i+a_2j+a_3k$ is non-zero show that $a=a_0^2+a_1^2+a_2^2+a_3^2\neq 0$ and that $\frac{1}{a}(a_0-a_1i-a_2j-a_3k)$ is the inverse in Q of $a_0+a_1i+a_2j+a_3k$. Deduce that Q forms a division ring (called the **ring of quaternions**).

§17. Arithmetic properties in commutative integral domains; unique factorisation domains; principal ideal domains; euclidean domains

This section is devoted to a study of particularly important types of commutative integral domain and their "arithmetic" properties (so-called since, loosely speaking, arithmetic begins when we have primes).

Suppose that A is a commutative ring. Given any non-empty subset X of A it is readily seen that the ideal generated by X (i.e., the intersection of all the ideals of A which contain X) is given by

$$\langle X\rangle=\left\{\sum_{i=1}^{m} p_i x_i+\sum_{i=1}^{n} a_i x_i;\ m, n\in \mathbf{N}\backslash\{0\},\ p_i\in \mathbf{Z},\ a_i\in A,\ x_i\in X\right\}.$$

In fact the expression given is clearly an ideal of A which contains X (for every $x\in X$ we have $x=1x$ where $1\in \mathbf{N}\backslash\{0\}$); moreover, every ideal which contains X also contains the given expression. In the case where A is unitary we can apply Theorem 15.2(4) and thus observe that, in this case,

$$\langle X\rangle=\left\{\sum_{i=1}^{n} a_i x_i;\ n\in \mathbf{N}\backslash\{0\},\ a_i\in A,\ x_i\in X\right\}.$$

In the particular case where $X=\{x\}$ we write $\langle x\rangle$ instead of $\langle\{x\}\rangle$ and call $\langle x\rangle$ the **principal ideal** generated by $\{x\}$. Thus, for every $x\in A$,

$$\langle x\rangle=\{px+ax;\ p\in \mathbf{Z},\ a\in A\}$$

and, when A is unitary,

$$\langle x\rangle=\{ax;\ a\in A\}=Ax.$$

If A is a commutative ring and $a, b\in A$ with $b\neq 0$ then we say that b **divides** a, and write $b\,|\,a$, if and only if there exists $c\in A$ such that

$bc=a$. In the case where A is unitary we therefore have

$$b \mid a \Leftrightarrow a \in \langle b \rangle \Leftrightarrow \langle a \rangle \subseteq \langle b \rangle.$$

We say that a is an **associate** of b if $a \mid b$ and $b \mid a$. By a **unit** of A we mean an associate of 1.

Theorem 17.1 *Let A be a commutative integral domain. Given non-zero $a, b \in A$, the following are equivalent:*

(1) *a is an associate of b;*
(2) $\langle a \rangle = \langle b \rangle$;
(3) *there is an invertible element $u \in A$ such that $a=bu$.*

Proof. It is clear that (1) and (2) are equivalent.

(1) $\Rightarrow$ (3): If $a=bu$ and $b=av$ then $a=avu$ whence $1=vu$ by the cancellation law and so u is invertible.

(3) $\Rightarrow$ (1): If $a=bu$ and u^{-1} exists then $b=au^{-1}$; thus $a \mid b$ and $b \mid a$.

Corollary. *If $u \in A \backslash \{0\}$ then the following are equivalent:*

(1) *u is a unit;*
(2) $\langle u \rangle = A$;
(3) *u is invertible.*

Proof. It suffices to take $b=1$ and $a=u$ in the theorem.

It is immediate from the above that the set of invertible elements of A forms a group under multiplication; we call this the **group of units** of A. Note that the relation R defined on $A \backslash \{0\}$ by

$$a \equiv b(R) \Leftrightarrow a \text{ is an associate of } b$$

is an equivalence relation; and equivalence classes modulo R are given by $x/R = Gx$ where G is the group of units of A.

Example 17.1. If F is a field then its group of units is $F \backslash \{0\}$.

Example 17.2. It is clear that 1 and -1 are units of the commutative integral domain $\mathbf{Z}$. We shall show that the group of units of $\mathbf{Z}$ is $\{1, -1\}$. Consider the mapping from $\mathbf{Z}$ to $\mathbf{N}$ given by $n \mapsto |n|$ where $|n| = \max\{n, -n\}$. We note that this mapping is a morphism with respect to multiplication. To see this, we remark that when $m, n \geqslant 0$ we have clearly $|mn| = |m|\,|n|$; and when at least one of m, n is <0 the required equality follows from

$$|(-m)(-n)| = |mn| = |-(mn)| = |(-m)n| = |m(-n)|.$$

Suppose now that u is a unit of $\mathbf{Z}$ so that there exists $v \in \mathbf{Z}$ with $uv=1$. Since $u \neq 0$ we have $|u|>0$ and so, by Theorem 8.12(2), $|u| \geqslant 1$. Now $|u|>1$ is impossible since this would give

$$1=|uv|=|u|\,|v|>|v|>0,$$

in contradiction to Theorem 8.12(2). Hence $|u|=1$ and so $u \in \{1, -1\}$.

The associates of any element x and the units of A are called **improper divisors** of x, any other divisors which x may have being called **proper divisors** of x. An element $b \in A$ is said to be **irreducible** if b is neither zero nor a unit but is such that if $a \mid b$ then a is either a unit or an associate of b (in other words, the divisors of b are all improper). An element $a \in A$ is said to be **prime** if it is neither zero nor a unit but is such that if $a \mid bc$ then $a \mid b$ or $a \mid c$.

Theorem 17.2 *Let A be a commutative integral domain. If $a \in A$ is prime then a is irreducible.*

Proof. Suppose that $c \mid a$. Then $cb=a$ for some $b \in A$ and since a is prime we have $a \mid c$ or $a \mid b$. If $a \mid c$ then we see that c is an associate of a; and if $a \mid b$ then $b=ad$ for some $d \in A$ whence $a=cb=cad$ which gives $1=cd$ and c is a unit.

Remark. The converse of Theorem 17.2 is false in general (see, for example, Exercise 17.16).

The first particular type of integral domain we shall consider is the following.

Definition. A **unique factorisation domain** is a commutative integral domain in which

(UFD 1) every element a which is neither zero nor a unit can be expressed as a product $a=\prod_{i=1}^{r} p_i$ of irreducible elements (with $r=1$ when a is itself irreducible);

(UFD 2) this decomposition is unique in that if a has two such decompositions, say $p_1 \ldots p_r=a=q_1 \ldots q_t$, then $r=t$ and there is a permutation σ on $\{1, \ldots, r\}$ such that p_i and $q_{\sigma(i)}$ are associates for all $i \in \{1, \ldots, r\}$.

Theorem 17.3 *Let D be a commutative integral domain in which* (UFD 1) *holds. Then* (UFD 2) *holds if and only if*

(UFD 2*) *every irreducible element of D is prime.*

Proof. Suppose that (UFD 1) and (UFD 2) hold. Let p be an irreducible element and suppose that $p \mid ab$. Then there exists $c \in D$ such that $ab = pc$. Now a and b cannot both be units; for if this were the case we would have $1 = pc(ab)^{-1}$ and p would be a unit, contrary to the hypothesis. If one of a, b is a unit, a say, then we have $b = pca^{-1}$ whence $p \mid b$. Suppose then that neither a nor b is a unit. Let

$$a = \prod_{i=1}^{r} a_i, \quad b = \prod_{j=1}^{s} b_j, \quad c = \prod_{k=1}^{t} c_k$$

be decompositions of a, b, c as products of irreducibles. By hypothesis we have

$$\prod_{i=1}^{r} a_i \prod_{j=1}^{s} b_j = p \prod_{k=1}^{t} c_k.$$

By (UFD 2), p is an associate of one of the elements a_i, b_j and so p must divide at least one of a, b.

Suppose conversely that (UFD 1) and (UFD 2*) hold. To establish (UFD 2) we argue by induction on r, the number of irreducible factors in the first decomposition of (UFD 2). The result clearly holds for $r = 1$ by the definition of irreducible element. Suppose, therefore, by way of using the second principle of induction (Theorem 8.15), that the result holds when the first decomposition has less than r factors ($r > 1$) and let $a = \prod_{i=1}^{r} p_i = \prod_{j=1}^{t} q_j$. The irreducible element p_1 divides $q_1 \prod_{j=2}^{t} q_j$ and so, by (UFD 2*), divides q_1 or $\prod_{j=2}^{t} q_j$. By repeated applications of this argument we see that p_1 divides at least one of the q_j. Suppose, without loss of generality, that p_1 and q_1 are associates. Then $q_1 = u_1 p_1$ where u_1 is a unit. Since $p_1 \neq 0$ is cancellative we then have $p_2 \ldots p_r = u_1 q_2 \ldots q_t = q_2^* \ldots q_t$ where $q_2^* = u_1 q_2$ is irreducible. We now apply the induction hypothesis and the result follows.

Suppose now that A is a unique factorisation domain and let

$$a = p_1^{\alpha_1} p_2^{\alpha_2} \ldots p_t^{\alpha_t} \quad (p_i \neq p_j, \ \alpha_i \geqslant 1)$$

be a non-zero non-unit element of A decomposed into a product of powers of *distinct* irreducible factors. If $a^* \in A$ divides a then clearly each of the irreducible factors of a^* divides a and so is associated with a certain p_i; and the corresponding exponent appearing in the decomposition of a^* is at most equal to α_i. It follows that all the divisors of a^* are obtained as associates of products of powers of irreducible factors p_i of a with exponents α_i^* satisfying $0 \leqslant \alpha_i^* \leqslant \alpha_i$.

Definition. Let $\{a_1, \ldots, a_n\}$ be a finite subset of the commutative integral domain A. Then $c \in A$ is a **greatest common divisor** of $\{a_1, \ldots, a_n\}$ if

(1) c divides every a_i;
(2) b divides every $a_i \Rightarrow b$ divides c.

Similarly, $d \in A$ is a **least common multiple** of $\{a_1, \ldots, a_n\}$ if

(3) d is a multiple of every a_i;
(4) b is a multiple of every $a_i \Rightarrow b$ is a multiple of d.

We denote by $\mathrm{GCD}\{a_1, \ldots, a_n\}$ and $\mathrm{LCM}\{a_1, \ldots, a_n\}$ respectively the sets of greatest common divisors and least common multiples of $\{a_1, \ldots, a_n\}$.

> *Remark.* It should be noted that greatest common divisors (also called highest common factors) and least common multiples, when they exist, are unique only to within association. Although the terminology is standard, it is misleading since the relation of divisibility on $A\backslash\{0\}$ is reflexive and transitive but not anti-symmetric and so is not an order.

Theorem 17.4 *Let D be a unique factorisation domain. If a, b are non-zero elements of D then* $\mathrm{GCD}\{a, b\} \neq \emptyset$ *and* $\mathrm{LCM}\{a, b\} \neq \emptyset$.

Proof. If one of a, b (say, b) is a unit then clearly $b \in \mathrm{GCD}\{a, b\}$ and $a \in \mathrm{LCM}\{a, b\}$. Suppose then that neither a nor b is a unit. We can write their decompositions in the form

$$a = p_1^{\alpha_1} p_2^{\alpha_2} \ldots p_t^{\alpha_t}, \quad b = p_1^{\beta_1} p_2^{\beta_2} \ldots p_t^{\beta_t} \quad (\alpha_i, \beta_i \geqslant 0)$$

in which the distinct irreducible factors p_i appear in at least one of a, b. [Note that here we allow the exponents to be zero; for example, if $a = p_1^2 p_2 p_3$ and $b = p_1 p_3 p_4^2$ then we can write $a = p_1^2 p_2 p_3 p_4^0$ and $b = p_1 p_2^0 p_3 p_4^2$.] If $\sim$ denotes "is an associate of" then for every common divisor h of a, b we have

$$h \sim p_1^{\lambda_1} p_2^{\lambda_2} \ldots p_t^{\lambda_t}$$

where $0 \leqslant \lambda_i \leqslant \min\{\alpha_i, \beta_i\} = \delta_i$. Clearly, every such divisor divides the common divisor

$$c = p_1^{\delta_1} p_2^{\delta_2} \ldots p_t^{\delta_t}$$

which therefore belongs to $\mathrm{GCD}\{a, b\}$. Similarly, for every common multiple k of a, b we have

$$k \sim p_1^{\mu_1} p_2^{\mu_2} \ldots p_t^{\mu_t}$$

where $\mu_i \geqslant \max\{\alpha_i, \beta_i\} = \gamma_i$ and every such multiple is a multiple of the common multiple

$$d = p_1^{\gamma_1} p_2^{\gamma_2} \dots p_t^{\gamma_t}$$

which therefore belongs to LCM $\{a, b\}$.

Corollary. *If $c \in$ GCD $\{a, b\}$ and $d \in$ LCM $\{a, b\}$ then cd and ab are associates.*

Proof. This is immediate since $\min\{\alpha_i, \beta_i\} + \max\{\alpha_i, \beta_i\} = \alpha_i + \beta_i$.

It is clear that a simple inductive argument will extend the result of Theorem 17.4 to any finite subset of D.

The second important type of integral domain which we shall consider is the following.

Definition. Let A be a commutative unitary ring. An ideal I of A is called **principal** if and only if it is of the form $I = \langle a \rangle = Aa$ for some $a \in A$. By a **principal ideal domain** we mean a commutative integral domain D in which every ideal is principal.

Example 17.3. It follows from Example 15.5 that every ideal of the ring $\mathbf{Z}$ is of the form $\langle n \rangle = n\mathbf{Z}$ for some $n \in \mathbf{Z}$. Hence $\mathbf{Z}$ is a principal ideal domain.

Suppose now that I, J are ideals of a commutative unitary ring A. It is readily seen that the ideal generated by $I \cup J$ (i.e., the smallest ideal to contain both I and J) is given (cf. Exercise 15.8) by

$$I + J = \{x + y;\ x \in I,\ y \in J\}$$

which is called the **sum** of the ideals I and J. In particular, for principal ideals, we have

$$Ax + Ay = \langle x \rangle + \langle y \rangle = \{\lambda x + \mu y;\ \lambda, \mu \in A\}.$$

Theorem 17.5 *Let A be a principal ideal domain and let p be a non-zero element of A. Then the following are equivalent:*

(1) *p is irreducible;*

(2) *p is prime;*

(3) *$\langle p \rangle$ is maximal;*

(4) *$\langle p \rangle$ is prime.*

Proof. (1) $\Rightarrow$ (3): Since p is not a unit we have $\langle p \rangle \subset A$ by the Corollary to Theorem 17.1. Now if $\langle p \rangle \subseteq \langle a \rangle \subset A$ then we have $a \mid p$ and a is not a unit. Since p is irreducible we deduce that a must be an associate of p where $\langle a \rangle = \langle p \rangle$ by Theorem 17.1. Thus the ideal $\langle p \rangle$ is maximal.

(3) ⇒ (2): Since $\langle p\rangle \subset A$ we see that p is not a unit. Suppose that $p \mid ab$ but that $p \nmid a$. Then $\langle a\rangle \nsubseteq \langle p\rangle$ and so $\langle a\rangle + \langle p\rangle$ properly contains $\langle p\rangle$. The maximality of $\langle p\rangle$ then gives $\langle a\rangle + \langle p\rangle = A$ and so there exist $s, t \in A$ such that $sp + ta = 1$. Now let $c \in A$ be such that $pc = ab$. Then we have

$$b = 1b = (sp+ta)b = spb + tab = spb + tpc = p(sb+tc)$$

and consequently $p \mid b$.

(2) ⇒ (1): Let a be a divisor of p with say $ab = p$. By (2) we have either $p \mid a$ or $p \mid b$. If $p \mid a$ then a and p are associates; and if $p \mid b$ with say $pc = b$ then $abc = pc = b$ whence $ac = 1$ and a is a unit. This then shows that p is irreducible.

(2) ⇔ (4): This is immediate from the fact that each is equivalent to

$$ab \in \langle p\rangle \Leftrightarrow a \in \langle p\rangle \text{ or } b \in \langle p\rangle.$$

Theorem 17.6 *Let A be a principal ideal domain. Then every non-empty family $\mathcal{F}$ of ideals of A ordered by set inclusion, has a maximal element.*

Proof. Given any $\langle x_0\rangle \in \mathcal{F}$ there are two possibilities: either $\langle x_0\rangle$ is maximal in $\mathcal{F}$, in which case the proof ends here, or $\langle x_0\rangle$ is not maximal in $\mathcal{F}$, in which case there exists $\langle x_1\rangle \in \mathcal{F}$ such that $\langle x_0\rangle \subset \langle x_1\rangle$. The same two possibilities are open to $\langle x_1\rangle$: either $\langle x_1\rangle$ is maximal, in which case the proof ends here, or $\langle x_1\rangle$ is not maximal, in which case there exists $\langle x_2\rangle \in \mathcal{F}$ such that $\langle x_1\rangle \subset \langle x_2\rangle$. Proceeding in this manner we obtain either an ideal $\langle m\rangle$ which is maximal in $\mathcal{F}$ or else an infinite chain of ideals $\langle x_0\rangle \subset \langle x_1\rangle \subset \langle x_2\rangle \subset \ldots$ such that $\langle x_n\rangle \subset \langle x_{n+1}\rangle$ for every n. By way of obtaining a contradiction, suppose that the latter situation arises and let $I = \bigcup_{n \in \mathbf{N}} \langle x_n\rangle$. Then I is an ideal of A since

(i) if $x, y \in I$ then there exist $i, j \in \mathbf{N}$ such that $x \in \langle x_i\rangle$, $y \in \langle x_j\rangle$ so that $x, y \in \langle x_k\rangle$ where $k = \max\{i, j\}$ whence $x - y \in \langle x_k\rangle \subseteq I$;

(ii) if $x \in I$ and $y \in A$ then there is an index i such that $x \in \langle x_i\rangle$ whence $xy = yx \in \langle x_i\rangle \subseteq I$.

Now since A is a principal ideal domain we must have $I = \langle a\rangle$ for some $a \in A$. By the definition of I there therefore exists $t \in \mathbf{N}$ such that $a \in \langle x_t\rangle$ whence $\langle a\rangle \subseteq \langle x_t\rangle$ and consequently $\langle x_t\rangle = \langle x_{t+1}\rangle = \langle x_{t+2}\rangle = \ldots$, a contradiction.

Theorem 17.7 *Every principal ideal domain is a unique factorisation domain.*

Proof. We show that every principal ideal domain A satisfies (UFD 1) and (UFD 2*); the result then follows by Theorem 17.3. That A satisfies (UFD 2*) follows from Theorem 17.5. To establish (UFD 1), we show first that *if $a \neq 0$ is not a unit then a is divisible by an irreducible element*. In fact, let $\mathscr{F}$ be the set of all proper ideals of A which contain $\langle a \rangle$. Since a is not a unit we have $\langle a \rangle \in \mathscr{F}$, so that $\mathscr{F} \neq \emptyset$. By Theorem 17.6 there exists $p \in A$ such that $\langle p \rangle$ is maximal in $\mathscr{F}$. This implies that $\langle p \rangle$ is a maximal ideal of A; for any proper ideal which contains $\langle p \rangle$ must also contain $\langle a \rangle$ so must belong to $\mathscr{F}$ and hence must coincide with $\langle p \rangle$. Applying Theorem 17.5, we see that p is irreducible; and since $\langle a \rangle \subseteq \langle p \rangle$ we have $p \mid a$. We shall now prove the theorem by showing that if $b \neq 0$ is not a unit then b is a product of irreducible elements [(UFD 1)]. Let Y be the set of all elements $y \neq 0$ in A such that $sy = b$ for some product s of irreducible elements and let $\mathscr{Y} = \{\langle y \rangle; y \in Y\}$. Since, as was shown above, b has an irreducible factor, we have $\mathscr{Y} \neq \emptyset$. By Theorem 17.6, there is a maximal element $\langle u \rangle$ in $\mathscr{Y}$. Let $p_1, \ldots, p_n$ be irreducible elements such that $p_1 \ldots p_n u = b$. If u were not a unit then there would exist an irreducible element q such that $q \mid u$; and if $qv = u$ then $p_1 \ldots p_n qv = b$ whence $\langle v \rangle \in \mathscr{Y}$ with $\langle u \rangle \subset \langle v \rangle$ as v divides, but is not an associate of, u. This contradicts the maximality of $\langle u \rangle$ in $\mathscr{Y}$. It follows that u is indeed a unit and so b is the product of the irreducible elements $p_1 u, p_2, \ldots, p_n$.

It is clear that the notion of the sum of two ideals can be extended to any finite number of ideals. Thus, A being unitary, we have

$$\langle a_1 \rangle + \langle a_2 \rangle + \ldots + \langle a_n \rangle = \{\lambda_1 a_1 + \lambda_2 a_2 + \ldots + \lambda_n a_n;\ \lambda_i \in A\}.$$

Theorem 17.8 *Let $a_1, \ldots, a_n$ be non-zero elements of a principal ideal domain. Then $d \in \mathrm{GCD}\{a_1, \ldots, a_n\}$ if and only if $\langle d \rangle = \langle a_1 \rangle + \ldots + \langle a_n \rangle$.*

Proof. As observed earlier, we have $\mathrm{GCD}\{a_1, \ldots, a_n\} \neq \emptyset$. Since $\langle a_1 \rangle + \ldots + \langle a_n \rangle$ is an ideal of A and since A is a principal ideal domain, there exists $e \in A$ such that $\langle e \rangle = \langle a_1 \rangle + \ldots + \langle a_n \rangle$. This implies that $\langle a_i \rangle \subseteq \langle e \rangle$ and so $e \mid a_i$ for every i. Now if x is any divisor of each of $a_1, \ldots, a_n$ then $\langle a_i \rangle \subseteq \langle x \rangle$ for each i and hence $\langle e \rangle \subseteq \langle x \rangle$, giving $x \mid e$. Thus we have $e \in \mathrm{GCD}\{a_1, \ldots, a_n\}$. Now since $d \in \mathrm{GCD}\{a_1, \ldots, a_n\}$ if and only if d is an associate of e; i.e., if and only if $\langle d \rangle = \langle e \rangle$ by Theorem 17.1, the result follows.

We say that $a_1, \ldots, a_n$ are **relatively prime** if $1 \in \mathrm{GCD}\{a_1, \ldots, a_n\}$. With this terminology we have the following important consequences of the above.

Corollary. **[Bezout]** *If $a_1, \ldots, a_n$ are non-zero elements of a principal ideal domain A then $a_1, \ldots, a_n$ are relatively prime if and only if there exist $x_1, \ldots, x_n \in A$ such that $a_1x_1 + \ldots + a_nx_n = 1$.*

Remark. In the case where $A = \mathbf{Z}$ we have the following important special case of the above: if m, n are non-zero integers then d is a greatest common divisor of m, n if and only if there exist $x, y \in \mathbf{Z}$ such that $mx + ny = d$; in particular, m and n are relatively prime if and only if there exist $x, y \in \mathbf{Z}$ such that $mx + ny = 1$.

We now come to the third important type of integral domain.

Definition. By a **euclidean valuation** on a commutative integral domain D we mean a mapping $\delta: D \backslash \{0\} \to \mathbf{N}$ such that, for all $a, b \in D \backslash \{0\}$,
(1) $b \mid a \Rightarrow \delta(b) \leqslant \delta(a)$;
(2) there exist $q, r \in D$ such that $a = bq + r$ with $r = 0$ or $\delta(r) < \delta(b)$. By a **euclidean domain** we mean a commutative integral domain D together with a euclidean valuation on D.

Example 17.4. $\mathbf{Z}$ together with the mapping $\delta: \mathbf{Z} \backslash \{0\} \to \mathbf{N}$ given by $\delta(n) = |n|$ is a euclidean domain. In fact, if $b \mid a$ then we have $a = bc$ and so $|a| = |bc| = |b||c| \geqslant |b|$. Moreover, if $a, b \neq 0$ then it follows by Theorem 13.5 that there exist $q, r \in \mathbf{Z}$ such that $a = |b|q + r$ with $0 \leqslant r < |b|$. If $r = 0$ then we have $b \mid a$ which we have already considered; and if $r \neq 0$ then $0 < r < b$ and so $|r| < |b|$.

Theorem 17.9 *Every euclidean domain is a principal ideal domain (and hence a unique factorisation domain).*

Proof. Let (D, δ) be a euclidean domain and let I be an ideal of D. Suppose that $I \neq \langle 0 \rangle$ so that there exists $b \neq 0$ in I. Choose b such that $\delta(b)$ is minimal in the set $\{\delta(x); x \in I \backslash \{0\}\}$, noting that such a choice is possible since $\mathbf{N}$ is well-ordered. We claim that $I = \langle b \rangle$. To prove that this is so, let $a \in I$; then there exist $q, r \in D$ such that $a = bq + r$ with either $r = 0$ or $\delta(r) < \delta(b)$. Now since I is an ideal we have $r = a - bq \in I$ and so $\delta(r) < \delta(b)$ is impossible by the choice of b. Hence we must have $r = 0$ and so $a = bq$. It therefore follows that $I = \langle b \rangle$. The statement in parentheses follows from Theorem 17.7.

If (D, δ) is a euclidean domain then every pair of non-zero elements a, b of D admits a greatest divisor which may be written in the form $ax + by$; this follows from Theorems 17.9, 17.7, 17.4 and 17.8. Another proof of the existence of a greatest common divisor (which also provides a convenient method of computing such an element in the case $D = \mathbf{Z}$)

is the following, known as the **euclidean algorithm**. Suppose that (D, δ) is a euclidean domain and that $a, b \in D\backslash\{0\}$. Then there exist $q_1, r_1 \in D$ such that $a=bq_1+r_1$ with either $r_1=0$ or $\delta(r_1)<\delta(b)$. If $r_1\neq 0$ then there exist $q_2, r_2 \in D$ such that $b=r_1q_2+r_2$ with either $r_2=0$ or $\delta(r_2)<\delta(r_1)$. Continuing in this way, we let in general q_{i+1}, r_{i+1} be such that $r_{i-1}=r_iq_{i+1}+r_{i+1}$ with either $r_{i+1}=0$ or $\delta(r_{i+1})<\delta(r_i)$. Since each $\delta(r_i)\in \mathbf{N}$ and since $\delta(r_i)<\delta(r_{i-1})$ it is clear that after a finite number of steps we arrive at some $r_i=0$. Now if $r_1=0$ we have $a=bq$ and clearly $b \in \mathrm{GCD}\{a, b\}$. Suppose then that $r_1\neq 0$. If $d|a$ and $d|b$ we have $d|(a-bq_1)$ and so $d|r_1$; but if $d_1|r_1$ and $d_1|b$ then $d_1|(bq_1+r_1)$ and so $d_1|a$. Thus we see that the set $\mathrm{CD}\{a, b\}$ of common divisors of a and b is the same as the set $\mathrm{CD}\{b, r_1\}$. Similarly, if $r_2\neq 0$ we have $\mathrm{CD}\{b, r_1\}=\mathrm{CD}\{r_1, r_2\}$. Continuing in this way we see that $\mathrm{CD}\{a, b\}=\mathrm{CD}\{r_{t-2}, r_{t-1}\}$ where r_t is the first r_i which is 0. It follows immediately that $\mathrm{GCD}\{a, b\}=\mathrm{GCD}\{r_{t-2}, r_{t-1}\}$. But since $r_t=0$ the equalities

$$r_{t-2}=q_tr_{t-1}+r_t=q_tr_{t-1}$$

show that $r_{t-1}\in \mathrm{GCD}\{r_{t-2}, r_{t-1}\}$ and so we have $r_{t-1}\in \mathrm{GCD}\{a, b\}$.

By way of illustration, let us determine $\mathrm{GCD}\{1561, 721\}$ in $\mathbf{Z}$. We have

$$\begin{aligned} 1561&=721.2+119 \qquad && \mathrm{CD}\{1561, 721\}=\mathrm{CD}\{721, 119\};\\ 721&=119.6+7 && \mathrm{CD}\{721, 119\}=\mathrm{CD}\{119, 7\};\\ 119&=17.7+0 \end{aligned}$$

Since the only units in $\mathbf{Z}$ are 1 and -1 (Example 17.2) it follows that $\mathrm{GCD}\{1561, 721\}=\{7, -7\}$. To illustrate Theorem 17.8, we read backwards in the above computation to obtain

$$7=721-119.6=721-(1561-721.2)6=13.721-6.1561.$$

Since, as we have seen above, $\mathbf{Z}$ is a euclidean domain, it is a principal ideal domain and hence a unique factorisation domain. It follows by Theorems 17.2 and 17.3 (or by Theorem 17.5) that the irreducible elements of $\mathbf{Z}$ and the prime elements of $\mathbf{Z}$ are the same. We therefore deduce the following result.

Theorem 17.10 **[Fundamental Theorem of Arithmetic]** *Every non-zero integer is either a unit (i.e., 1 or -1) or can be written as a product of powers of primes which is unique in that if $p_1^{i_1}p_2^{i_2}\ldots p_n^{i_n}=q_1^{j_1}q_2^{j_2}\ldots q_m^{j_m}$ where each p_i and each q_j is a prime with $p_i\neq p_j$ and $q_i\neq q_j$ when $i\neq j$ then $n=m$ and there is a permutation σ on $\{1, \ldots, n\}$ such that for $k=1, \ldots, n$, $i_k=j_{\sigma(k)}$ and $p_k, q_{\sigma(k)}$ are associates (i.e., $p_k=\pm q_{\sigma(k)}$).*

It is clear that, using the above prime decomposition, we can find GCD$\{a, b\}$ for a, b non-zero integers, just as in the proof of Theorem 17.4. However, it is often very difficult to factorise a given integer (e.g., try 912673) so it is better to use the euclidean algorithm.

Exercises for §17

1. Let p be an integer greater than 1. Prove that the following are equivalent:
(1) p is a prime;
(2) $\mathbf{Z}/p\mathbf{Z}$ is an integral domain;
(3) $\mathbf{Z}/p\mathbf{Z}$ is a field.

2. If p, k are positive integers with p a prime and $1 \leqslant k \leqslant p-1$ prove that p divides the binomial coefficient $\binom{p}{k}$.

[*Hint.* Look at $k!\binom{p}{k}\Big/p\mathbf{Z}$ and use Exercise 1.]

3. If $p \in \mathbf{Z}$ is a prime define $R_p = \left\{\frac{n}{m} \in \mathbf{Q};\, p \nmid m\right\}$. Show that R_p is a unitary ring and that the units of R_p are the elements of the form $\frac{n}{m}$ where neither n nor m is divisible by p. Hence show that every element of R_p can be written uniquely in the form αp^r where α is a unit and $r \in \mathbf{N}$. Define αp^r to have *order r*. Given any positive integer k, prove that the elements of order $\geqslant k$ form an ideal of R_p. Show further that all ideals of R_p are of this form.

4. Let S_n be the subset of $\mathbf{C}$ given by

$$S_n = \{x + y\sqrt{n};\ x, y, n \in \mathbf{Z},\ \sqrt{n} \notin \mathbf{Z}\}.$$

Show that S_n is a subdomain of $\mathbf{C}$ and that

$$x + y\sqrt{n} \mid p + q\sqrt{n} \Rightarrow x^2 - ny^2 \mid p^2 - nq^2.$$

Hence show that $x + y\sqrt{n}$ is a unit of S_n if and only if x, y satisfy **Pell's equation** $x^2 - ny^2 = \pm 1$. Use this result to determine the group of units of S_{-1} (called the **domain of gaussian integers**).

5. Let D be a commutative integral domain and consider the following conditions:

(A) D contains no infinite sequences $(a_i)_{i \geqslant 1}$ with the property that a_{i+1} is a proper factor of a_i for each i;

(B) $\mathrm{GCD}\{a, b\} \neq \emptyset$ for all $a, b \in D \backslash \{0\}$.

Prove that if (A) holds then every non-zero element of D is either a unit or a product of irreducible elements. Prove also that, writing $\mathrm{GCD}\{a, X\}$ for $\bigcup_{x \in X} \mathrm{GCD}\{a, x\}$, condition (B) implies that

(a) $\mathrm{GCD}\{a, \mathrm{GCD}\{b, c\}\} = \mathrm{GCD}\{\mathrm{GCD}\{a, b\}, c\}$;
(b) $\mathrm{GCD}\{ca, cb\} = c.\mathrm{GCD}\{a, b\}$;
(c) if $1 \in \mathrm{GCD}\{a, b\}$ and $1 \in \mathrm{GCD}\{a, c\}$ then $1 \in \mathrm{GCD}\{a, bc\}$.

Deduce from the above that every commutative integral domain which satisfies (A) and (B) is a unique factorisation domain.

6. By a **norm** on a commutative integral domain D we mean a mapping $N: D \to \mathbf{Z}$ such that

(1) $(\forall x \in D) \quad N(x) \geqslant 0$;
(2) $N(x) = 0 \Leftrightarrow x = 0$;
(3) $(\forall x, y \in D) \quad N(xy) = N(x)N(y)$.

If N is a norm on D prove that $N(u) = 1$ for every unit u of D. If, furthermore, every $x \in D$ which satisfies $N(x) = 1$ is a unit, show that every $y \in D$ such that $N(y)$ is a prime is an irreducible element of D.

7. Let $G = \mathrm{Gp}\{g\}$ be a multiplicative cyclic group of order n. Prove that the order of g^m is n/d where d is the (positive) greatest common divisor of m, n.

[*Hint.* Let $n = da$, $m = db$ and observe that a, b are relatively prime. Show that $(g^m)^a = 1$ so that the order of g^m divides a. Show also that if $(g^m)^t = 1$ then a divides t.]

8. In the ring $\mathbf{Z}/m\mathbf{Z}$ of integers modulo $m > 1$ show that the units are the elements $n/m\mathbf{Z}$ where m and n are relatively prime.

9. If $m > 1$ prove that $n/m\mathbf{Z}$ is a generator of the group $\mathbf{Z}/m\mathbf{Z}$ if and only if $n/m\mathbf{Z}$ is a unit in the ring $\mathbf{Z}/m\mathbf{Z}$.

[*Hint.* Observe that $n/m\mathbf{Z}$ is a generator if and only if $1/m\mathbf{Z} \in \mathrm{Gp}\{n/m\mathbf{Z}\}$. Use Exercise 8.]

10. The **Euler φ-function** is defined as follows: for every integer $n \geqslant 1$, $\phi(n)$ is the number of generators of a cyclic group of order n. Observe that, by Exercises 17.8 and 17.9, $\phi(1) = 1$ and, for $n > 1$, $\phi(n)$ is the number of invertible elements in the ring $\mathbf{Z}/n\mathbf{Z}$. Show that if $p > 1$ is prime then $\phi(p) = p - 1$.

[*Hint.* Use Exercise 1.]

Given any integer $n \geqslant 1$ show that $\sum_{d|n} \phi(d) = n$, the summation being over all positive divisors of n.

[*Hint.* Let G be a cyclic group of order n. By Theorem 14.8, for every positive divisor d of n there is a unique subgroup of order d. Show that G contains exactly $\phi(d)$ elements of order d. Now use the fact that every element has order d for some divisor d of n.]

11. If A is a commutative unitary ring then A is said to

(a) be **noetherian** if and only if every ascending chain of ideals is finite (in that for every chain $I_1 \subseteq I_2 \subseteq I_3 \subseteq \dots$ of ideals of A, there is an integer p such that $I_n = I_p$ for all $n \geqslant p$);

(b) satisfy the **maximum condition** if every non-empty family of ideals has a maximal element.

Prove that the following are equivalent:

(1) A is noetherian;
(2) A satisfies the maximum condition;
(3) every ideal of A is finitely generated (in that for every ideal I there is a finite subset X such that I is the ideal generated by X).

[*Hint.* (2) $\Rightarrow$ (3): For every ideal I consider the collection C of all ideals of I which are finitely generated; $C \neq \emptyset$ since clearly $\langle 0 \rangle \in C$. Let J be a maximal element of C and for every $x \in I$ consider the ideal $J + \langle x \rangle$.]

Deduce that if A is an integral domain then the following are equivalent:

(a) A is a principal ideal domain;
(b) A is noetherian and the sum of any two principal ideals of A is also a principal ideal.

12. If (D, δ) is a euclidean domain and $u \in D$ prove that the following are equivalent:

(1) u is a unit;
(2) $\delta(u) = \delta(1)$;
(3) $(\forall x \in D) \quad \delta(ux) = \delta(x)$.

13. If (D, δ) is a euclidean domain and $f : \mathbf{N} \to \mathbf{N}$ is strictly increasing [in that $m < n \Rightarrow f(m) < f(n)$] show that $(D, f \circ \delta)$ is also euclidean.

14. Let G be the domain of gaussian integers (Exercise 4). Show that $\delta : G \to \mathbf{N}$ described by $\delta(m + in) = m^2 + n^2$ is a euclidean valuation.

[*Hint.* Show that δ is a norm on G (Exercise 6). Given non-zero elements $g, h \in G$, to find $q, r \in G$ such that $g = hq + r$ let $gh^{-1} = \lambda + i\mu$

where $\lambda, \mu \in \mathbf{Q}$. Let $a, b \in \mathbf{N}$ be such that $|a-\lambda| \leqslant \frac{1}{2}$ and $|b-\mu| \leqslant \frac{1}{2}$ and let $\epsilon=\lambda-a$, $\eta=\mu-b$. Show that $q=a+ib$ and $r=h(\epsilon+i\eta)$ satisfy the required equality. Use the norm property of δ to complete the proof.] Prove that in G the ideals $\langle 1+i\rangle$, $\langle 3\rangle$, $\langle 2+i\rangle$, $\langle 7\rangle$ are maximal.

15. Prove that the subset $\{m+n\sqrt{2};\, m, n \in \mathbf{Z}\}$ of $\mathbf{R}$ forms a euclidean domain under the mapping δ given by $\delta(m+n\sqrt{2})=m^2-2n^2$.

16. For every $n \in \mathbf{N}$ let $G_n=\{a+b\sqrt{-n};\, a, b \in \mathbf{Z}\}$. Show that if $n<3$ then (G_n, δ) is a euclidean domain where $\delta(a+b\sqrt{-n})=a^2+nb^2$. Show that 2 and $1 \pm \sqrt{-3}$ are irreducible elements of G_3 and deduce from the equality $2.2=(1+\sqrt{-3})(1-\sqrt{-3})$ that G_3 cannot be a euclidean domain under any valuation. Show also that in G_3 the irreducible element 2 is not prime.

17. Let G_3^* be the set of all numbers of the form $a+b\sqrt{-3}$ where a, b are either both integers or both halves of odd integers. Show that G_3^* can be made into a euclidean domain.

[*Hint.* Proceed as in Exercise 14 but take integer approximations in modulus $\leqslant \frac{1}{4}$.]

18. If D is a principal ideal domain and if $a, b, c \in D$ are such that $a \mid bc$ with a, b relatively prime, prove that $a \mid c$.

19. Let D be a principal ideal domain. If $a, b, c \in D$ show that there exist $x, y \in D$ such that $c=ax+by$ if and only if c is divisible by every greatest common divisor of x, y. If $u, v \in D$ are such that $c=au+bv$, call the pair (u, v) a *solution* of the equation $c=ax+by$. If (u, v) is a solution, prove that every solution (x, y) is of the form $\left(u-\dfrac{nb}{d}, v-\dfrac{na}{d}\right)$ where $n \in D$ and $d \in \mathrm{GCD}\{a, b\}$.

[*Hint.* Observe that $\dfrac{a}{d}$ and $\dfrac{b}{d}$ are relatively prime. If $ax+by=c=au+bv$ then $b(y-v)=a(u-x)$; divide by d and use Exercise 18.]

20. Show that the "integer congruence equation" $ax \equiv b \pmod{n}$ has a solution x if and only if b is divisible by every $d \in \mathrm{GCD}\{a, n\}$. If x_1 is a particular solution, show that every solution is of the form $x_1+\dfrac{mn}{d}$.

[*Hint.* Use Exercise 19.]

21. **[Chinese Remainder Theorem]** Let A be a commutative unitary ring

and let $I_1, \ldots, I_k$ be ideals of A such that

$$(*) \qquad (j=1, \ldots, k) \quad A = I_j + \bigcap_{t \neq j} I_t.$$

Define $a \equiv b (\text{mod } I_j) \Leftrightarrow a - b \in I_j$. Prove that the system of congruences

$$(j=1, \ldots, k) \quad x \equiv r_j \ (\text{mod } I_j)$$

has a solution.

[*Hint.* Proceed by induction. In the case $k=2$ we have $A = I_1 + I_2$ so that $r_1 = r_{11} + r_{12}$ and $r_2 = r_{21} + r_{22}$ where $r_{11}, r_{21} \in I_1$ and $r_{12}, r_{22} \in I_2$. Show that $x = r_{12} + r_{21}$ is a solution of the pair of congruences. Now assume the result holds for systems of $n-1$ congruences so that we know a solution x^* of any system $(j=1, \ldots, n-1)$ $x \equiv r_j \ (\text{mod } I_j)$. We are looking for a solution to the system

$$\begin{cases} (j=1, \ldots, n-1) & x \equiv x^* \ (\text{mod } I_j) \\ & x \equiv r_n \ (\text{mod } I_n). \end{cases}$$

Observe that this is equivalent to the system

$$\begin{cases} x \equiv x^* \left(\text{mod } \bigcap_{j=1}^{n-1} I_j \right) \\ x \equiv r_n \ (\text{mod } I_n) \end{cases}$$

and use the result for $k=2$ whilst remembering (*).]

Deduce that if D is a principal ideal domain and $a_1, \ldots, a_k$ are elements of D such that a_i is relatively prime to a_j whenever $i \neq j$ then the system of congruences

$$(j=1, \ldots, k) \quad x \equiv r_j \ (\text{mod } a_j)$$

has a solution.

[*Hint.* For each i, a_i is relatively prime to $\prod_{j \neq i} a_j$.]

§18. Field of quotients of a commutative integral domain; Q; characteristic of a ring; ordered integral domains

We shall now turn our attention towards a definition of the field **Q** of rational numbers. Intuitively, we consider this as the set of objects of the form a/b where $a, b \in \mathbf{Z}$ with $b \neq 0$, endowed with the laws of composition $(a/b)(c/d) = ac/bd$ and $a/b + c/d = (ad + bc)/bd$. In this set we identify the objects a/b and ma/mb for every $m \neq 0$ and likewise the objects $a/1$ and a. The latter identification means that **Z** becomes a subdomain of **Q**, whence the number system **Q** is an extension of **Z**.

In order to construct **Q** in a rigorous algebraic manner, we have to be quite explicit in our aims. Basically, the situation is as follows. In the integral domain **Z** equations of the form $bx=a$ $(b\neq 0)$ do not in general have a solution (but when they do, a solution is unique) and what we are seeking is an enriched number system which contains an isomorphic copy of **Z** and in which every equation of the above type has a unique solution. In other words, the structure we are seeking is precisely a field which is an extension of **Z**. However, in order to be as economic as possible we wish this field to be as small as possible; for example, our intuitive notions of **R** and **C** show that they are also fields containing a copy of **Z**. Our aim, therefore, is to establish the existence and uniqueness (up to isomorphism) of a smallest field F which contains an isomorphic copy of **Z**. Now it transpires that the process we shall use to obtain this field is useful in connection with integral domains other than **Z**, so we shall formulate the theory in terms of an arbitrary (non-trivial) commutative integral domain D. The reader may at this stage imagine that the procedure for constructing groups of quotients as given in §13 might be applicable to obtain such a field F. The discussion which follows shows that this is precisely the case.

If F is a field and X is a subset of F then it is clear that the intersection of all the subfields of F which contain X is also a subfield of F which contains X. It is called the **subfield generated by** X.

Example 18.1. If F is a field then the subfield generated by $\{1_F\}$ is $\{m1_F/n1_F;\ m, n\in \mathbf{Z}, n1_F\neq 0\}$. It is clear that this is a subfield of F containing 1_F. Moreover, every subfield of F which contains 1_F contains every element of the form $m1_F/n1_F$ where $m, n\in\mathbf{Z}$ with $n1_F\neq 0$. Thus it is the smallest subfield of F containing 1_F.

Definition. Let D be a (non-trivial) commutative integral domain. By a **field of quotients** of D we shall mean a field F together with a ring monomorphism $f: D\to F$ such that, for every division ring X and every ring monomorphism $g: D\to X$, there is a unique ring monomorphism $h: F\to X$ such that the diagram

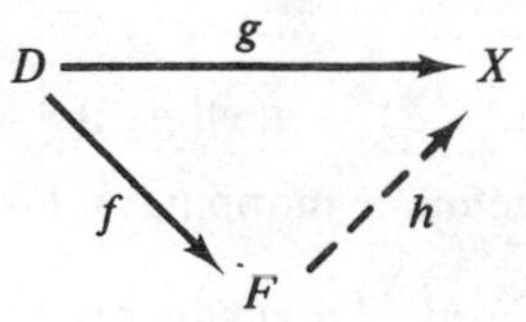

is commutative.

Theorem 18.1 *If D is a commutative integral domain then there exists, to within isomorphism, a unique field of quotients (F, f) of D. Moreover, F is generated by* Im *f.*

Proof. Let $D^* = D\backslash\{0\}$ and define the binary relation R on $D \times D^*$ by

$$(x_1, y_1) \equiv (x_2, y_2)(R) \Leftrightarrow x_1y_2 = x_2y_1.$$

Just as in the proof of Theorem 13.4, it can be shown that R is an equivalence relation on $D \times D^*$. Let $F = (D \times D^*)/R$ and denote the equivalence class $(x, y)/R$ by $\frac{x}{y}$. If we define laws of composition on $D \times D^*$ by

$$(x_1, y_1) + (x_2, y_2) = (x_1y_2 + x_2y_1, y_1y_2), \quad (x_1, x_2)(y_1, y_2) = (x_1x_2, y_1y_2)$$

(both of which are commutative and associative) then it is readily seen that R is compatible with each of these laws and so F is endowed with the induced laws

$$\frac{x_1}{y_1} + \frac{x_2}{y_2} = \frac{x_1y_2 + x_2y_1}{y_1y_2}, \quad \frac{x_1}{y_1} \cdot \frac{x_2}{y_2} = \frac{x_1x_2}{y_1y_2}.$$

It is now readily verified that F thus forms a commutative unitary ring with identity element $\frac{x}{x}$ for every $x \in D^*$ and zero element $\frac{0}{y}$ for every $y \in D^*$. To show that F is a field, let $\frac{x}{y}$ be any non-zero element of F. Then $x \neq 0$ and so $\frac{y}{x}$ is also a non-zero element of F. Since F is commutative and since $\frac{x}{y} \cdot \frac{y}{x} = \frac{xy}{xy}$ = the identity element, it follows that F is a field.

Now define $f: D \to F$ by the prescription $(\forall x \in D)\ f(x) = \frac{x}{1}$. It is clear that f is a ring morphism. Moreover, f is injective and so is a monomorphism. We shall show that (F, f) is a field of quotients of D. For this purpose, consider an arbitrary monomorphism $g: D \to X$ of D into any division ring X. Precisely as in the proof of Theorem 13.4 it can be shown that, in $D \times D^*$,

$$(x, y) \equiv (x^*, y^*)(R) \Rightarrow g(x)[g(y)]^{-1} = g(x^*)[g(y^*)]^{-1}.$$

Consequently we can define a mapping $h: F \to X$ by setting

$$h\left(\frac{x}{y}\right) = g(x)[g(y)]^{-1}.$$

This mapping h is a monomorphism and is unique with respect to the property $h \circ f = g$. The proof of this is exactly as in Theorem 13.4. As the rest of the proof is similar, we leave it to the reader to supply the details.

The above result shows that every (non-trivial) commutative integral domain D has a field of quotients and that all such fields are isomorphic. We can therefore identify all fields of quotients of D and thus talk of *the* field of quotients of D, taking as a model of this the field F constructed in the above proof. If (F, f) is the field of quotients of D then since f is a monomorphism it is common practice to identify D with the subdomain $\operatorname{Im} f$ of F. In particular, in the case where $D = \mathbf{Z}$ the field of quotients of $\mathbf{Z}$ is written $\mathbf{Q}$ and is called the **field of rational numbers**. The identification of $\mathbf{Z}$ as a subdomain of $\mathbf{Q}$ is then obtained by identifying n with $\frac{n}{1}$.

We shall show later in this section how the total order of $\mathbf{Z}$ can be extended to a unique total order on $\mathbf{Q}$.

If R is a unitary ring then the subring $\mathrm{Rg}\{1\}$ of R generated by the identity element is given by $\{n1\,;\, n \in \mathbf{Z}\}$ where, for every $m \in \mathbf{Z}$ with $m \geqslant 0$, $m1 = 1 + 1 + \ldots + 1$ (m terms). The mapping $(\bullet 1): \mathbf{Z} \to \mathrm{Rg}\{1\}$ given by $n \mapsto n1$ is then a ring epimorphism. Since $\mathbf{Z}$ is a principal ideal domain and since $\operatorname{Ker}(\bullet 1)$ is an ideal of $\mathbf{Z}$ we then have $\operatorname{Ker}(\bullet 1) = p\mathbf{Z}$ for some $p \in \mathbf{Z}$.

Definition. If R is a unitary ring then the **characteristic** of R is the natural number n such that $n\mathbf{Z} = \operatorname{Ker}(\bullet 1)$.

Theorem 18.2 *If R is a (non-trivial) commutative integral domain then the characteristic of R is either* 0 *(in which case* $\mathrm{Rg}\{1\}$ *is isomorphic to* $\mathbf{Z}$*) or a prime p (in which case* $\mathrm{Rg}\{1\}$ *is isomorphic to* $\mathbf{Z}/p\mathbf{Z}$*).*

Proof. Let the characteristic of R be n. Since $(\bullet 1): \mathbf{Z} \to \mathrm{Rg}\{1\}$ is a ring epimorphism we have $\mathbf{Z}/n\mathbf{Z} = \mathbf{Z}/\operatorname{Ker}(\bullet 1) \simeq \operatorname{Im}(\bullet 1) = \mathrm{Rg}\{1\}$. If $n = 0$ then clearly $\mathrm{Rg}\{1\} \simeq \mathbf{Z}$. Suppose then that $n \neq 0$. Since R has no zero divisors neither does $\mathrm{Rg}\{1\}$ and so neither does $\mathbf{Z}/n\mathbf{Z}$ which is then a non-trivial integral domain. By Theorem 16.4 the ideal $n\mathbf{Z}$ is a prime ideal of $\mathbf{Z}$ and, by Theorem 17.5, n is a prime.

Theorem 18.3 *A (non-trivial) commutative integral domain R is of characteristic* 0 *if and only if every non-zero element of R is of infinite order in the additive group R; and is of (prime) characteristic p if and only if every non-zero element is of order p in the additive group R.*

Proof. For every non-zero $x \in R$ we have $nx = n(1x) = (n1)x$ and so, since R has no zero divisors, $nx = 0$ if and only if $n1 = 0$. It follows that the order of every non-zero element of R is the order of the identity element 1. Now the order of 1 is infinite if and only if Rg{1} is infinite and hence, by Theorem 18.2, isomorphic to **Z**, in which case the characteristic of R is 0; and the order of 1 is finite if and only if Rg{1} is finite and hence, by Theorem 18.2, isomorphic to $\mathbf{Z}/p\mathbf{Z}$, in which case the characteristic of R (and the order of every non-zero element of R) is the prime p.

If X is a division ring then the **centre** of X is given by

$$Z(X) = \{x \in X;\ (\forall y \in X)\ xy = yx\}.$$

$Z(X)$ is a division subring of X; in fact, it is clearly a subring of X, contains 1 and contains x^{-1} whenever it contains x (since from $xy = yx$ we obtain $yx^{-1} = x^{-1}xyx^{-1} = x^{-1}yxx^{-1} = x^{-1}y$). Now $Z(X)$ is clearly commutative and so we see that $Z(X)$ is a subfield of X. Now let P be the intersection of all the division subrings of X. It is clear that P is a division subring of X and, being contained in the intersection of all subfields of X (since every subfield is a division subring), is contained in the subfield $Z(X)$. Thus P is also commutative and so is also a subfield of X. In summary, P is the intersection of all the subfields of X and so is the smallest subfield of X. We call P the **prime subfield** of X. It is readily seen that P is none other than the subfield generated by $\{1\}$, namely (Example 18.1)

$$P = \{m1/n1\,;\ m, n \in \mathbf{Z},\ n1 \neq 0\}.$$

Theorem 18.4 *Let X be a division ring. If the characteristic of X is zero then the prime subfield of X is isomorphic to the field* **Q**; *and if the characteristic of X is the prime p then the prime subfield of X is isomorphic to the field* $\mathbf{Z}/p\mathbf{Z}$.

Proof. Suppose that X is of characteristic zero. Then, by Theorem 18.3, $n1 = 0$ if and only if $n = 0$, and so the mapping $g: \mathbf{Z} \to X$ given by $g(n) = n1$ is a monomorphism. Since **Q** is the field of quotients of **Z** there exists a unique monomorphism $h: \mathbf{Q} \to X$ such that the following diagram is commutative, where ι denotes the canonical embedding of **Z** in **Q**.

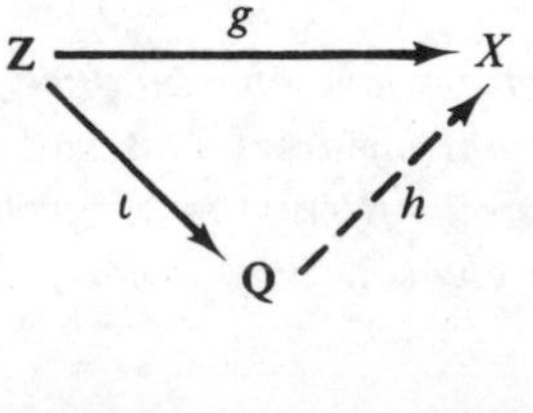

From the proof of Theorem 18.1 the monomorphism h is given by

$$h\left(\frac{m}{n}\right)=g(m)[g(n)]^{-1}=m1/n1.$$

We thus see that the prime subfield of X is isomorphic to $\mathbf{Q}$.

Suppose now that X is of (prime) characteristic p. Then by Theorem 18.2 the subring generated by $\{1\}$ is isomorphic to the ring $\mathbf{Z}/p\mathbf{Z}$. But p is a prime and so it follows by Theorems 17.5 and 16.4 that $\mathbf{Z}/p\mathbf{Z}$ is a field.

Corollary. *Every field of characteristic zero contains a copy of* $\mathbf{Q}$.

We shall now proceed to show how the usual order on $\mathbf{Z}$ can be extended to a unique total order on $\mathbf{Q}$. This arises as a particular case of a more general situation. By way of preparation for this, we require the following results.

Recall from §13 the notions of an *ordered group* and the *positive cone* of an ordered group. If $(G, \top)$ is an ordered group then for every non-empty subset X of G we write $X^{-1}=\{x^{-1}; x \in X\}$. We also write, for non-empty subsets X, Y of G, $X \top Y=\{x \top y; x \in X, y \in Y\}$.

Theorem 18.5 *A non-empty subset P of a group $(G, \top)$ is the positive cone relative to some order on G if and only if*

(1) $P \cap P^{-1}=\{e_G\}$;

(2) $P \top P \subseteq P$;

(3) $(\forall x \in G)\quad x \top P \top x^{-1} \subseteq P$.

Furthermore, G is totally ordered with respect to P if and only if, in addition,

(4) $P \cup P^{-1}=G$.

Proof. Suppose that $(G, \top, \leqslant)$ is an ordered group and that $G_+=\{x \in G; x \geqslant e_G\}$ is its positive cone. Given $x \in G_+$ we have

$$e_G=x^{-1} \top x \geqslant x^{-1} \top e_G=x^{-1}$$

and so we see that $G_+ \cap G_+^{-1}=\{e_G\}$. If now $z \in G_+ \top G_+$ then $z=x \top y$ where $x, y \in G_+$ so that $z \geqslant e_G \top e_G=e_G$ and hence $z \in G_+$. Thus we see that $G_+ \top G_+ \subseteq G_+$. Also, if x is any element of G we have

$$z \in x \top G_+ \top x^{-1} \Rightarrow (\exists y \in G_+)\ z=x \top y \top x^{-1} \geqslant x \top e_G \top x^{-1}=e_G \Rightarrow z \in G_+$$

and so $x \top G_+ \top x^{-1} \subseteq G_+$. Consequently the positive cone of G satisfies (1), (2) and (3).

Conversely, suppose that G has a non-empty subset P which satisfies the properties (1), (2), (3). Define a relation $\leqslant$ on G by

$$a \leqslant b \Leftrightarrow b \top a^{-1} \in P.$$

For every $a \in G$ we have $a \top a^{-1} = e_G \in P$ by (1) and so $\leqslant$ is reflexive on G. That $\leqslant$ is antisymmetric follows from the fact that, by (1),

$$\left.\begin{array}{l} a \leqslant b \Rightarrow b \top a^{-1} \in P \\ b \leqslant a \Rightarrow a \top b^{-1} \in P \end{array}\right\} \Rightarrow b \top a^{-1} = (a \top b^{-1})^{-1} \in P \cap P^{-1} = \{e_G\} \Rightarrow b = a.$$

That $\leqslant$ is transitive follows from (2):

$$\left.\begin{array}{l} a \leqslant b \Rightarrow b \top a^{-1} \in P \\ b \leqslant c \Rightarrow c \top b^{-1} \in P \end{array}\right\} \Rightarrow c \top a^{-1} = c \top b^{-1} \top b \top a^{-1} \in P \top P \subseteq P \Rightarrow a \leqslant c.$$

Thus $\leqslant$ is an order on G. To show that G is an ordered group we still have to show that if $a \leqslant b$ then $(\forall c \in G)$ $a \top c \leqslant b \top c$ and $c \top a \leqslant c \top b$. Now we have

$$\begin{aligned} a \leqslant b \Rightarrow b \top a^{-1} \in P &\Rightarrow (\forall c \in G) \quad b \top c \top (a \top c)^{-1} = b \top a^{-1} \in P \\ &\Rightarrow (\forall c \in G) \quad a \top c \leqslant b \top c, \end{aligned}$$

and the other inequality is established similarly using (3). We note also that P is the positive cone of $(G, \top, \leqslant)$; for

$$e_G \leqslant a \Leftrightarrow a = a \top e_G = a \top e_G^{-1} \in P.$$

Suppose now that (4) holds. Then for all $x, y \in G$ we have $x \top y^{-1} \in P$ or $x \top y^{-1} \in P^{-1}$ so that $x \top y^{-1} \geqslant e_G$ or $x \top y^{-1} \leqslant e_G$ whence $x \geqslant y$ or $x \leqslant y$ and consequently G is totally ordered. Conversely, if G is totally ordered then for every $x \in G$ we have either $x \geqslant e_G$ or $x \leqslant e_G$ so that $x \in P$ or $x \in P^{-1}$ whence we see that $G = P \cup P^{-1}$.

Definition. By an **ordered ring** we mean a ring $(R, +, \cdot)$ which is endowed with an order $\leqslant$ such that $(R, +, \leqslant)$ is an ordered group and its positive cone R_+ is a subsemigroup of $(R, \cdot)$.

Remark. Note that by saying that R_+ is a subsemigroup of $(R, \cdot)$ we mean that $(\forall x, y \in R)$ $x, y \geqslant 0 \Rightarrow xy \geqslant 0$.

Theorem 18.6 *A non-empty subset P of an ordered ring R is the positive cone relative to some order on R if and only if*

(1) $P \cap (-P) = \{0\}$;

(2) $P + P \subseteq P$;

(3) $PP \subseteq P$.

Furthermore, R is totally ordered with respect to P if and only if, in addition,

(4) $P \cup (-P) = R$.

Proof. Suppose that P is the positive cone relative to some order $\leqslant$ on R. Then by Theorem 18.5 the properties (1) and (2) are satisfied. As for (3), this is clearly equivalent to the property $(\forall x, y \in R)\ x, y \geqslant 0 \Rightarrow xy \geqslant 0$.

Conversely, let $P \neq \emptyset$ satisfy (1), (2) and (3). Then, as in the proof of Theorem 18.5, (1) and (2) show that the relation $\leqslant$ defined on R by $a \leqslant b \Leftrightarrow b - a \in P$ is an order and that $(R, +, \leqslant)$ is an ordered group with positive cone P. As the property (3) is equivalent to $(\forall x, y \in P)\ xy \in P$ we see that R is an ordered ring. Finally, as in Theorem 18.5, (4) is equivalent to saying that this order is a total order.

Example 18.2. The ring $\mathbf{Z}$ of integers is an ordered ring which is in fact totally ordered. The order in question is the usual order, namely that given by $m \leqslant n \Leftrightarrow n - m \geqslant 0$ and indeed this is the *only* total order possible on the ring $\mathbf{Z}$. For, suppose that some total order is defined by the positive cone P. Then since $\mathbf{Z} = P \cup (-P)$ we have either $1 \in P$ or $1 \in -P$, the latter being equivalent to $-1 \in P$. Now we cannot have both $1 \in P$ and $1 \in -P$ since $P \cap (-P) = \{0\}$; and since P is stable under multiplication we cannot have $-1 \in P$, for $(-1)(-1) = 1$ which would imply that also $1 \in P$. Hence we must have $1 \in P$. Since P is stable under addition, it now follows that $\mathbf{Z}_+ \subseteq P$ whence $-\mathbf{Z}_+ \subseteq -P$. Now we observe that $\{\{0\}, \mathbf{Z}_+\backslash\{0\}, -\mathbf{Z}_+\backslash\{0\}\}$ and $\{\{0\}, P\backslash\{0\}, -P\backslash\{0\}\}$ are each partitions of $\mathbf{Z}$ from which we conclude that $\mathbf{Z}_+ = P$ and so only one total order is possible.

Theorem 18.7 *Let $(D, \leqslant)$ be a totally ordered commutative integral domain and let F be its field of quotients. Then there is one and only one total order $\preccurlyeq$ on F which extends that of D (in that, regarding D as a subset of F, $a \preccurlyeq b \Leftrightarrow a \leqslant b$ whenever $a, b \in D$).*

Proof. Consider the subset F_+ of F given by

$$F_+ = \{x/y;\ x \in D_+,\ y \in D_+\backslash\{0\}\}.$$

We show that F_+ is the positive cone of a total order on F. Since D_+ is stable under both addition and multiplication in R, the laws of composition in F show that F_+ is stable under addition and multiplication in F so that F_+ satisfies (2) and (3) of Theorem 18.6. As for (1), let

$a \in F_+ \cap (-F_+)$. Then there exist $x, p \in D_+$ and $y, q \in D_+\backslash\{0\}$ such that $a = \frac{x}{y} = -\frac{p}{q}$ whence $0 \leqslant xq = -yp \leqslant 0$, giving $xq = 0$; and since D is an integral domain we deduce that $x = 0$, whence $a = 0$. To establish (4), we note that if $\frac{x}{y} \in F$ then there is no loss in generality if we assume that $y \in D_+\backslash\{0\}$. This being the case, if $x \in D_+$ then we have $a \in F_+$; and if $x \notin D_+$ then, as D is totally ordered, $-x \in D_+$ so that $-a = \frac{-x}{y} \in F_+$ whence $a = -(-a) \in F_+$. We conclude that $F_+ \cup (-F_+) = F$.

Now it is readily seen that $D \cap F_+ = D_+$. Hence the total order on F defined by F_+ is an extension of that defined on D by D_+.

Finally, let $\preccurlyeq$ be any order on F with respect to which F forms a totally ordered field in which $\preccurlyeq$ extends the total order $\leqslant$ on D. Let $F_\oplus = \{x \in F;\ 0 \preccurlyeq x\}$ be the positive cone of F relative to $\preccurlyeq$. We show that $F_\oplus = F_+$ whence the uniqueness follows. Now both $\{\{0\}, F_\oplus\backslash\{0\}, -F_\oplus\backslash\{0\}\}$ and $\{\{0\}, F_+\backslash\{0\}, -F_+\backslash\{0\}\}$ are partitions of F. But if $a \in F_\oplus$, say $a = \frac{x}{y}$ where $x \in D$ and (with no loss of generality) $y \in D_+\backslash\{0\}$, then we have $x = ay$ where $a \succcurlyeq 0$ and, since $D_+ = D \cap F_\oplus$, $y \succcurlyeq 0$. Since $(F, \preccurlyeq)$ is an ordered ring we obtain $x = ay \succcurlyeq 0$ and so $x \geqslant 0$. Consequently we have $a = \frac{x}{y} \in F_+$. This shows that $F_\oplus \subseteq F_+$ and consequently we have $F_\oplus\backslash\{0\} \subseteq F_+\backslash\{0\}$ and $-F_\oplus\backslash\{0\} \subseteq -F_+\backslash\{0\}$. We conclude from the above partitions that $F_\oplus = F_+$.

As we know, **Z** is a totally ordered commutative integral domain. It follows from the above result that there is a unique total order on **Q** which extends that of **Z**. We shall show presently that every totally ordered commutative integral domain contains an isomorphic copy of the totally ordered commutative integral domain **Z** and that every totally ordered field contains an isomorphic copy of the totally ordered field **Q**. Before proceeding with this, we prove some elementary facts concerning ordered rings (which are often used without explicit reference).

Theorem 18.8 *Let R be an ordered ring. Then for all $x, y \in R$,*

(1) $x \leqslant y \Rightarrow (\forall z \geqslant 0)\quad xz \leqslant yz$ *and* $zx \leqslant zy$;

(2) $x \leqslant y \Rightarrow (\forall z \leqslant 0)\quad xz \geqslant yz$ *and* $zx \geqslant zy$.

Moreover, if R is totally ordered,

(3) $(\forall x \in R) \quad x^2 \geqslant 0$;

(4) *if* $x \in R$ *is invertible then* $x>0 \Leftrightarrow x^{-1}>0$.

Proof. (1) If $x \leqslant y$ then $y-x \in R_+$ so for all $z \in R_+$ we have $zy-zx=z(y-x) \in R_+R_+ \subseteq R_+$ and likewise $yz-xz \in R_+$.

(2) If $x \leqslant y$ then $y-x \in R_+$ so for all $z \leqslant 0$ we have $-z \in R_+$ and consequently $-zy+zx=-z(y-x) \in R_+R_+ \subseteq R_+$ and likewise $-yz+xz \in R_+$.

(3) Given any $x \in R$ we have $x \geqslant 0$ or $x \leqslant 0$. If $x \geqslant 0$ then $x \in R_+$ and so $x^2 \in R_+R_+ \subseteq R_+$; and if $x \leqslant 0$ then $-x \in R_+$ whence again $x^2=(-x)(-x) \in R_+R_+ \subseteq R_+$. Thus in all cases $x^2 \in R_+$.

(4) If x is invertible then we have $x=x^2x^{-1}$. Thus if $x^{-1} \in R_+$ we have $x \in R_+$ by (3). Conversely, $x^{-1}=(x^{-1})^2x$ and so if $x \in R_+$ we have $x^{-1} \in R_+$ again by (3).

Theorem 18.9 *Every totally ordered commutative integral domain D is of characteristic zero.*

Proof. Given any $x \in D$ we have $x^2 \geqslant 0$ by Theroem 18.3 (3). It follows that $x^2>0$ for all $x \in D \backslash \{0\}$ since D has no zero divisors. In particular, $1=1^2>0$. Since $D_+ + D_+ \subseteq D_+$ it follows by induction that $n1>0$ for all positive integers n. Thus we see that the identity element of D is of infinite order. It follows from the proof of Theorem 18.3 that every non-zero element of D is of infinite order and D is of characteristic zero.

Our aim now is to show that every totally ordered commutative integral domain contains a copy of **Z** and that every totally ordered field contains a copy of **Q**. For this purpose, we require the following elementary result.

Theorem 18.10 *In any totally ordered field X we have, for n, q of the same parity,*

$$\frac{m}{n} \leqslant \frac{p}{q} \Leftrightarrow mq \leqslant np.$$

Proof. Choosing $n>0$ and $q>0$ without loss of generality, we have $nq \in X_+X_+ \subseteq X_+$ and so $nq>0$ (equality being excluded since X has no zero divisors). Thus if $\frac{m}{n} \leqslant \frac{p}{q}$ we have, by Theorem 18.8(1), $mq=nq\left(\frac{m}{n}\right) \leqslant nq\left(\frac{p}{q}\right)=np$. Conversely, suppose that $mq \leqslant np$. Then since $nq>0$ we have $\frac{1}{nq}>0$ by Theorem 18.8(4), so that $\frac{m}{n}=\frac{1}{nq}(mq)\leqslant \frac{1}{nq}(np)=\frac{p}{q}$.

Theorem 18.11 *If D is a totally ordered commutative integral domain then there is an order-preserving monomorphism $f:\mathbf{Z}\to D$. If F is a totally ordered field then there is an order-preserving monomorphism $f:\mathbf{Q}\to F$.*

Proof. By Theorem 18.9, D is of characteristic zero and so the mapping $(\bullet 1):\mathbf{Z}\to D$ is a ring monomorphism. Since

$$m\leqslant n \Leftrightarrow n-m\geqslant 0 \Leftrightarrow n1-m1=(n-m)1\geqslant 0 \Leftrightarrow m1\leqslant n1$$

it follows that $(\bullet 1)$ is also order-preserving.

If F is a totally ordered field then the monomorphism $(\bullet 1):\mathbf{Z}\to F$ extends to a unique monomorphism $h:\mathbf{Q}\to F$ since $\mathbf{Q}$ is the field of quotients of $\mathbf{Z}$. The monomorphism h is given by $h\left(\frac{m}{n}\right)=\frac{m1}{n1}$ and we have to show that it is order-preserving. Applying Theorem 18.10 to both $\mathbf{Q}$ and F we obtain, n and q being greater than 0,

$$\begin{aligned}\frac{m1}{n1}\leqslant\frac{p1}{q1} &\Leftrightarrow (mq)1=(m1)(q1)\leqslant(n1)(p1)=(np)1\\ &\Leftrightarrow mq\leqslant np\\ &\Leftrightarrow \frac{m}{n}\leqslant\frac{p}{q}\end{aligned}$$

from which the result follows.

We end this section with another surprise:

Theorem 18.12 $\mathbf{Q}$ *is denumerable.*

Proof. Our model for $\mathbf{Q}$ as constructed in Theorem 18.1 is given by $(\mathbf{Z}\times\mathbf{Z}\backslash\{0\})/R$ where R is the equivalence relation

$$(x_1, y_1)\equiv(x_2, y_2)(R) \Leftrightarrow x_1y_2=x_2y_1.$$

Since, by Theorem 13.8, $\mathbf{Z}$ is denumerable it follows by Theorem 10.2 that so also is $\mathbf{Z}\times\mathbf{Z}\backslash\{0\}$. The diagram

$$\mathbf{Z}\xrightarrow{\ i\ }\mathbf{Q}=(\mathbf{Z}\times\mathbf{Z}\backslash\{0\})/R\ \underset{\iota}{\overset{\natural_R}{\leftrightarrows}}\ \mathbf{Z}\times\mathbf{Z}\backslash\{0\}$$

in which i is the canonical embedding of $\mathbf{Z}$ into $\mathbf{Q}$ and ι is an injection associated with the canonical surjection $\natural_R$ (Theorem 5.6), together with the Schröder–Bernstein Theorem now shows that $\mathbf{Q}$ is denumerable.

Exercises for §18

1. Show that $E=\{m+n\sqrt{2};\ m, n \in \mathbf{Z}\}$ is a subdomain of $\mathbf{R}$ and determine its field of quotients.

2. In Exercise 17.14 determine the characteristic of the field $G/\langle 1+i\rangle$. Determine similarly the characteristics of the other quotient fields.

3. Show that for every positive integer n there is a ring of characteristic n.

4. If D is a commutative integral domain of prime characteristic p prove that $f: D \to D$ given by $f(x)=x^p$ is a ring monomorphism.

[*Hint.* Use Exercises 15.3 and 17.2.]

5. Given any positive rational p/q prove that there exist positive integers $a_1, \ldots, a_n, \alpha_1, \ldots, \alpha_n$ and distinct primes $p_1, \ldots, p_n$ such that

$$p/q=a_1/p_1^{\alpha_1}+\ldots+a_n/p_n^{\alpha_n}.$$

[*Hint.* Consider the prime factorisation of q and use induction.]

6. Prove that if $q \in \mathbf{Q}$ then there is a unique $n \in \mathbf{Z}$ such that $n \leqslant q < n+1$.

[*Hint.* Use euclidean division.]

7. Prove that there is no $x \in \mathbf{Q}$ such that $x^2=2$.

[*Hint.* Use Exercise 1.4.]

8. Show that $\sup\{x \in \mathbf{Q};\ 0 < x^2 < 2\}$ does not exist in $\mathbf{Q}$.

[*Hint.* Let $t \in \mathbf{Q}$ be the supremum and consider $t-\dfrac{t^2-2}{2t}$.]

§19. Archimedean, Cauchy complete and Dedekind complete ordered fields; R

Definition. By an (additive abelian) **archimedean ordered group** we shall mean a totally ordered group $(G, +, \leqslant)$ such that, whenever $0 < a < b$ in G, there exists a positive integer n such that $na > b$; and by an

archimedean ordered field we shall mean a totally ordered field $(F, +, \cdot, \leqslant)$ such that $(F, +, \leqslant)$ is an archimedean ordered group.

Example 19.1. By Theorem 8.18, **Z** is an archimedean ordered group.

Example 19.2. **Q**, together with the unique total order inherited from **Z** as in Theorem 18.8, is an archimedean ordered field. In fact, if we have $0 < \frac{m}{n} < \frac{p}{q}$ in **Q** with $n, q > 0$ then

$$(n^2p+1)\frac{m}{n} > n^2p\frac{m}{n} = npm \geqslant np \geqslant \frac{np}{nq} = \frac{p}{q}.$$

Since every totally ordered field contains a copy of **Q** as its prime subfield, we see by Example 19.2 that **Q** is, to within isomorphism, the smallest archimedean ordered field. In this section we shall show the existence of a greatest archimedean ordered field; i.e., one which contains an isomorphic copy of every archimedean ordered field. We begin by characterising an archimedean ordered field in terms of its prime subfield.

Definition. A non-empty subset S of a totally ordered field F is said to be **dense** in F if, whenever $a<b$ in F, there exists $x \in S$ such that $a<x<b$.

Theorem 19.1 *A totally ordered field F is archimedean ordered if and only if its prime subfield is dense in F.*

Proof. Since the prime subfield of F is isomorphic to **Q** we shall identify it with **Q** for convenience. Suppose then that **Q** is dense in F and let $a, b \in F$ be such that $0<a<b$. By the density there exist integers $m, n>0$ such that $0<\frac{m}{n}<\frac{a}{b}$. Consequently

$$na=(bn)\frac{a}{b}>(bn)\frac{m}{n}=bm>b$$

and so F is archimedean ordered.

Conversely, let F be archimedean ordered and let $a, b \in F$ be such that $a<b$. There are several cases to consider:

(1) $a<0<b$: in this case $0 \in$ **Q** satisfies the requirements.

(2) $0<a<b$: in this case $b-a>0$ and so $\frac{1}{b-a}>0$ whence there exists

$n>0$ such that $n1>\dfrac{1}{b-a}$ and hence $\dfrac{1}{n}<b-a$. Since F is archimedean ordered and since **N** is well-ordered, there is a smallest $m \in \mathbf{N}$ such that $b \leqslant m\left(\dfrac{1}{n}\right)=\dfrac{m}{n}$. We then have $b>\dfrac{m-1}{n}=\dfrac{m}{n}-\dfrac{1}{n}>b-(b-a)=a$ and so $\dfrac{m-1}{n}$ is the required element of **Q**.

(3) $a<b<0$: in this case we have $0<-b<-a$ and so by case (2) there exists $x \in \mathbf{Q}$ with $-b<x<-a$ whence $a<-x<b$.

Definition. If F is a totally ordered field and $a \in F$ then we define the **modulus** of a by

$$|a|=\begin{cases} a & \text{if } a \geqslant 0; \\ -a & \text{if } a \leqslant 0. \end{cases}$$

In other words, $|a|=\max\{a, -a\}$.

Theorem 19.2 *Let F be a totally ordered field. Then for all $a, b \in F$,*

(1) $|a|=0 \Leftrightarrow a=0$;

(2) $a \leqslant |a|$;

(3) $|a|=|-a|$;

(4) $|ab|=|a|\,|b|$;

(5) $|a+b| \leqslant |a|+|b|$;

(6) $||a|-|b|| \leqslant |a-b|$;

(7) $|a|<b \Leftrightarrow -b<a<b$;

(8) *if* $a \neq 0$ *then* $|a^{-1}|=|a|^{-1}$.

Proof. (1), (2), (3), (4) are obvious from the definition. To prove (5) we note that if $a+b \geqslant 0$ then $|a+b|=a+b \leqslant |a|+|b|$; and if $a+b \leqslant 0$ then $|a+b|=-(a+b)=(-a)+(-b) \leqslant |-a|+|-b|=|a|+|b|$. As for (6), we have, from (5), $|a|=|a-b+b| \leqslant |a-b|+|b|$ and so $|a|-|b| \leqslant |a-b|$ whence (6) follows. Property (7) follows immediately from (2) and (3); and finally (8) follows from (4) on taking $b=a^{-1}$.

We recall now that if X is any non-empty set then a **sequence of elements of X indexed by N** is a mapping $a:\mathbf{N} \to X$.

Definition. Let F be a totally ordered field. We say that a sequence a of elements of F is

(1) a **bounded sequence** if there exists $b \in F$ such that $|a(n)| \leqslant b$ for every $n \in \mathbf{N}$;

(2) a **Cauchy sequence** if, for every $\epsilon>0$ in F, there is a positive integer a_ϵ such that $|a(p)-a(q)|<\epsilon$ for all $p,q\geqslant a_\epsilon$;

(3) a **null sequence** if, for every $\epsilon>0$ in F, there is a positive integer a_ϵ such that $|a(p)|<\epsilon$ for all $p\geqslant a_\epsilon$.

Theorem 19.3 *If $B(F)$, $C(F)$ and $N(F)$ denote respectively the sets of all bounded, Cauchy and null sequences in a totally ordered field F then $N(F)\subset C(F)\subset B(F)$.*

Proof. If $a\in N(F)$ then for any $\epsilon>0$ in F we have, for $p,q\geqslant a_{\epsilon/2}$,

$$|a(p)-a(q)|\leqslant|a(p)|+|-a(q)|=|a(p)|+|a(q)|<\frac{\epsilon}{2}+\frac{\epsilon}{2}=\epsilon$$

whence $a\in C(F)$ and hence $N(F)\subseteq C(F)$. That $N(F)\subset C(F)$ follows from the fact that the sequence a such that $a(n)=1$ for every $n\in\mathbf{N}$ belongs to $C(F)$ but not to $N(F)$.

If $a\in C(F)$ then given in particular $1>0$ we have $|a(p)-a(q)|<1$ for all $p,q\geqslant a_1$. Let $b=1+\max\{|a(1)|,\ldots,|a(a_1)|\}$; then $|a(p)|\leqslant b$ for $p\leqslant a_1$ and for $p>a_1$ we have

$$|a(p)|=|a(p)-a(a_1)+a(a_1)|\leqslant|a(p)-a(a_1)|+|a(a_1)|<1+|a(a_1)|\leqslant b.$$

Hence we have $a\in B(F)$ and so $C(F)\subseteq B(F)$. That $C(F)\subset B(F)$ follows from the fact that $a(n)=(-1)^n$ defines a sequence in $B(F)$ which is not in $C(F)$.

Our next result is of fundamental importance.

Theorem 19.4 *The set $C(F)$ of all Cauchy sequences in a totally ordered field F forms a commutative unitary ring under the laws of composition given by $(a+b)(n)=a(n)+b(n)$ and $(ab)(n)=a(n)b(n)$. In this ring the set $N(F)$ of all null sequences is a proper ideal and the quotient ring $C(F)/N(F)$ is a field.*

Proof. If $a,b\in C(F)$ then for any given $\epsilon>0$ in F there exist positive integers a_ϵ, b_ϵ such that $|a(p)-a(q)|<\epsilon$ and $|b(p)-b(q)|<\epsilon$ for all $p,q\geqslant\max\{a_\epsilon,b_\epsilon\}$. Thus if $p,q\geqslant\max\{a_{\epsilon/2},b_{\epsilon/2}\}$ we have

$$|a(p)+b(p)-[a(q)+b(q)]|\leqslant|a(p)-a(q)|+|b(p)-b(q)|<\frac{\epsilon}{2}+\frac{\epsilon}{2}=\epsilon.$$

Hence $a+b\in C(F)$. Now since $C(F)\subset B(F)$ by Theorem 19.3, there exists $d\in F$ such that $|a(n)|,|b(n)|<d$ for all $n\in\mathbf{N}$. If $p,q\geqslant$

$\max\{a_{\epsilon/2d}, b_{\epsilon/2d}\}$ we have

$$\begin{aligned}|a(p)b(p)-a(q)b(q)| &= |a(p)[b(p)-b(q)]+[a(p)-a(q)]b(q)| \\ &\leqslant |a(p)|\,|b(p)-b(q)|+|a(p)-a(q)|\,|b(q)| \\ &< d(\epsilon/2d)+(\epsilon/2d)d=\epsilon.\end{aligned}$$

Hence $ab \in C(F)$. It is readily seen that $C(F)$, equipped with the above laws of composition, forms a commutative unitary ring, the identity element being the sequence a such that $a(n)=1$ for every $n \in \mathbf{N}$.

To show that $N(F)$ is a proper ideal of $C(F)$ we note first that if $a, b \in N(F)$ then $a-b \in N(F)$. Suppose now that $a \in N(F)$ and $b \in C(F)$. Since b is bounded there exists $d \in F$ such that $|b(n)| \leqslant d$ for every $n \in \mathbf{N}$. Given any $\epsilon > 0$ in F, for $p \geqslant a_{\epsilon/d}$ we have

$$|a(p)b(p)| = |a(p)|\,|b(p)| \leqslant |a(p)|\,d < \frac{\epsilon}{d}d = \epsilon$$

and consequently $ab \in N(F)$. Thus we see that $N(F)$ is an ideal of $C(F)$. That it is a proper ideal is clear from the fact that the constant sequence $\mathbf{1}$ all of whose terms are 1 belongs to $C(F)$ but not to $N(F)$.

To show that $C(F)/N(F)$ is a field we have to show that for every $a \in C(F)$ with $a \notin N(F)$ there exists $x \in C(F)$ such that $ax/N(F) = \mathbf{1}/N(F)$. Now to say that a is a null sequence we mean that

$$(\forall \epsilon > 0)(\exists a_\epsilon > 0)(\forall n \geqslant a_\epsilon) \qquad |a(n)| < \epsilon.$$

But by hypothesis a is not a null sequence. We therefore have the existence of an $\epsilon > 0$ such that, for every positive integer t, there is some integer $m \geqslant t$ for which $|a(m)| \geqslant \epsilon$. Let $m \geqslant t = a_{\epsilon/2}$ be such that $|a(m)| \geqslant \epsilon$. Then for all $p \geqslant a_{\epsilon/2}$ we have

$$\epsilon \leqslant |a(m)| = |a(m)-a(p)+a(p)| \leqslant |a(m)-a(p)|+|a(p)| < \frac{\epsilon}{2}+|a(p)|$$

and hence $|a(p)| > \dfrac{\epsilon}{2}$. This shows that $a(p) \neq 0$. Now define the sequence x by

$$x(n) = \begin{cases} 1/a(n) & \text{if } n \geqslant a_{\epsilon/2}; \\ 1 & \text{otherwise.} \end{cases}$$

To show that $x \in C(F)$ we observe that, for any given $\delta > 0$ and all $p, q \geqslant \max\{a_{\epsilon/2}, a_{\epsilon^2\delta/4}\}$,

$$|x(p)-x(q)| = \left|\frac{1}{a(p)} - \frac{1}{a(q)}\right| = \left|\frac{a(q)-a(p)}{a(p)a(q)}\right| < \frac{\epsilon^2\delta/4}{(\epsilon/2)^2} = \delta,$$

and so $x \in C(F)$. Finally, we have

$$a(n)x(n)-1=\begin{cases} 0 & \text{if } n \geqslant a_{\epsilon/2}; \\ a(n)-1 & \text{otherwise,} \end{cases}$$

which shows that $ax-\mathbf{1} \in N(F)$ and thus completes the proof.

Corollary. *The ideal* N(F) *is a maximal ideal of* C(F).

Proof. This is immediate from Theorem 16.4.

Definition. A sequence a of elements of a totally ordered field F is said to be a **positive sequence** if

$$(\exists \epsilon_a \in F_+\backslash\{0\})(\exists n_a \in \mathbf{N})(\forall n \geqslant n_a) \qquad a(n) > \epsilon_a.$$

Theorem 19.5 *If F is a totally ordered field then $C(F)/N(F)$ can be given the structure of a totally ordered field which contains an isomorphic copy of F.*

Proof. Suppose first that $a, b \in C(F)$ are such that $b-a$ is a positive sequence; i.e.,

$$(*) \quad (\exists \epsilon_{a,b} \in F_+\backslash\{0\})(\exists n_{a,b} \in \mathbf{N})(\forall n \geqslant n_{a,b}) \quad b(n)-a(n) > \epsilon_{a,b}.$$

If $a^* \in C(F)$ is such that $a^*/N(F)=a/N(F)$ then $a^*-a \in N(F)$ and so

$$(\exists n_{a,a^*} \in \mathbf{N})(\forall n \geqslant n_{a,a^*}) \quad |a^*(n)-a(n)| < \epsilon_{a,b}/4.$$

Likewise, if $b^* \in C(F)$ is such that $b^*/N(F)=b/N(F)$ then $b-b^* \in N(F)$ and

$$(\exists n_{b,b^*} \in \mathbf{N})(\forall n \geqslant n_{b,b^*}) \quad |b(n)-b^*(n)| < \epsilon_{a,b}/4.$$

Let $n^*=\max\{n_{a,b}, n_{a,b^*}, n_{b,b^*}\}$; then for all $n \geqslant n^*$ we have

$$\begin{aligned} b^*(n)-a^*(n) &= b^*(n)-b(n)+b(n)-a(n)+a(n)-a^*(n) \\ &\geqslant -|b^*(n)-b(n)|+b(n)-a(n)-|a(n)-a^*(n)| \\ &> -\epsilon_{a,b}/4+\epsilon_{a,b}-\epsilon_{a,b}/4 \\ &= \epsilon_{a,b}/2. \end{aligned}$$

We thus see that a^*, b^* also satisfy (*) and so b^*-a^* is a positive sequence. It follows from this that we can define a binary relation $\prec$ on $C(F)/N(F)$ by writing $A \prec B$ if and only if there exist $a \in A$ and $b \in B$ such that $b-a$ is a positive sequence.

We shall now show that $\prec$ is transitive. Suppose that $A \prec B$ and $B \prec C$; then there exist $a \in A$, $b \in B$, $c \in C$ such that a, b and b, c

satisfy (*). Let $\epsilon=\epsilon_{a,b}+\epsilon_{b,c}$ and let $n^*=\max\{n_{a,b}, n_{b,c}\}$; then for all $n\geqslant n^*$ we have

$$a(n)-c(n)=a(n)-b(n)+b(n)-c(n)>\epsilon_{a,b}+\epsilon_{b,c}=\epsilon$$

which shows that a, c also satisfy (*) and hence that $A\prec C$.

To show that the set $C(F)/N(F)$ is totally ordered under the order $\preccurlyeq$ it clearly suffices to show that for any A, B precisely one of the statements $A\prec B$, $A=B$, $B\prec A$ holds. Clearly no two of these can hold simultaneously and so it will suffice to show that if $A\prec B$ and $B\prec A$ are each false then we must have $A=B$. Now given $a\in A$ and $b\in B$ we have, since each is a Cauchy sequence,

$$(\forall\epsilon\in F_+\backslash\{0\})(\exists n^*\in\mathbf{N})(\forall m, n\geqslant n^*)\quad |a(m)-a(n)|<\frac{\epsilon}{3}, |b(m)-b(n)|<\frac{\epsilon}{3}.$$

Since $B\prec A$ is false we obtain, from the negation of (*), the existence of an integer $p\geqslant n^*$ such that $a(p)-b(p)\leqslant\frac{\epsilon}{3}$. It follows that, for all $n\geqslant n^*$.

$$\begin{aligned}a(n)-b(n)&=a(n)-a(p)+a(p)-b(p)+b(p)-b(n)\\&\leqslant|a(n)-a(p)|+a(p)-b(p)+|b(p)-b(n)|\\&<\frac{\epsilon}{3}+\frac{\epsilon}{3}+\frac{\epsilon}{3}=\epsilon.\end{aligned}$$

Since $A\prec B$ is also false we can deduce similarly that $b(n)-a(n)<\epsilon$ for all $n\geqslant n^*$. We deduce that $|a(n)-b(n)|<\epsilon$ for all $n\geqslant n^*$ whence we see that $a-b\in N(F)$ and consequently $A=B$.

It follows readily from the property (*) that if $A\preccurlyeq B$ then for any C we have $A+C\preccurlyeq B+C$; and for any D with $0\preccurlyeq D$ we have $AD\preccurlyeq BD$. We thus see that $C(F)/N(F)$ is a totally ordered field. That it contains a copy of F follows by observing that if, to every $x\in F$, we associate the constant sequence $\mathbf{x}$ (every element of which is x) then the assignment $x\mapsto\mathbf{x}/N(F)$ is a monomorphism with

$$x<y\Leftrightarrow y-x>0\Leftrightarrow\mathbf{x}/N(F)\prec\mathbf{y}/N(F).$$

Definition. If F is a totally ordered field then we say that an element $t\in F$ is a **limit** of a sequence a of elements of F if

$$(\forall\epsilon\in F_+\backslash\{0\})(\exists m\in\mathbf{N})(\forall n\geqslant m)\quad |a(n)-t|<\epsilon.$$

We say that a sequence a of elements of F is **convergent** if it has a limit in F.

Theorem 19.6 *Let a be a convergent sequence in a totally ordered field F. Then a has a unique limit and is a Cauchy sequence.*

Proof. Suppose that t_1 and t_2 are each limits of a. We show that $t_1 < t_2$ is impossible. Suppose in fact that $t_1 < t_2$; then, taking $\epsilon = (t_2 - t_1)/2$ there exist $m, p \in \mathbf{N}$ such that $(\forall n \geqslant m)\ |a(n) - t_1| < (t_2 - t_1)/2$ and $(\forall n \geqslant p)\ |a(n) - t_2| < (t_2 - t_1)/2$. If $q = \max\{m, p\}$ we then have the contradiction

$$t_2 - t_1 = t_2 - a(n) + a(n) - t_1 < |a(n) - t_2| + |a(n) - t_1| < t_2 - t_1.$$

Similarly we see that $t_2 < t_1$ is impossible. Hence $t_2 = t_1$ and so the sequence has a unique limit which we shall denote by t.

Now by definition there exists $m \in \mathbf{N}$ such that, for all $n \geqslant m$, $|a(n) - t| < \epsilon/2$. If $n, p \geqslant m$ we then have

$$|a(n) - a(p)| = |a(n) - t + t - a(p)| \leqslant |a(n) - t| + |a(p) - t| < \frac{\epsilon}{2} + \frac{\epsilon}{2} = \epsilon$$

and so a is indeed a Cauchy sequence.

> *Remark.* The unique limit of a convergent sequence a is often written $\lim_{n \to \infty} a(n)$. We shall write simply $\lim a$.

We have just seen that every convergent sequence is a Cauchy sequence. The converse of this is not in general true (see Exercise 19.4). However, ordered fields in which the converse does hold are of especial importance as we shall see.

Definition. A totally ordered field F is said to be **Cauchy complete** if every Cauchy sequence in F is convergent.

An important example of a Cauchy complete ordered field which is also archimedean ordered is given in the following result (in which the order referred to is that given in the proof of Theorem 19.5).

Theorem 19.7 *The totally ordered field $C(\mathbf{Q})/N(\mathbf{Q})$ is Cauchy complete and archimedean ordered.*

Proof. For convenience we write $C(\mathbf{Q})/N(\mathbf{Q})$ as R and the elements of R in the form $[a]$ where $a \in C(\mathbf{Q})$. Moreover, $\mathbf{q}$ will denote the constant sequence each term of which is the rational q.

We show first that R is archimedean. For this purpose, we shall show that the prime subfield of R is dense in R and appeal to Theorem 19.1. Now the prime subfield of R consists of the elements $[\mathbf{q}]$ where $q \in \mathbf{Q}$. Our aim, therefore, is to show that for any $A, B \in R$ with $A \prec B$ there exists $q \in \mathbf{Q}$ such that $A \prec [\mathbf{q}] \prec B$. Now given $a \in A$ and $b \in B$ we have, since $b-a$ is a positive sequence,

$$(\exists \epsilon_{a,\,b} \in \mathbf{Q}_+\backslash\{0\})(\exists n_{a,\,b} \in \mathbf{N})(\forall n \geqslant n_{a,\,b}) \quad |b(n)-a(n)| > \epsilon_{a,\,b}.$$

Since a and b are Cauchy sequences we also have

$$(\exists n^*_{a,\,b} \in \mathbf{N})(\forall m, n \geqslant n^*_{a,\,b}) \quad |a(m)-a(n)|, |b(m)-b(n)| < \epsilon_{a,\,b}/3.$$

Let $n^* = \max\{n_{a,\,b}, n^*_{a,\,b}\}$ and let $p = a(n^*+1)+b(n^*+1) \in \mathbf{Q}$. Then for all $n \geqslant n^*$,

$$\begin{cases} \dfrac{p}{2}-a(n)=\dfrac{1}{2}[b(n^*+1)-a(n^*+1)]+a(n^*+1)-a(n) \\ \qquad\qquad\qquad\qquad\qquad\qquad > \epsilon_{a,\,b}/2-\epsilon_{a,\,b}/3=\epsilon_{a,\,b}/6; \\ b(n)-\dfrac{p}{2}=\dfrac{1}{2}[b(n^*+1)-a(n^*+1)]+b(n)-b(n^*+1) \\ \qquad\qquad\qquad\qquad\qquad\qquad > \epsilon_{a,\,b}/2-\epsilon_{a,\,b}/3=\epsilon_{a,\,b}/6 \end{cases}$$

and consequently, taking $q=p/2$, we have $A \prec [\mathbf{q}] \prec B$.

Let us now show that R is Cauchy complete. For this purpose, we show first that if $a \in C(\mathbf{Q})$ then $\lim [\mathbf{a(n)}] = [a]$. Now given any $[\delta] \succ [0]$ there exists $\epsilon \in \mathbf{Q}$ such that $[0] \prec [\epsilon] \prec [\delta]$ as we have just seen. Since a is a Cauchy sequence we have

$$(\forall m, n \geq a_{\epsilon/2}) \qquad |a(m)-a(n)| < \epsilon/2.$$

Let $n \geq a_{\epsilon/2}$ be fixed. Then for $p \geq a_{\epsilon/2}$ we have

$$a(p)-a(n) < \epsilon/2 \qquad \text{and} \qquad a(n)-a(p) < \epsilon/2$$

so that

$$\epsilon-(a(p)-a(n)) > \epsilon/2 \qquad \text{and} \qquad \epsilon-(a(n)-a(p)) > \epsilon/2.$$

It follows that, in R,

$$[a]-[\mathbf{a(n)}] \prec [\epsilon] \prec [\delta],$$
$$[\mathbf{a(n)}]-[a] \prec [\epsilon] \prec [\delta].$$

Consequently,

$$(\forall n \geq a_{\epsilon/2}) \qquad |[a]-[\mathbf{a(n)}]| \prec [\delta]$$

and so $\lim[\mathbf{a(n)}] = [a]$.

Suppose now that A^* is a Cauchy sequence in R, so that the elements of A^* are of the form $A^*(m) = [A(m)]$ where $A(m) \in C(\mathbf{Q})$ for each m. As there is nothing to prove if A^* is a constant sequence, we can assume without loss of generality that all the terms of A^* are distinct. By the above, for every p we can choose $a(p) \in \mathbf{Q}$ such that

$$|[\mathbf{a(p)}]-[A(p)]| < |[A(p+1)]-[A(p)]|.$$

The inequality

$$|[\mathbf{a(p)}]-[\mathbf{a(q)}]| \leqslant |[\mathbf{a(p)}]-[A(p)]| + |[A(p)]-[A(q)]| + |[A(q)]-[\mathbf{a(q)}]|$$

and the order-preserving isomorphism $[\mathbf{x}]\mapsto x$ then show that the sequence given by $p\mapsto a(p)$ belongs to $C(\mathbf{Q})$. We complete the proof by showing that $[a]=\lim\ [A(n)]$. This follows immediately from the fact that $[a]=\lim\ [\mathbf{a(n)}]$ and the inequality

$$|[A(n)]-[a]| \leqslant |[A(n)]-[\mathbf{a(n)}]|+|[\mathbf{a(n)}]-[a]|.$$

We shall now introduce what appears to be (but, as we shall see, is not) a different type of ordered field. For this purpose we ask the reader to recall that the *supremum* of a subset of an ordered set is, when it exists, the least upper bound of the subset.

Definition. A totally ordered field is said to be **Dedekind complete** if every non-empty subset which is bounded above has a supremum.

We shall soon see that there is, to within isomorphism, only one Dedekind complete ordered field and only one Cauchy complete archimedean ordered field; moreover, they are isomorphic.

As an example of a totally ordered field which is not Dedekind complete, we mention the field **Q** (see Exercise 18.8).

Definition. By a **universally archimedean field** we shall mean a totally ordered field R which is archimedean and such that, if X is any archimedean ordered field, there is a unique monomorphism $h: X\to R$ such that the diagram

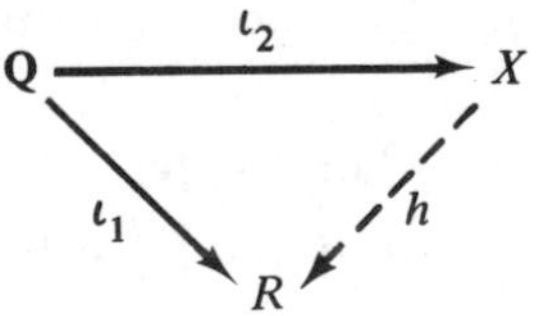

is commutative where ι_1 and ι_2 are the canonical embeddings.

We now prove the main result of this section.

Theorem 19.8 *There exists, to within isomorphism, a unique universally archimedean field. Moreover, the following conditions on a totally ordered field F are equivalent:*

(1) *F is a universally archimedean field;*

(2) *F is Dedekind complete;*

(3) *F is archimedean and Cauchy complete.*

Proof. We shall show that (3) $\Rightarrow$ (2), (2) $\Rightarrow$ (1) and (1) $\Rightarrow$ (3), the first sentence being established in (1) $\Rightarrow$ (3).

(3) $\Rightarrow$ (2): If the non-empty subset A of F is bounded above in F let $\bar{A}$ denote its set of upper bounds. For every positive integer p define

$$E_p = \{k \in \mathbf{N};\ 2^{-p}k \in \bar{A}\}.$$

That $E_p \neq \emptyset$ follows from the fact that if $\bar{a} \in \bar{A} \cap \mathbf{N}$ then $2^p\bar{a} \in E_p$. Now given any $a \in A$ we see that if $k \leqslant 2^p(a-1)$ then $2^{-p}k \leqslant a-1 < a$ and so $k \notin E_p$. Thus E_p is bounded below and hence has a smallest element $k(p)$. Define a sequence in F by writing $a(p) = 2^{-p}k(p)$ for each p. Then we have $a(p) \in \bar{A}$ and $a(p) - 2^{-p} = 2^{-p}[k(p)-1] \notin \bar{A}$; i.e., we have $2^{-(p+1)}[2k(p)] \in \bar{A}$ and $2^{-(p+1)}[2k(p)-2] \notin \bar{A}$. It follows from this that either (i) $k(p+1) = 2k(p)$ or (ii) $k(p+1) = 2k(p)-1$. Consequently

$$a(p+1) = 2^{-(p+1)}k(p+1) = \begin{cases} 2^{-p}k(p) = a(p) & \text{in case (i);} \\ 2^{-(p+1)}[2k(p)-1] = a(p) - 2^{-(p+1)} & \\ & \text{in case (ii).} \end{cases}$$

Thus the sequence a is such that, for every p, $0 \leqslant a(p) - a(p+1) \leqslant 2^{-(p+1)}$. We leave to the reader (see Exercise 19.3) the task of showing that a is a Cauchy sequence in F. By the standing hypothesis that F is Cauchy complete, it then follows that $\lim a = \alpha$ exists in F. We prove that $\alpha = \sup A$.

We note first that $\alpha \leqslant a(p)$ for every p. [In fact, if there existed p such that $\alpha > a(p)$ then $\alpha - a(p+1) > a(p) - a(p+1) \geqslant 0$ whence, by induction, $\alpha > a(t) \geqslant a(t+1)$ for all $t \geqslant p$; and this contradicts the fact that $\alpha = \lim a$.] We also note that $\alpha \in \bar{A}$. [In fact, if $\alpha \notin \bar{A}$ then there exists $x \in A$ such that $x > \alpha$ and so, for some p, $a(p) - \alpha = |a(p) - \alpha| < x - \alpha$ whence $a(p) < x$ and this contradicts the fact that $a(p)$ belongs to $\bar{A}$.] Suppose now that $\beta \in \bar{A}$ with $\beta < \alpha$. Since F is archimedean ordered we can choose p such that $2^{-p} < \alpha - \beta$ whence $a(p) - 2^{-p} \geqslant \alpha - 2^{-p} > \beta$ and consequently $a(p) - 2^{-p} \in \bar{A}$. But we have shown above that $a(p) - 2^{-p} \notin \bar{A}$. We conclude, therefore, that every $\beta \in \bar{A}$ is such that $\beta > \alpha$ and hence that $\alpha = \sup A$.

(2) $\Rightarrow$ (1): Suppose now that F is Dedekind complete. If $0 < a < b$ in F let $E = \{na;\ n \in \mathbf{N}\backslash\{0\}\}$. If now we had $na \leqslant b$ for all $n \in \mathbf{N}$ then b would be an upper bound for E and so, by hypothesis, E would have a supremum α, say. Since $0 < a$ we have $\alpha - a < \alpha$ whence there exists $n > 0$ such that $\alpha - a < na$ [for if no such n existed $\alpha - a$ would be an upper bound which is strictly less than the least upper bound α, which is

impossible]. We then have $\alpha=\alpha-a+a<na+a=(n+1)a<\alpha$. This contradiction therefore shows that, for some n, we must have $na>b$. Hence F is archimedean.

Suppose now that X is any archimedean ordered field. For every $x\in X$ define the subset L_x of F by

$$L_x=\{q1_F;\ q\in\mathbf{Q},\ q1_X<x\}.$$

Since X is archimedean there exists $n>0$ such that $n1_X>-x$ whence $(-n)1_X<x$ and so $L_x\neq\emptyset$. Again since X is archimedean, there exists $p>0$ such that $p1_X>x$ and so $p1_F$ is an upper bound for L_x. Since, by hypothesis, F is Dedekind ordered, L_x has a supremum. For every $x\in X$ define also the subset G_x of F by

$$G_x=\{q1_F;\ q\in\mathbf{Q},\ x<q1_X\}.$$

By a similar argument to the above, the set $-G_x$ has a supremum whence G_x clearly has an infimum. We now note that

$$\sup L_x=\inf G_x.$$

In fact, it is clear that every element of G_x is strictly greater than every element of L_x and so every element of G_x is an upper bound for L_x whence $\inf G_x\geqslant\sup L_x$. Now we must have equality for otherwise, F being archimedean, there would exist $t\in\mathbf{Q}$ such that $\sup L_x<t1_F<\inf G_x$ and likewise $r, s\in\mathbf{Q}$ such that $\sup L_x<r1_F<t1_F<s1_F<\inf G_x$. We would then have $r1_X\geqslant x$ and $s1_X\leqslant x$ whence $r\geqslant s$ and $r1_F\geqslant s1_F$, a contradiction.

We now define a mapping $h: X\to F$ by the prescription $h(x)=\sup L_x$. We note that, for every $q\in\mathbf{Q}$,

$$(h\circ\iota_2)(q)=h(q1_X)=\sup L_{q1_X}=q1_F=\iota_1(q)$$

and so $h\circ\iota_2=\iota_1$. We now show that h is a morphism.

Suppose first that $x, y\in X$ are such that $0<x, y$. Then for all positive $p1_F\in L_x$ and all positive $q1_F\in L_y$ we have $pq1_F=p1_Fq1_F<xy$ and so

$$pq1_F=\sup L_{pq1_X}\leqslant\sup L_{xy}=h(xy).$$

It follows that $p1_F\leqslant h(xy)(q1_F)^{-1}$ and so $h(x)\leqslant h(xy)(q1_F)^{-1}$ whence $q1_F\leqslant h(xy)[h(x)]^{-1}$ and so $h(y)\leqslant h(xy)[h(x)]^{-1}$ whence $h(x)h(y)\leqslant h(xy)$. Since we also have $h(x)=\inf G_x$ for every $x\in X$ we can argue similarly to deduce that $h(xy)\leqslant h(x)h(y)$. Thus, for $x, y>0$, $h(xy)=h(x)h(y)$.

A similar argument, using sums instead of products, shows that $h(x+y)=h(x)+h(y)$, irrespective of whether $x, y>0$ or not; and in particular we have $h(-x)=-h(x)$.

Suppose now that one of x, y is 0. Clearly, we have $h(xy)=h(x)h(y)$. In the final case where, say $x<0$ and $y>0$, we have $x=-z$ where $z>0$ and so $h(x)h(y)=h(-z)h(y)=-h(z)h(y)=-h(zy)=-h(-xy)=-[-h(xy)]=h(xy)$. This then completes the proof that h is a morphism.

Now suppose that $x<y$ in X. Since X is archimedean there is an integer n such that $\frac{1}{n}1_X<y-x$ whence $\frac{1}{n}1_F\in L_{y-x}$. We thus have

$h(y)=h(y-x+x)=h(y-x)+h(x)\geqslant\frac{1}{n}1_F+h(x)>h(x)$ from which we deduce that h is injective.

It remains to establish the uniqueness of h. For this purpose, suppose that $g:X\to F$ is also an order-preserving monomorphism such that $g\circ\iota_2=\iota_1$. Suppose that $(\exists x\in X)\; g(x)<h(x)$. Then since X is archimedean there exists $q\in\mathbf{Q}$ such that $g(x)<q1_F<h(x)$. It then follows from the definition of h that $q1_F\in L_x$ and so $q1_X<x$. Hence $q1_F=\iota_1(q)=g[\iota_2(q)]=g(q1_X)<g(x)<q1_F$ and this contradiction shows that $(\forall x\in X)\; g(x)\geqslant h(x)$. Suppose now that $(\exists x\in X)\; g(x)>h(x)$. Again there exists $q\in\mathbf{Q}$ such that $h(x)<q1_F<g(x)$. In this case $q1_F\notin L_x$ and so $q1_X\geqslant x$ whence we deduce similarly the contradiction $q1_F>q1_F$. We conclude from these observations that $(\forall x\in X)\; g(x)=h(x)$ and so $g=h$.

(1) $\Rightarrow$ (3): We note first that, by Theorem 19.7, the ordered field $C(\mathbf{Q})/N(\mathbf{Q})$ is archimedean and Cauchy complete. By (3) $\Rightarrow$ (2) $\Rightarrow$ (1) it is therefore universally archimedean. Now if F is any universally archimedean field then the uniqueness of the order-preserving monomorphisms, h, k in the diagrams

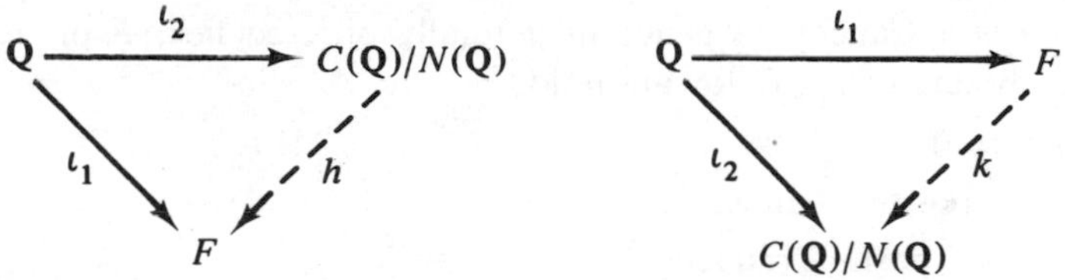

show that $h\circ k:F\to F$ is an order-preserving monomorphism such that $h\circ k\circ\iota_1=\iota_1$. But since F is universally archimedean, the only such map possible is clearly id_F. Hence $h\circ k=id_F$ whence h is also surjective. Thus we see that h is an order-preserving isomorphism whose inverse is the order-preserving isomorphism k. This then shows the existence and, up to isomorphism, the uniqueness of a universally archimedean field.

It also establishes (1) $\Rightarrow$ (3) since then F is isomorphic to $C(\mathbf{Q})/N(\mathbf{Q})$ which satisfies (3) by Theorem 19.7.

Remark. The above result shows that there is an essentially unique universally archimedean field. As a model of this we can choose $C(\mathbf{Q})/N(\mathbf{Q})$. This essentially unique field is called the **field of real numbers** and is denoted by $\mathbf{R}$. The fact that $\mathbf{R}$ is Dedekind complete is of immense importance in analysis. We are in fact now equipped with enough machinery to begin a course of real analysis!

Exercises for §19

1. Prove that every totally ordered field is dense in itself.
2. Prove that for every $r \in \mathbf{R}$ there is a unique $n \in \mathbf{Z}$ such that $n \leqslant r < n+1$.

[*Hint.* Use Exercise 18.6 and the fact that $\mathbf{Q}$ is dense in $\mathbf{R}$.]

3. Prove that if F is a totally ordered field then the sequence b given by $b(n) = 2^{-n}$ is a null sequence in F. Deduce that any sequence a which is such that $0 \leqslant a(n) - a(n+1) \leqslant 2^{-n-1}$ is a Cauchy sequence in F.

4. For every positive integer n let $a(n)$ be the unique integer such that $[a(n)]^2 \leqslant 2.10^{2n} < [a(n)+1]^2$. Show that, for each n,

$$2 - \left(\frac{a(n)}{10^n}\right)^2 < \frac{4}{10^n} + \frac{1}{10^{2n}}$$

and hence that the sequence b given by $b(n) = (a(n)/10^n)^2$ is convergent with $\lim b = 2$. Deduce that the sequence c given by $c(n) = a(n)/10^n$ is a Cauchy sequence in $\mathbf{Q}$ but that it does not converge in $\mathbf{Q}$.

5. If a is a Cauchy sequence in a totally ordered field F prove that precisely one of the following holds:

(1) $\lim a = 0$;

(2) a is a positive sequence;

(3) $-a$ is a positive sequence.

[*Hint.* To show at least one holds, show that if (1) is false then either (2) or (3) is true. To show that no more than one holds, show that if (1) is true then both (2) and (3) are false and that (2) and (3) cannot hold simultaneously.]

6. Let $q \in \mathbf{Q}$ be such that $|q| < 1$. Define a sequence a by $a(n) = \sum_{m=1}^{n} q^m$. Prove that a converges to $1/(1-q)$.

[*Hint*. Observe that $(1-q)a(n)=1-q^n$.]

7. If a is a sequence of real numbers then by the **series** defined by a we mean the sequence Σ_a which is given by $\Sigma_a(n)=\sum_{i=0}^{n} a(i)$. If the sequence Σ_a converges then we say that the sequence a is **summable** and write $\lim \Sigma_a$ as $\sum_{n\in\mathbf{N}} a(n)$. Prove that if the sequences a and b are summable then so also is the sequence $\lambda a+\mu b$ for all $\lambda, \mu \in \mathbf{R}$ and that

$$\sum_{i\in\mathbf{N}} (\lambda a+\mu b)(i)=\lambda \sum_{i\in\mathbf{N}} a(i)+\mu \sum_{i\in\mathbf{N}} b(i).$$

Let a, b be sequences such that $0\leqslant b(n)\leqslant a(n)$ for each n. Prove that if a is summable then so also is b and that $\sum_{i\in\mathbf{N}} b(i)\leqslant \sum_{i\in\mathbf{N}} a(i)$.

8. Let b be an integer with $b>1$. Let a be a sequence of integers such that $0\leqslant a(n)\leqslant b-1$ for each n. Prove that the sequence t given by $t(n)=a(n)/b^n$ is summable with $0\leqslant \sum_{i\in\mathbf{N}} a(i)/b^i\leqslant b$.

9. Let $x\in\mathbf{R}$ be such that $0\leqslant x<1$ and let b be a fixed integer with $b>1$. For every $y\in\mathbf{R}$ let $[y]$ denote the greatest integer which is less than or equal to y (Exercise 19.2). Show that if c is the sequence given by $c(n)=[xb^n]/b^n$ then $\lim c=x$. Deduce that the sequence a given by

$$a(n)=[xb^n]-[xb^{n-1}]b$$

is such that

(1) $(\forall n\in\mathbf{N})\quad 0\leqslant a(n)\leqslant b-1$;

(2) $(\forall n\in\mathbf{N})(\exists m\geqslant n)\quad a(m)<b-1$;

(3) $x=\sum_{i\in\mathbf{N}} a(i)b^{-i}$.

Show further that a is the only sequence satisfying (1), (2) and (3). [The equality (3) is known as the **expansion of x to the base** b. When $b=10$ we obtain the **decimal** expansion of x and when $b=2$ we obtain the **binary** expansion of x.]

10. Use binary expansions (Exercise 9) to prove that the interval $]0, 1[$ of $\mathbf{R}$ is equipotent to $\{0, 1\}^{\mathbf{N}}$. Deduce from Exercises 11.5 and 8.11 that

$$\text{Card } \mathbf{R}=2^{\aleph_0}=\text{Card } \mathbf{P}(\mathbf{N}).$$

Finally, use Theorem 10.10 to show that $\mathbf{R}$ is not denumerable.

11. Prove that if $x\in\mathbf{R}$ is such that $x>0$ and if $n\in\mathbf{Z}$ is such that $n>0$

then there is a unique $y \in \mathbf{R}$ such that $y^n = x$.

[*Hint*. Let $E_x = \{t \in \mathbf{R};\ t^n < x\}$. Show that $\dfrac{x}{1+x} \in E_x$ so that $E_x \neq \emptyset$; and that $1+x$ is an upper bound for E_x so that $\alpha = \sup E_x$ exists. Show as follows that $\alpha^n = x$. Suppose that $\alpha^n < x$ and choose h such that

$$0 < h < \frac{x - \alpha^n}{(\alpha+1)^n - \alpha^n};$$

use the binomial expansion of $(\alpha+h)^n$ to deduce the contradiction $\alpha + h \in E_x$. Suppose now that $x < \alpha^n$ and choose h such that

$$0 < h < \min\left\{1, \alpha, \frac{\alpha^n - x}{(\alpha+1)^n - \alpha^n}\right\};$$

use the binomial expansion of $(\alpha-h)^n$ to deduce the contradiction $(\alpha-h)^n > x$, that $\alpha - h$ is an upper bound of E_x.]

12. Show that it is impossible to give $\mathbf{C}$ the structure of a totally ordered field.

[*Hint*. Consider the consequences of saying that the complex number i belongs to a positive cone of $\mathbf{C}$.]

§20. Polynomials; C

Our aim in this section is to make precise the notion of a polynomial and to show how this can be used to extend the number system to include complex numbers.

Let us begin by supposing that S is a ring, that R is a subring of S and that t is any element of S. Since the intersection of any family of subrings of S is also a subring of S it is clear that the intersection of all the subrings of S which contain $R \cup \{t\}$ is also a subring of S containing $R \cup \{t\}$ and is the smallest subring to do so. We call this the **subring generated by** $R \cup \{t\}$ and denote it by $\mathrm{Rg}(R \cup \{t\})$. If we define, for every $t \in S$,

$$R[t] = \{x \in S;\ (\exists a_0, \ldots, a_n \in R)\ x = a_0 + a_1 t + \ldots + a_n t^n\}$$

then it is clear that we have $R \subseteq R[t] \subseteq \mathrm{Rg}(R \cup \{t\})$. Moreover, these three sets coincide whenever $t \in R$. We shall therefore consider the case where $t \notin R$. If we suppose that R has an identity element 1_R then we have $t = 1_R t \in R[t]$ and so $R \cup \{t\} \subseteq R[t]$. Now it is not in general true that the additive (abelian) subgroup $R[t]$ of S is a subring of S; for, given $t \in R[t]$ and $a \in R \subseteq R[t]$, the product ta may not be of the form $a_0+a_1t+\ldots+a_nt^n$ so that $R[t]$ may not be stable under multiplication. However, if we impose on t the condition that it commutes with every element of R, so that $(\forall a \in R)\ ta=at$, then it is readily seen that $R[t]$ does become stable under multiplication (and hence a subring of S) with products given by

$$(a_0+a_1t+\ldots+a_nt^n)(b_0+b_1t+\ldots+b_mt^m)$$
$$=a_0b_0+(a_1b_0+a_0b_1)t+\ldots+\left(\sum_{i=0}^{r} a_{r-i}b_i\right)t^r+\ldots+a_nb_mt^{n+m}.$$

Under these conditions we thus see that $\text{Rg}(R \cup \{t\})=R[t]$, so that the subring of S generated by $R \cup \{t\}$ is of a particularly interesting form.

The above discussion rests on the fact that we are working in a ring S which is given in advance. The problem we shall be interested in solving now is of a slightly more general nature, namely: given a unitary ring R and an object $t \notin R$ does there exist a smallest unitary ring A which contains $R \cup \{t\}$ and in which t commutes with every element of R? More precisely, given R and t does there exist a unitary ring A such that

(1) A has a subring R^* and an element $t^* \notin R^*$ such that t^* commutes with every element of R^*;
(2) there is a mapping $f: R \cup \{t\} \to A$ with $f(t)=t^*$ such that $f^*: R \to R^*$ given by $f^*(r)=f(r)$ for every $r \in R$ is an isomorphism;
(3) A is generated by $R^* \cup \{t^*\}$?

Before proceeding, let us note that if A, B are unitary rings and if $f: A \to B$ is a ring morphism then, although $f(1_A)$ is always the identity element of the subring $\text{Im} f$ of B [for, given any $x \in A$ we have $f(x)= f(1_Ax)=f(1_A)f(x)$ and $f(x)=f(x1_A)=f(x)f(1_A)$], this need not be the identity element of B itself; for example, the mapping f given by $f(x)=0_B$ for all $x \in A$ is a ring morphism and $0_B \neq 1_B$ if B is non-trivial. We shall therefore consider the following type of ring morphism.

Definition. If A, B are unitary rings and if $f: A \to B$ is a ring morphism then we shall say that f is **1-preserving** if $f(1_A)=1_B$.

We now introduce, in a similar fashion to the notions of group and field of quotients, the concept of a polynomial ring over a given unitary ring and show how this may be used to solve the problem in hand.

Definition. Let R be a (non-trivial) unitary ring. By a **polynomial ring** on R we shall mean a unitary ring P together with a 1-preserving ring monomorphism $f: R \to P$ and an element $t \in P \backslash \operatorname{Im} f$ which commutes with every element of $\operatorname{Im} f$ such that, for every unitary ring X, every 1-preserving ring monomorphism $g: R \to X$ and every $u \in X \backslash \operatorname{Im} g$ which commutes with every element of $\operatorname{Im} g$, there is a unique 1-preserving ring monomorphism $h: P \to X$ such that $h(t) = u$ and the diagram

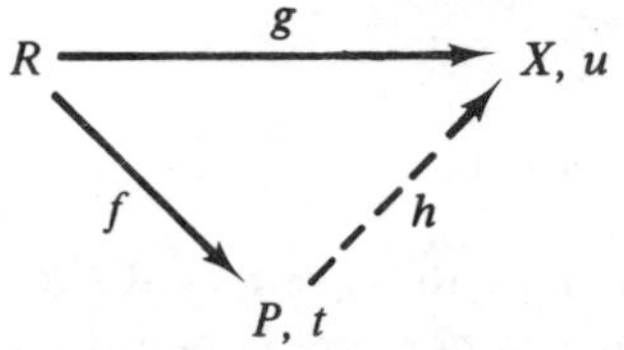

is commutative. We shall often write such a polynomial ring as (P, f, t).

Theorem 20.1 *For every non-trivial unitary ring R there exists, to within isomorphism, a unique polynomial ring (P, f, t) on R. Moreover, P is generated by* $\operatorname{Im} f \cup \{t\}$.

Proof. Let P be the set of all mappings $\alpha: \mathbf{N} \to R$ (i.e., sequences of elements of R) which are such that $\alpha(n) = 0$ for all but a finite number of natural numbers n. We leave to the reader the task of showing that the prescriptions

$$(\alpha + \beta)(n) = \alpha(n) + \beta(n), \quad (\alpha\beta)(n) = \sum_{i=0}^{n} \alpha(n-i)\beta(i)$$

define laws of composition on P relative to which P forms a unitary ring, the identity element of which is the mapping $\delta: \mathbf{N} \to R$ given by

$$\delta(n) = \begin{cases} 1 & \text{if } n = 0; \\ 0 & \text{otherwise.} \end{cases}$$

Given any $x \in R$, define a mapping $f_x: \mathbf{N} \to R$ by setting

$$f_x(n) = \begin{cases} x & \text{if } n = 0; \\ 0 & \text{otherwise.} \end{cases}$$

Clearly each mapping f_x belongs to P and so we can define a mapping $f: R \to P$ by the prescription $f(x) = f_x$. It is readily verified that f is a ring monomorphism and that $f_1 = \delta$, the identity element of P, so that f is 1-preserving. Now let $t \in P$ be given by

$$t(n) = \begin{cases} 1 & \text{if } n = 1; \\ 0 & \text{otherwise.} \end{cases}$$

It is readily verified that t commutes with every f_x. We shall now show that (P, f, t) is a polynomial ring on R.

For this purpose, consider an arbitrary 1-preserving ring monomorphism $g: R \to X$ of R into a unitary ring X and any element $u \in X \setminus \operatorname{Im} g$ which commutes with every element of Im g. Define a mapping $h: P \to X$ by setting

$$(\forall \alpha \in P) \qquad h(\alpha) = g[\alpha(0)] + \sum_{n \geqslant 1} g[\alpha(n)]u^n$$

noting that the summation is in fact finite since $\alpha(n) = 0$ for all but a finite number of n. Using the ring properties of X as well as the commutativity of u with every element of Im g, it is readily verified that h is a 1-preserving ring monomorphism which is such that $h(t) = u$ and $h \circ f = g$. To establish the uniqueness of h, we note first that, from the definition of multiplication in P, t^m is given by

$$t^m(n) = \begin{cases} 1 & \text{if } n = m; \\ 0 & \text{otherwise.} \end{cases}$$

We leave to the reader the easy task of showing that each $\alpha \in P$ can be written uniquely in the form

$$\alpha = f_{\alpha(0)} + \sum_{n \geqslant 1} f_{\alpha(n)} t^n.$$

Suppose then that $k: P \to X$ is an arbitrary 1-preserving ring monomorphism such that $k(t) = u$ and $k \circ f = g$. For every $\alpha \in P$ we have

$$\begin{aligned} k(\alpha) &= k\Big(f_{\alpha(0)} + \sum_{n \geqslant 1} f_{\alpha(n)} t^n\Big) \\ &= (k \circ f)[\alpha(0)] + \sum_{n \geqslant 1} (k \circ f)[\alpha(n)] k(t^n) \\ &= g[\alpha(0)] + \sum_{n \geqslant 1} g[\alpha(n)] u^n \\ &= h(\alpha). \end{aligned}$$

This then shows that $k = h$ as required and establishes the existence of a polynomial ring (P, f, t) on R. There remains the proof that such a polynomial ring is unique and that P is generated by $\operatorname{Im} f \cup \{t\}$. The latter statement is clear; and the former is established in a way which is entirely similar to the last part of the proof of Theorem 13.4, so we leave the details to the reader.

The above result shows that there is an essentially unique polynomial ring on R. As a model for this, we choose that constructed in the above existence theorem. Since the mapping f therein is a 1-preserving monomorphism, we shall agree to identify R with the subring Im f of P. In so doing, the mapping f becomes the inclusion monomorphism. We note from the above proof that the element t plays a distinguished rôle in that every element α of P can be expressed uniquely in the form

$$\alpha=\alpha_0+\alpha_1 t+\ldots+\alpha_n t^n$$

where, for each i, $\alpha_i=f_{\alpha(i)}=f[\alpha(i)]=\alpha(i)\in R\subseteq P$ with $\alpha_n\neq 0$. This gives rise to the following standard terminology: an element $\alpha\in P$ is called a **polynomial in t with coefficients in R**, the elements α_i being the coefficients; by the **degree** $\deg\alpha$ of a non-zero $\alpha\in P$ we mean the greatest integer n such that $\alpha_n\neq 0$; the **leading coefficient** of α is the coefficient α_n where $n=\deg\alpha$; the **constant term** of α is the coefficient α_0. The distinguished element t of P is traditionally called the **indeterminate** (although there is nothing indeterminate about it). Since it will be confusing later to retain a lower-case letter for this, we shall henceforth denote it by X and denote the polynomial ring on R by $R[X]$.

Theorem 20.2 *Let f and g be non-zero polynomials over a unitary ring R. If either the leading coefficient of f or that of g is not a zero divisor in R then $fg\neq 0$ and* $\deg fg=\deg f+\deg g$.

Proof. If a_n and b_m denote respectively the leading coefficients of f, g then a_nb_m is the coefficient of X^{n+m} in the product fg. If either a_n or b_m is not a zero divisor then $a_nb_m\neq 0$ and so a_nb_m is the leading coefficient of fg. It follows that $fg\neq 0$ in $R[X]$ and that $\deg fg=n+m=\deg f+\deg g$.

Corollary. *If R is an integral domain so also is $R[X]$.*

Proof. Since R has no zero divisors, it follows from the above that the product of two non-zero elements of $R[X]$ is also non-zero in $R[X]$.

Definition. We shall say that a polynomial $f\in R[X]$ is **monic** (or that f is a **monomial**) if its leading coefficient is the identity element 1_R of R.

We now establish a division algorithm for $R[X]$.

Theorem 20.3 *Let R be a unitary ring and let f, g be polynomials over R with g monic. Then there exist polynomials q, r over R such that $f=gq+r$ with either $r=0$ or* $\deg r<\deg g$.

Proof. If $\deg f<\deg g$ then taking $q=0$ and $r=f$ we obtain the result. Suppose then that $\deg f\geqslant\deg g$. We establish the result in this case by induction on $\deg f-\deg g$. Suppose that $f=a_0+a_1X+\ldots+a_mX^m$ and $g=b_0+b_1X+\ldots b_{n-1}X^{n-1}+X^n$ with $\deg f=m$. If we have $\deg f-$

$\deg g=0$ then $m=n$ and f can be written in the form

$$f=ga_n+[(a_{n-1}-b_{n-1}a_n)X^{n-1}+\ldots+(a_0-b_0a_n)]$$

in which case the property holds. Suppose then, by way of induction, that the result holds for all f^*, g^* with $0\leqslant \deg f^*-\deg g^*<\deg f-\deg g$. We have $f=ga_mX^{m-n}+h$ where $h=(a_{m-1}-b_{n-1}a_m)X^{m-1}+\ldots+(a_0-b_0a_m)$ and $\deg h\leqslant m-1<m=\deg f$. If $\deg h<\deg g$ no more proof is required; if $\deg h\geqslant \deg g$ then $0\leqslant \deg h-\deg g<\deg f-\deg g$ and so, by the induction hypothesis, $h=gk+r$ where $r=0$ or $\deg r<\deg g$. Hence

$$f=ga_mX^{m-n}+gk+r=g[a_mX^{m-n}+k]+r$$

and the proof is complete.

An immediate consequence of this result is the following.

Theorem 20.4 *If F is a field then F[X] is a euclidean domain.*

Proof. By the Corollary to Theorem 20.2, $F[X]$ is an integral domain. We show that the mapping $\delta: F[X]\backslash\{0\}\to \mathbf{N}$ given by $\delta(f)=\deg f$ is a euclidean valuation. It is clear that if f divides g then $\deg f\leqslant \deg g$. If now f and g are non-zero elements of $F[X]$ and if the leading coefficient of g is b_n then the polynomial $g^*=b_n^{-1}g$ is monic with $\deg g^*=\deg g$. Applying Theorem 20.3 we obtain $f=g^*q+r$ where $r=0$ or $\deg r<\deg g^*=\deg g$. This can be written $f=gb_n^{-1}q+r$, which is what we had to prove.

Corollary. *If F is a field then F[X] is a principal ideal domain and a unique factorisation domain.*

Proof. This is immediate from Theorems 17.9 and 17.7.

Remark. Note that when F is a field the decomposition $f=gq+r$ is unique in the sense that if also $f=gq_1+r_1$ with $\deg r_1<\deg g=n$ then $q=q_1$ and $r=r_1$. In fact we have $g(q-q_1)=r_1-r$ and if $q-q_1\neq 0$ we deduce that $r_1-r\neq 0$ and, by Theorem 20.2,

$$\deg\,[g(q-q_1)]=\deg g+\deg\,(q-q_1)\geqslant \deg g=n>\deg\,(r_1-r),$$

a contradiction. Hence $q-q_1=0$ whence also $r_1-r=g(q-q_1)=0$.

We shall now investigate the process of substituting elements for the indeterminate of a polynomial and some of the consequences.

In the proof of the next result we require the identity

$$\sum_{i=0}^{n}\sum_{j=0}^{m} a_ib_jr^{i+j}=\sum_{k=0}^{n+m}\left(\sum_{j=0}^{k} a_{k-j}b_j\right)r^k$$

where a_i, b_j, r are elements of a given commutative unitary ring and

n, m, k are positive integers. Perhaps the easiest way to see that this result holds is to note that the left hand side is the sum of all the elements in the following array whereas the right hand side is the same sum taken diagonally.

$$\begin{array}{cccccc}
a_0b_0r^0 & a_1b_0r^1 & a_2b_0r^2 & \ldots\ldots & \ldots\ldots & a_nb_0r^n \\
a_0b_1r^1 & a_1b_1r^2 & \ldots\ldots & \ldots\ldots & \ldots\ldots & a_nb_1r^{n+1} \\
a_0b_2r^2 & \ldots\ldots & \ldots\ldots & \ldots\ldots & \ldots\ldots & a_nb_2r^{n+2} \\
\ldots\ldots & \ldots\ldots & \ldots\ldots & \ldots\ldots & \ldots\ldots & \ldots\ldots \\
\ldots\ldots & \ldots\ldots & \ldots\ldots & \ldots\ldots & \ldots\ldots & \ldots\ldots \\
a_0b_mr^m & \ldots\ldots & \ldots\ldots & \ldots\ldots & \ldots\ldots & a_nb_mr^{n+m}
\end{array}$$

Theorem 20.5 *Let R be a commutative unitary ring and let S be a subring of R containing 1_R. For any given polynomial*

$$f = a_0 + a_1X + \ldots + a_nX^n \subset S[X]$$

and any $r \in R$ define the element $f(r) \in R$ by

$$f(r) = a_0 + a_1r + \ldots + a_nr^n.$$

Then for every $r \in R$ the mapping ζ_r given by the prescription $\zeta_r(f) = f(r)$ is an epimorphism from the ring $S[X]$ onto the subring of R generated by $\{1_R, r\}$.

Proof. It is clear that $\zeta_r(f+g) = \zeta_a(f) + \zeta_r(g)$. If f and g are given by $f = a_0 + a_1X + \ldots + a_nX^n$ and $g = b_0 + b_1X + \ldots + b_mX^m$ then, defining X^0 to be the identity element of $S[X]$ and r^0 to be that of R, we have

$$\begin{aligned}
\zeta_r(f)\zeta_r(g) = f(r)g(r) = \left(\sum_{i=0}^{n} a_ir^i\right)\left(\sum_{j=0}^{m} b_jr^j\right) &= \sum_{i=0}^{n}\sum_{j=0}^{m} a_ib_jr^{i+j} \\
&= \sum_{k=0}^{n+m}\left(\sum_{j=0}^{k} a_{k-j}b_j\right)r^k \\
&= (fg)(r) \\
&= \zeta_r(fg).
\end{aligned}$$

Thus each ζ_r is a ring morphism. Clearly $\zeta_r(X^0)=r^0$, so that ζ_r is 1-preserving, and $\zeta_r(X)=r$. Hence $\operatorname{Im}\zeta_r$ is a subring of R containing $\{1_R, r\}$. It is easily seen by induction that any subring of R which contains $\{1_R, r\}$ also contains $f(r)=\zeta_r(f)$. Hence we see that $\operatorname{Im}\zeta_r$ coincides with the subring generated by $\{1_R, r\}$.

Definition. The morphism ζ_r in the above is called the **substitution morphism** determined by r and we say that the element $f(r)$ is obtained by **substituting** r for the indeterminate X in f. As before, we write the subring generated by $\{1_R, r\}$ as $R[r]$, noting that this is not to be confused with $R[X]$. If $f(r)=0$ then we say that r is a **root** of the polynomial f. We say that r is a **root of multiplicity** k of f whenever k is the greatest integer such that f is divisible by $(X-r)^k$.

When F is a field, it is clear from the unique factorisation property of $F[X]$ that if f has distinct roots $a_1, \ldots, a_t$ where a_i is of multiplicity k_i for each i then

$$f=(X-a_1)^{k_1}(X-a_2)^{k_2}\ldots(X-a_t)^{k_t}g$$

for some polynomial g.

It should be noted that a given polynomial may factorise over one field but not over another. For example, the polynomial X^2+1 factorises over **C** with $X^2+1=(X+i)(X-i)$ but does not factorise over **R** (for any factorisation over **R** must also be a factorisation over **C** and the above factorisation over **C** is unique). Note also that X^2+1, when considered as a polynomial over the field $\mathbf{Z}/2\mathbf{Z}$, admits the factorisation $X^2+1=(X+1)(X+1)$ so that in this field 1 is a root of multiplicity two.

We shall now prove one of the basic theorems of field theory. For this purpose we shall introduce the following terminology. By an extension field of a given field F we shall mean a field E which contains F as a subfield. There are many types of field extensions, a proper study of which belongs to a course on field theory.[1] We shall restrict our attention to the following. If E is an extension field of a field F and if S is any subset of E then the smallest subfield of E which contains $F\cup S$ is called the **extension field of** F **generated by** S and will be denoted by $F(S)$. In the case where $S=\{r\}$ we write this as $F(r)$. Thus $F(r)$ is the smallest *subfield* of E containing $F\cup\{r\}$. Note that the notation $F[r]$ used previously denotes the smallest *subdomain* of E containing $F\cup\{r\}$ and so $F(r)$ is the field of quotients of $F[r]$.

Our aim is to prove that if $f\in F[X]$ is any non-zero polynomial over the field F then there is an extension field E of F which contains a root

[1] See, for example, *Introduction to Field Theory* by I. T. Adamson (Oliver & Boyd).

of f. Since $F[X]$ is a unique factorisation domain, it is clearly sufficient to establish this result whenever f is a monic irreducible polynomial.

Definition. If f is a monic irreducible polynomial over a field F and if E is an extension field of F containing a root r of f such that $E=F(r)$ then we say that E is a field **obtained by adjoining a root of** f **to** F.

Let us temporarily suppose that the field E is obtained by adjoining to the field F a root r of the monic irreducible polynomial f over F. Since $f(r)=0$ we see that f belongs to the kernel of the substitution morphism $\zeta_r:F[X]\rightarrow F[r]$. As $F[X]$ is a principal ideal domain Ker ζ_r is a principal ideal of $F[X]$ and so is of the form $\langle g\rangle$ where $g\in F[X]$. Since F is a field we can divide g by its leading coefficient and hence suppose without loss of generality that g is monic. This monic generator of Ker ζ_r is called the **minimum polynomial** of r. Since $f\in$ Ker $\zeta_r=\langle g\rangle$ we see that g divides f. But since f is monic and irreducible and g is non-constant and monic we deduce that $g=f$. Hence Ker $\zeta_r=\langle f\rangle$ and so, since $\zeta_r:F[X]\rightarrow F[r]$ is an epimorphism, we deduce that $F[r]\simeq F[X]/\langle f\rangle$. But since f is irreducible the ideal $\langle f\rangle$ is maximal (Theorem 17.5) and so $F[X]/\langle f\rangle$ is a field (Theorem 16.4). Thus $F[r]$ is a field and so this suggests that to construct an extension field of F containing a root of the monic irreducible polynomial f one should construct a field isomorphic to $F[X]/\langle f\rangle$. This is the basis of the following fundamental result.

Theorem 20.6 [Kronecker] *Let F be a field and let $f\in F[X]$ be monic and irreducible. Then there exists an extension field E of F which is obtained by adjoining a root of f to F.*

Proof. We know that $F[X]/\langle f\rangle$ is a field. Consider the mapping $\theta:F\rightarrow F[X]/\langle f\rangle$ defined by $\theta(x)=x/\langle f\rangle$. It is clear that θ is a morphism. Since the only constant polynomial which belongs to $\langle f\rangle$ is the zero polynomial, we see that θ is not the zero morphism. It then follows from Theorem 16.3 that θ is a monomorphism. We can therefore identify F with Im θ; in this way θ becomes the identity map on F and $F[X]/\langle f\rangle$ becomes an extension field of F. If now $g\in F[X]$ is given by $g=a_0X^0+a_1X+\ldots+a_nX^n$ then by substituting the element $X/\langle f\rangle$ of $F[X]/\langle f\rangle$ for X in g we obtain

$$\begin{aligned}g(X/\langle f\rangle)&=a_0(X/\langle f\rangle)^0+a_1(X/\langle f\rangle)+\ldots+a_n(X/\langle f\rangle)^n\\&=(a_0X^0+a_1X+\ldots+a_nX^n)/\langle f\rangle\\&=g/\langle f\rangle.\end{aligned}$$

In particular, we have $f(X/\langle f\rangle)=f/\langle f\rangle$. Since $f\in\langle f\rangle$ we see that $f/\langle f\rangle$ is the zero element of $F[X]/\langle f\rangle$; hence $X/\langle f\rangle$ is a root of f in $F[X]/\langle f\rangle$. Since, for every element $g/\langle f\rangle$ of $F[X]/\langle f\rangle$, we have shown above that $g/\langle f\rangle=g(X/\langle f\rangle)$ it follows that $F[X]/\langle f\rangle=F[X/\langle f\rangle]$. Thus $F[X/\langle f\rangle]$ is a field and so coincides with $F(X/\langle f\rangle)$. We conclude that $F[X]/\langle f\rangle$ is an extension field of F obtained by adjoining a root of f to F.

Let us now apply the above theorem to the case where $F=\mathbf{R}$ and f is the polynomial X^2+1. Clearly f is monic and, as is easily seen, is irreducible over $\mathbf{R}$. The quotient $R[X]/\langle X^2+1\rangle$ is thus an extension field of $\mathbf{R}$ obtained by adjoining a root of X^2+1 to $\mathbf{R}$. We denote this field by $\mathbf{C}$ and call it the **field of complex numbers**. Writing $X/\langle X^2+1\rangle$ as simply i, we see that $i^2+1=0$ in $\mathbf{C}$. It follows that, as in the above,

$$\mathbf{C}=\mathbf{R}[i]=\{x+yi;\ x, y\in\mathbf{R},\ i^2=-1\}.$$

Speaking somewhat more loosely, we see that $\mathbf{C}$ is a field which contains $\mathbf{R}$ and a square root of -1.

Remark. It is a truly remarkable fact that the field $\mathbf{C}$ enjoys the property that *every* non-constant polynomial $f\in\mathbf{C}[X]$ admits a root in $\mathbf{C}$, a consequence of which is that every such f factorises completely over $\mathbf{C}$ (i.e., into a product of linear factors, of the form $aX+b$). This is often expressed in the following form:

Theorem [Fundamental Theorem of Algebra] *The field* $\mathbf{C}$ *of complex numbers is algebraically closed.*

Despite its name, this is not really a theorem of algebra since its proof (which is considerably beyond the scope of this development) requires techniques from analysis.

Exercises for §20

1. **[Remainder Theorem]** Let F be a field. If $f\in F[X]$ and $r\in F$ prove that there exists $q\in F[X]$ such that $f=(X-r)q+f(r)$.

[*Hint.* Use Theorem 20.3.]

Deduce that $r\in F$ is a root of a non-constant polynomial $f\in F[X]$ if and only if $X-r$ divides f in $F[X]$.

2. If F is a field and if $f\in F[X]$ is of degree greater than or equal to 1 prove that f has at most $\deg f$ distinct roots in F.

[*Hint.* Use induction and Exercise 1.]

Deduce that if $f, g \in F[X]$ are such that $\deg f \leqslant n$ and $\deg g \leqslant n$ and if $a_1, \ldots, a_{n+1} \subseteq F$ are distinct and such that $f(a_i) = g(a_i)$ for each i then $f = g$. If $f \in F[X]$ then the **polynomial function defined by** f is the mapping $\theta_f : F \to F$ given by $\theta_f(x) = f(x)$. Prove that, if F^F denotes the ring of all mappings from F to itself, then the mapping $\theta : F[X] \to F^F$ given by $\theta(f) = \theta_f$ is a morphism. In the case where F is infinite, show that θ is a monomorphism.

3. If $a_1, \ldots, a_n$ are n distinct elements of a field F and if $b_1, \ldots, b_n$ are any n elements of F prove that there is a unique polynomial $f \in F[X]$ such that $\deg f \leqslant n-1$ and $f(a_i) = b_i$ for each i.

[*Hint*. Use Exercise 2 and consider the polynomial

$$f = \sum_{i=1}^{n} \left(b_i \prod_{j \neq i} (X - a_j) \Big/ \prod_{j \neq i} (a_i - a_j) \right).]$$

4. Prove that $\mathbf{Z}[X]/\langle X \rangle \simeq \mathbf{Z}$. Deduce that the ideal $\langle X \rangle$ is prime in $\mathbf{Z}[X]$ but is not maximal.

5. Let $\zeta_{\sqrt{2}}$ be the substitution morphism on $\mathbf{Q}[X]$. Determine $\operatorname{Im} \zeta_{\sqrt{2}}$ and $\operatorname{Ker} \zeta_{\sqrt{2}}$. Deduce that the set of real numbers $L = \{a + b\sqrt{2}; a, b \in \mathbf{Q}\}$, equipped with the usual laws of addition and multiplication, is an extension field of $\mathbf{Q}$ isomorphic to $\mathbf{Q}[X]/\langle X^2 - 2 \rangle$.

6. If D is an integral domain and Q is its field of quotients prove that the field of quotients $D(X)$ of $D[X]$ and the field of quotients $Q(X)$ of $Q[X]$ are isomorphic.

7. A non-zero polynomial $f \in \mathbf{Z}[X]$ is called **primitive** if the highest common factor of the non-zero coefficients of f is 1. Prove that the product of two primitive polynomials in $\mathbf{Z}[X]$ is again primitive.

[*Hint*. Suppose that f, g are primitive but that fg is not. Let p be a prime dividing all the coefficients of fg. Let a_r be the first coefficient of f not divisible by p and let b_s be the first coefficient of g not divisible by p. Examine the coefficient of X^{r+s} in fg.]

Deduce that if a non-zero $h \in \mathbf{Z}[X]$ is irreducible in $\mathbf{Z}[X]$ then it is irreducible in $\mathbf{Q}[X]$.

[*Hint*. Suppose that h is irreducible in $\mathbf{Z}[X]$ but that $h = fg$ in $\mathbf{Q}[X]$. Show that $f = af^*$ and $g = bg^*$ where f^*, g^* are primitive in $\mathbf{Z}[X]$ and deduce the contradiction that h is reducible over $\mathbf{Z}$.]

8. **[Eisenstein's Irreducibility Criterion]** Let $f \in \mathbf{Z}[X]$ be given by $f = a_0 + a_1X + \ldots + a_nX^n$. If p is a prime such that

(1) p does not divide a_n;

(2) p divides a_i for $i = 0, \ldots, n-1$;

(3) p^2 does not divide a_0,

prove that f is irreducible over $\mathbf{Q}$.

[*Hint.* Show as follows that f is irreducible over $\mathbf{Z}$ and use the previous exercise. Suppose f reducible, say $f = gh$ where $g = b_0 + b_1X + \ldots + b_rX^r$ and $h = c_0 + c_1X + \ldots + c_sX^s$. Suppose that $p \mid b_0$ and that c_j is the first coefficient of h which is divisible by p. Achieve the contradiction by examining $a_j = b_0c_j + b_1c_{j-1} + \ldots + b_jc_0$.]

9. If (D, δ) is a euclidean domain prove that the following are equivalent:

(1) either D is a field or D is isomorphic to a ring of polynomials over a field;

(2) for all non-zero $a, b \in D$ there exist *unique* elements $q, r \in D$ such that $a = bq + r$ with either $r = 0$ or $\delta(r) < \delta(b)$;

(3) for all non-zero $a, b \in D$,

$$a + b \neq 0 \Rightarrow \delta(a+b) \leqslant \max\{\delta(a), \delta(b)\}.$$

[*Hint.* (2) $\Rightarrow$ (3): Divide $a+b$ into $a^2 - b^2 + b$ in two ways.

(3) $\Rightarrow$ (1): If D^* denotes the group of units of D show that $K = D^* \cup \{0\}$ is a field. If $D \neq K$ then the set $D \backslash K$ of non-zero non-invertible elements of D is not empty; let $\xi \in D \backslash K$ be such that $\delta(\xi)$ is minimum in $\{\delta(x); x \in D \backslash K\}$. Show that D coincides with the set of elements of the form $a_0 + a_1\xi + \ldots + a_n\xi^n$ where each $a_i \in K$.]

10. **[Polynomials in several variables]** Let R be a unitary ring. Define recursively $R^{(n+1)} = R^{(n)}[X_n]$ where $R^{(0)} = R$, so that $R^{(1)} = R[X_0]$, $R^{(2)} = R^{(1)}[X_1] = (R[X_0])[X_1]$, etc. Establish the chain

$$R^{(n)} \supset R^{(n-1)} \supset \ldots \supset R^{(1)} \supset R^{(0)}$$

and show that every $x \in R^{(n+1)}$ can be expressed uniquely in the form

$$x = \sum_{i_0, \ldots i_n} a_{i_0 \ldots i_n} X_0^{i_0} X_1^{i_1} \ldots X_n^{i_n}$$

where each $a_{i_0 \ldots i_n}$ belongs to R.

[*Hint.* Use induction.]

11. **[Partial fractions]** Let F be a field and let f_1, f_2 be non-zero polynomials over F which are relatively prime. Prove that there exists a

unique pair of polynomials h_1, h_2 over F such that

$$\frac{1}{f_1 f_2} = \frac{h_1}{f_1} + \frac{h_2}{f_2}.$$

[*Hint.* Use the Corollary to Theorem 17.8.]

Hence show that every **rational function** over f (i.e., every element of the field $F(X)$) can be written in the form

$$f/g = h_1/p_1^{i_1} + \ldots + h_n/p_n^{i_n} + H$$

where $p_1, \ldots, p_n$ are distinct irreducible monic polynomials and, for each j, $\deg h_j < \deg p_j^{i_j}$ with p_j not a divisor of h_j. Show also that the polynomials $h_1, \ldots, h_n$ and H are uniquely determined.

[*Hint.* Use induction.]

12. For every $z = x + iy \in \mathbf{C}$ define the **complex conjugate** of z to be $\bar{z} = x - iy$. Define the **trace** and **norm** of z respectively by $T(z) = z + \bar{z}$ and $N(z) = z\bar{z}$. Prove that if $f \in \mathbf{R}[X]$ is monic then f is irreducible over $\mathbf{R}$ if and only if f is either a first degree polynomial or a second degree polynomial of the form $X^2 - T(z)X + N(z)$ for some $z \in \mathbf{C}\backslash\mathbf{R}$.

[*Hint.* Show first that $X^2 - T(z)X + N(z)$ is irreducible in $\mathbf{R}[X]$ for every $z \in \mathbf{C}\backslash\mathbf{R}$. Now observe that $\overline{f(z)} = f(\bar{z})$, so that if z is a root of f over $\mathbf{C}$ then so also is $\bar{z}$, whence $X^2 - T(z)X + N(z) = (X - z)(X - \bar{z})$ is a divisor of f.]

Index